GRID COMPUTING

Infrastructure, Service, and Applications

GRID COMPUTING

Infrastructure, Service, and Applications

Edited by

Lizhe Wang • Wei Jie • Jinjun Chen

CRC Press
Taylor & Francis Group
Boca Raton London New York

CRC Press is an imprint of the
Taylor & Francis Group, an **informa** business

CRC Press
Taylor & Francis Group
6000 Broken Sound Parkway NW, Suite 300
Boca Raton, FL 33487-2742

First issued in paperback 2017

ISBN-13: 978-1-4200-6766-8 (hbk)
ISBN-13: 978-1-138-11225-4 (pbk)

Library of Congress Cataloging-in-Publication Data

Grid computing : infrastructure, service, and applications / editors, Lizhe Wang, Wei Jie, Jinjun Chen.
    p. cm.
Includes bibliographical references and index.
ISBN 978-1-4200-6766-8 (hardcover : alk. paper)
1. Computational grids (Computer systems) I. Wang, Lizhe, 1974- II. Jie, Wei. III. Chen, Jinjun. IV. Title.

QA76.9.C58G694 2009
004'.36--dc22   2009003681

Visit the Taylor & Francis Web site at
http://www.taylorandfrancis.com

and the CRC Press Web site at
http://www.crcpress.com

Contents

Foreword

Grid computing, one of the buzzwords in the ICT industry, has emerged as a new paradigm for Internet-based parallel and distributing computing. Grids are aimed at exploiting synergies that result from cooperation of autonomous distributed entities. The synergies that develop as a result of grids include the sharing, exchange, selection, and aggregation of geographically distributed resources for solving large-scale problems in science, engineering, and commerce. At the same time, the grid community has embraced the integration of commodity Web services and grid technologies, and has adopted a utility-oriented computing model. Grids have become enablers for the creation of cyberinfrastructure for e-science and e-business applications. The recent widespread interest in grid computing from commercial organizations is pushing it toward mainstream computing and grid services to become valuable economic commodities.

Grids leverage existing IT infrastructure to optimize compute resources, and manage data and computing workloads. The developers of grids and grid applications need to address numerous challenges: security, heterogeneity, dynamicity, scalability, reliability, service creation and pricing, resource discovery, resource management, application decomposition and service composition, and quality of services. A number of projects around the world have developed technologies that help address one or more of these challenges. Some prominent projects include: Globus from Argonne National Laboratory (USA), Gridbus from the University of Melbourne (Australia), EGEE (Enabling Grids for E-Science) from Europe, and Educational Grid from China. Some of these have been prominent in developing open-source grid middleware technologies.

The grid community worldwide has made many advances recently with its focus on creation of innovative applications and their deployments on large-scale grids spanning multiple continents. This edited book, by Dr. Lizhe Wang, Dr. Wei Jie, and Dr. Jinjun Chen, does a great job in identifying the recent technological developments in grid computing worldwide and putting them together as a single coherent collection. The book is organized in bottom-up fashion, starting with infrastructure and ending with application and the rest; that is, grid middleware, in between! Chapters included in the book are authored by some of the world's leading experts in grid computing originating from Asia, Europe, and the United States. I am confident that readers, especially grid technology and application developers including the next-generation cyberinfrastructure developers, will find the book useful.

Dr. Rajkumar Buyya
Director, Grid Computing and
Distributed Systems Laboratory,
The University of Melbourne,
Australia

Preface

Computer architectures, networking, computer applications as well as computing paradigms have strong impacts on each other and evolve together to build modern computing infrastructures. The recent several decades have been characterized by rapid advancement of computing technologies, such as the single-chip microprocessor, high-performance networking, and capacity for large-size data storage. It could be therefore declared that we are facing an exciting epoch: geographically distributed computing resources as well as storage resources, connected by high-speed networking, providing a wide-area computing environment for innovative applications. Several terms have been identified for such computing paradigms; for example, metacomputing, high-performance distributed computing (HPDC), global computing, Internet computing, wide-area computing, and grid computing. This book uses the term grid computing.

Grid computing is one of the most innovative aspects of computing techniques in recent years. It mainly focuses on "resource sharing among geographically distributed sites and the development of innovative, high-performance oriented applications." From the middle of the last decade, when the first several grid projects were carried out, grid computing practice has lasted more than ten years. During these years, research on grid computing has made great progress, for example, grid computing standards and definitions, grid infrastructures, virtualization, middleware, and innovative applications. It is therefore an appropriate time to conclude these research results and look at the next step for grid computing, which is facing more challenging issues such as petascale data management, QoS-driven computing environment provision, global e-infrastructure, and advances in virtualization technologies.

Advances in grid computing are described in this book, which is structured into three parts: grid infrastructure and middleware, grid services, and grid applications. In the first part, several national and international grid systems are presented with the focus on grid infrastructure and middleware. Chapter 1 presents CROWN—China Research and Development environment Over Wide-area Network, one of the main projects in NSFCGrid program, which is lead by Beihang University (BUAA), with several partners in China such as the National University of Defense Technology of China (NUDT), Peking University, Computer Network Information Center of China Academy of Science (CNIC, CAS), and Tsinghua University. Chapter 2 discusses several ongoing cyberinfrastructure efforts in New York State. The major focus is on techniques for Web and grid service performance optimization, component frameworks for grids, instruments and sensors for grid environments, adaptive information dissemination protocols across grid overlays, emulation framework for grid computation on multicore processors, and secure grid data transfer. The EGEE project is elaborated in Chapter 3. Enabling Grids for E-sciencE (EGEE) represents the world's largest multidisciplinary grid infrastructure today. Co-funded by the European Commission, it brings together more than 250 resource centers from 48 countries to produce a reliable and scalable computing resource available to the European and global research community. The ChinaGrid project, funded by the Ministry of Education of China, is an attempt to achieve the above goals by exploring various resources on existing and well-developed Internet infrastructures. Chapter 4 is devoted to offer ChinaGrid project details. Chapter 5 presents a workflow enactment engine from Australia Grid, which utilizes tuple spaces to provide an event-driven mechanism for workflow execution entities.

In the second part, the book discusses recent grid service advances. The UK National Grid Service (NGS), which is discussed in Chapter 6, currently provides researchers within the United Kingdom and their collaborators with access to a wide range of computing and infrastructure resources using grid technologies that have been developing in the United Kingdom since 2001. Chapter 7 introduces the basic idea of resource allocation in a grid environment as well as major approaches to efficient resource selections. Chapter 8 contributes the work of grid services orchestration with OMIIBPEL. Chapter 9 analyzes the possibility of treating scientific workflow issues using techniques from the data stream community and proposes a data stream view for scientific workflow. Chapter 10 describes an SLA model that derives all other SLA-related classes (e.g., SLA proposals, SLA templates) and allows the clients and the service providers to specify the QoS requirements, economic considerations, service responsibilities, and so on, in a structured way. Chapter 11 reviews portal and workflow technologies, and discusses using them in grid environment. A development model of using AJAX in portal applications is proposed. Chapter 12 presents an overview of PKIs and their limitations, and highlights how recent work in UK academia allows user-oriented security models that are aligned with access to Internet resources more generally through the UK. Access Management Federation is based upon the Internet2 Shibboleth technologies. Chapter 13 presents PIndex, a peer-to-peer model for grid information services. PIndex builds on Globus MDS4, but introduces peer groups to dynamically split one large grid information search space into many small sections to enhance its scalability and resilience.

The third part of the book focuses on innovative grid applications. Chapter 14 discusses the WISDOM initiative, which is one of the main accomplishments in the use of grids for biomedical sciences achieved on grid infrastructures in Europe. Chapter 15 looks at incorporating flow-level networking models into grid simulators, in order to improve the scalability and speed of grid simulations by reducing the overhead of data and network intensive experiments, and improving their accuracy. Chapter 16 employs system-level virtualization and provides virtual machines resources for grid applications. Chapter 17 describes the usage of the grid in the high-energy physics environment in the LHC project. Service Oriented HLA RTI (SOHR) framework implements the functionalities of a HLA RTI using grid services and enables distributed simulations to be conducted on a heterogeneous grid environment. Chapter 18 elaborates on the design, implementation, and evaluation of the SOHR framework.

Contributors

Gabrielle Allen
Center for Computation and
    Technology
Louisiana State University
Baton Rouge, Louisiana

Jonathan J. Bednasz
Center for Computational
    Research
The State University of New
    York–Buffalo
Buffalo, New York

Vincent Breton
Corpusculaire Physics Laboratory
    of Clermont-Ferrand
Campus Cézeaux
Aubière, France

James Broberg
Department of Computer Science
    and Software Engineering
The University of Melbourne
Melbourne, Australia

Rajkumar Buyya
Department of Computer
    Science and Software
    Engineering
The University of Melbourne
Melbourne, Australia

Wentong Cai
School of Computer Engineering
Nanyang Technological
    University
Singapore

Jinjun Chen
Centre for Information
    Technology Research
Swinburne University of
    Technology–Hawthorn
Melbourne, Australia

Liang Chen
University College London
London, United Kingdom

Kenneth Chiu
Department of Computer Science
Binghamton University
Binghamton, New York

Cong Du
Department of Computer
   Science
Illinois Institute of Technology
Chicago, Illinois

Wolfgang Emmerich
University College London
London, United Kingdom

Steven M. Gallo
Center for Computational
   Research
The State University of New
   York–Buffalo
Buffalo, New York

Madhu Govindaraju
Department of Computer Science
Binghamton University
Binghamton, New York

Chunming Hu
Beihang University
Beijing, China

Jinpeng Huai
Beihang University
Beijing, China

Subu Iyer
Hewlett-Packard Labs
Palo Alto, California

Hai Jin
Cluster and Grid Computing Lab
Huazhong University of
   Science and Technology
Wuhan, China

Bob Jones
CERN
European Organization for
   Nuclear Research
Geneva, Switzerland

Doman Kim
School of Biological Science and
   Technology
Chonnam National University
Gwangju, Korea

Massimo Lamanna
CERN
European Organization for
   Nuclear Research
Geneva, Switzerland

Erwin Laure
CERN
European Organization for
   Nuclear Research
Geneva, Switzerland

Zhou Lei
Center for Computation and
   Technology
Louisiana State University
Baton Rouge, Louisiana

Michael Lewis
Department of Computer Science
Binghamton University
Binghamton, New York

Maozhen Li
Department of Electronic and
   Computer Engineering
Brunel University
Uxbridge, United Kingdom

Xiaorong Li
Institute of High Performance
   Computing
Singapore

Zengxiang Li
School of Computer Engineering
Nanyang Technological
    University
Singapore

Russ Miller
Hauptman-Woodward Medical
    Research Institute
Department of Computer
    Science and Engineering
The State University of
    New York–Buffalo
Buffalo, New York

Henry Palit
Institute of High Performance
    Computing
Singapore

Ke Pan
School of Computer Engineering
Nanyang Technological
    University
Singapore

Giulio Rastelli
Department of Pharmaceutical
    Sciences
University of Modena and
    Reggio Emilia
Modena, Italy

Andrew Richards
e-Science
Science and Technology
    Facilities Council
Rutherford Appleton Laboratory
Oxfordshire, United Kingdom

Catherine L. Ruby
Department of Computer Science
    and Engineering
The State University of the
    New York–Buffalo
Buffalo, New York

Vijay Sahota
Department of Electronic and
    Computer Engineering
Brunel University
Uxbridge, United Kingdom

Xuanhua Shi
IRISA
Rennes, France

Prerak Shukla
Department of Computer Science
Illinois Institute of Technology
Chicago, Illinois

Gillian M. Sinclair
University of Manchester
Manchester, United Kingdom

Richard O. Sinnott
National e-Science Centre
University of Glasgow
Glasgow, United Kingdom

Xian-He Sun
Department of Computer Science
Illinois Institute of Technology
Chicago, Illinois

Stephen John Turner
School of Computer
    Engineering
Nanyang Technological
    University
Singapore

Bruno Wassermann
University College London
London, United Kingdom

Charles M. Weeks
Hauptman-Woodward Medical
    Research Institute
Buffalo, New York

Chi Yang
Centre for Information
    Technology Research
Swinburne University of
    Technology–Hawthorn
Melbourne, Australia

Xiaoyu Yang
Earth Sciences Department
University of Cambridge
Cambridge, United Kingdom

Jia Yu
Department of Computer Science
    and Software Engineering
University of Melbourne
Melbourne, Australia

Zhifeng Yun
Center for Computation and
    Technology
Louisiana State University
Baton Rouge, Louisiana

Part I

Grid Infrastructure and Middleware

1

CROWN: A Service Grid Middleware for e-Science

Chunming Hu and Jinpeng Huai

CONTENTS

1.1 Background

In the year 2004, with the development of grid technologies in both academia
and industry, the Natural Science Foundation Committee of China (NSFC),
which is one of the main funding parties in China, announced its e-Science
Program named Network-based e-Science Environment. As one of the grid-
related research programs, this program is also referred to as the NSFCGrid.
The program started in 2004 and was completed at the end of 2007. The
main goal of this program is to build up a virtual scientific and experimental
environment to enable wide-area research such as large-scale computing
and distributed data processing. The research projects are organized into
three layers: basic theory and principles, general testbed, and pilot applications. In contrast to CNGrid and ChinaGrid, NSFGrid pays more attention
to the fundamental research of grid-related technologies.

    CROWN is the brief name for China Research and Development environ-
ment Over Wide-area Network [1], one of the main projects in the NSFCGrid
program, which is led by Beihang University (BUAA), with several partners

in China, such as the National University of Defense Technology of China (NUDT), Peking University, the Computer Network Information Center of China Academy of Science (CNIC, CAS), and Tsinghua University. The output of the CROWN project falls into three parts: the middleware, set to build a service-oriented grid system; the testbed to enable the evaluation and verification of grid-related technologies; and the applications.

The CROWN service grid middleware is the kernel for building an application service grid. The basic features of CROWN are listed as follows: first, it adopts an Open Grid Services Architecture/Web Services Resource Framework (OGSA/WSRF) compatible architecture [2]; second, considering the application requirements and the limitations of the security architecture of OGSA/WSRF, more focus is put on the grid resource management and dynamic management mechanism in the design stage, and a new security architecture with a distributed access control mechanism and trust management mechanism, which are proposed to support the sharing and collaborating of resources in a loosely coupled environment, is proposed in CROWN.

Under the framework of the CROWN project, we also created the CROWN testbed, integrating 41 high-performance servers or clusters distributed among 11 institutes in 5 cities (as of April 2007). They are logically arranged in 16 domains of 5 regions by using the CROWN middleware. The testing environment is growing continuously and has become very similar to the real production environment. The CROWN testbed will eventually evolve into a wide-area grid environment for both research and production.

CROWN has become one of the important e-Science infrastructures in China. We have developed and deployed a series of applications from different disciplines, which include the Advanced Regional Eta-coordinate Numerical Prediction Model (AREM), the Massive Multimedia Data Processing Platform (MDP), gViz for visualizing the temperature field of blood flow [3], the Scientific Data Grid (SDG), and the Digital Sky Survey Retrieval (DSSR) for virtual observation. These applications are used as test cases to verify the technologies in CROWN.

1.2 CROWN Middleware

1.2.1 Design Motivation

Since many researchers are focusing on technologies and applications of grids, the interoperability and loosely coupled integration problem between different grid systems are now becoming a hot topic. At the same time, the application and standardization of Web services technology are developing rapidly, and service-oriented architectures (SOAs) have become an important trend in building a distributed computing environment for wide-area networks, which helps the merging of grid and Web services. Recently, the OGSA [2] and WSRF were proposed and have become two of the fundamental technologies in grid competing. SOA and related standardization work provide an important methodology to the research and application of grid technology. First, the resources are encapsulated into services with standardized interfaces, supporting the unified service management protocol, which helps to solve the problem caused by the heterogeneity of resources; second, the resources are utilized through a service discovery and dynamic binding procedure, which helps to set up a loosely coupled computing environment. But the current resource management mechanism is not enough for all the grid application scenarios because of the distributed and autonomic resource environment, and the existing security mechanism cannot provide features such as privacy protection and dynamic trust relationship establishment, which hinders the further application of grid technology.

In fact, it is not only grid computing but also peer-to-peer computing and ubiquitous computing that try to explore the Internet-oriented distributed computing paradigm. The common issue in these computing paradigms is how to use the capability of resources efficiently in a trustworthy and coordinated way in an open and dynamic network environment. As we know, the Internet (especially the wireless mobile networks) is growing rapidly, while it is deficient in effective and secure mechanisms to manage resources, especially when the resource environment and relationships between different autonomic systems are changing constantly. At this point, three basic problems such as cooperability, manageability, and trustworthiness are proposed. The cooperability problem is how to make the resources in different domains work in a coordinated way to solve one big problem. The manageability problem is how to manage heterogamous resources and integrate the resources on demands in a huge network environment, which is a basic condition of building an Internet-oriented distributed computing environment. The trustworthiness problem is how to set up a reliable trust relationship between cross-domain resources when they are sharing and collaborating.

Based on our previous work on Web service supporting environments, and an OGSA/OGSI-compatible service grid middleware WebSASE4G, a WSRF-compatible CROWN grid middleware was proposed in 2002. According to the three basic problems, several key issues have been explored, such as resource management of the service grid, distributed access control, cross-domain trust management, grid service workflow, and service orchestration-based software development.

1.2.2 Architecture

Grid computing started from metacomputing in the 1990s. In recent years, grid computing has changed from metacomputing to computing grids and service grids, but the basic architecture of such a computing paradigm has not changed much. In 2001, the five-layer sand–glass architecture was proposed and generally accepted. With the application and standardization of grid computing, the architecture and its supporting technology became an important research issue. OGSA is a service-oriented architecture, which adopts the service as a unified resource encapsulation format to provide better extensibility and interoperability between grid resources. WSRF refines the service interface and interoperating protocols of OGSA, and makes OGSA a Web service-compatible implementation framework, which helps the merging of grid and Web service technology proceed more smoothly.

In fact, the five-layer sand–glass architecture simply proposed an abstract functionality structure for the service grid. OGSA/WSRF provided an implementation framework based on the service concept and a set of grid service interfaces, neither of which discussed the design principles, middleware component definitions, and the detailed solutions for access control and trust management in the service grid. In this chapter, we analyze the requirements of typical grid applications, and provide a detailed introduction of CROWN middleware and its architecture, design principles, and kernel technologies based on OGSA/WSRF service grid architecture.

Generally, there are three kinds of services in an OGSA/WSRF service grid: general services, resource encapsulating services, and application-specific services. General services, such as the grid job broker service, metascheduling service, and grid information services (GISs), are an important part of a service grid middleware. In a typical application grid scenario (see Figure 1.1), a user first submits a job to the metascheduling service, and then gets the job status and result from the job broker service; the metascheduling service retrieves the resource requirements from the

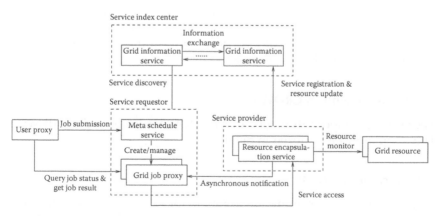

FIGURE 1.1 CROWN-based service grid application pattern.

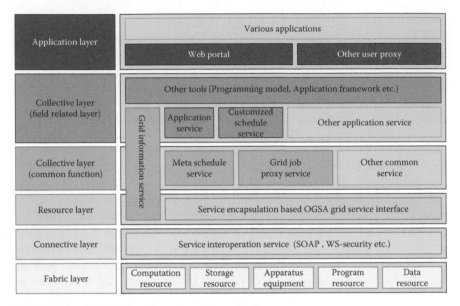

FIGURE 1.2   CROWN-based service grid application.

job description language, queries GIS to discover the necessary service, submits the job to the service, and traces the status of the job execution.

Based on the above analysis, we provide a layered architecture for CROWN and compare it with the five-layer sand–glass architecture (see Figure 1.2). In Figure 1.2, the service grid middleware covers the resource layer, the collective layer, and the application layer. In our system design, services in the resource layer (e.g., resource encapsulation service), collective layer (e.g., grid information service, metascheduling service, and job broker service), and part of the services in the application layer are OGSA-compatible grid services, which uses the information service for registration and dynamic discovery. Grid application support tools are used to enable the interoperation between the collective layer and the application layer (e.g., the application-specific developing framework, routines, and Web-based grid application portals).

1.2.3 Components in CROWN

There are 11 components in total in the CROWN service grid middleware. These will now be analyzed in detail.

1.2.3.1 Node Server [4]

This provides a basic runtime environment for grid service. Using the node server, the underlying resources can be encapsulated into grid services.

The node server provides all the generic functionalities when running a service instance, such as SOAP message processing, service instance management, instance/invoke lifecycle management, and notification mechanisms. Based on the Globus Toolkit 4.0, the node server adds many features such as remote and hot service deployment, resource status monitoring and reporting, logging, remote control, and management. By adding security modules, the node server can provide features like PKI/ Kerberos-based authentication, fine-gained authorization, trust management, and automatic trust negotiation (ATN), which could guarantee security and privacy effectively when resources are used by remote users or cross-domain resources.

1.2.3.2 Resource Locating and Description Service (RLDS)

This is a distributed information service system for service registration and discovery. Multiple RLDS instances use information exchange and the topology maintenance protocol to build a hierarchical architecture and overlay the network at runtime as needed to get better resource management and service discovery performance.

1.2.3.3 CROWN Scheduler

This is a metascheduling service in CROWN that queues and schedules users' jobs according to a set of predefined strategies, interoperates with the RLDS to get current service deployment information and job status, uses a predefined scheduling policy (random policy, load balancing policy, etc.) to do the matchmaking, and performs the service invocation. The CROWN scheduler supports two types of jobs: POSIX application invocation and grid service invocation. A job submission description language (JSDL) is used to describe the QoS requirements and security demands of the job.

1.2.3.4 CROWN CommSec

This is a plug-in for the node server and a generic Web service to provide basic security communication features such as building and verifying certificate chains. Administrators can edit the predefined policy file according to complex security requirements to provide independent, extensible, and feasible security solutions.

1.2.3.5 CROWN AuthzService

This is a generic service that uses an XACML-based authorization policy description language and provides authorization decision and policy management capabilities. It supports the multigranularity access control policy and the domain access control policy.

1.2.3.6 CROWN CredMan

This is used to manage user credentials. Through the agent certificate issue, the identified subjects can be managed, especially when the job is submitted from a Web portal or in mobile networks.

1.2.3.7 CROWN CredFed

This contains a plug-in for the node server and a credential mapping service. It can be used to map credentials from different security infrastructures (such as PKI and Kerberos) to enable identity mapping between two security domains. Administrators can modify the mapping policy to control the behavior of CredFed.

1.2.3.8 CROWN ATN

This contains a plug-in for the node server and a set of generic services. The ATN establishes the trust relationship between strangers on the Internet, protecting the privacy (e.g., the information on attributes-based certificates and the negotiation policies) of both sides. It provides a security decision and trust management mechanism for an open network environment.

1.2.3.9 CROWN Portal and Rich Client Framework

These two tools provide a unified user interface to the service grid to support the job parameter configuration, job submitting, JSDL generating, and result demonstration. The CROWN Portal provides a Web-based interaction model for the applications, and the Rich Client Framework provides a Java-based application framework for applications that have extensive visualization and interaction demands (such as complex visualization). The framework can be customized according to the application scenario to speed up the application development.

1.2.3.10 CROWN Designer

This is a grid service developing and deploying tool based on the Eclipse platform. A set of wizards and dialogs provided make the development and deployment of a grid service easier. By using the remote and hot deploy features of the node server, the designer provides drag-and-drop features to deploy the GAR file. In the near future, more service orchestration tools will be integrated into the CROWN Designer.

1.2.3.11 CROWN Monitor

This is an Eclipse RCP-based client tool written in Java. It is used to retrieve, store, and analyze events/information from different grid entities, and to show current runtime information using maps and charts. We can also adjust the parameters of the tool to change the monitor behavior of the target service grid systems.

Based on these middleware modules, the CROWN middleware is designed as shown in Figure 1.3. There are three layers in a service grid. CROWN middleware connects resources in a resource layer. The application grid uses Web portals or other customized client interfaces to submit jobs and solve the user's problem. First, the node server should be deployed on each resource to support service deployment and runtime management; second, all the grid resources should be divided into multiple domains, at least one RLDS instance should be deployed into each domain, and all the RLDS instances have to be configured into a predefined heretical architecture to form a distributed information systems; third, the CROWN scheduler will be deployed into the grid, to get the job request from the user and to find the proper services for each job; finally, monitoring and developing tools simplify the building of a service grid and its applications.

1.2.4 Main Features

CROWN adopts an OGSA/WSRF-compatible architecture, with the following features.

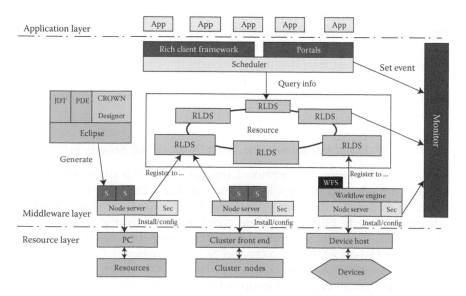

FIGURE 1.3 Service grid design principle based on CROWN middleware.

1.2.4.1 Overlay-Based Distributed Resource Management

The overlay technique is an effective way of supporting new applications as well as protocols without any changes in the underlying network layer. The basic idea of the overlay network is to build a new network over the existing physical network nodes according to some selected logical rules. In CROWN, resources are managed by an information service overlay consisting of a set of RLDS services [5]. These RLDS instances are linked with each other according to a tree-based topology with carefully selected shortcuts and exchange resource and request information with their logical neighbors. Such a fully decentralized structure can provide better performance while avoiding a single point of failure in the information systems.

1.2.4.2 Remote and Hot Deployment with Trust (ROST)

Traditionally, remote service deployment is supported in a cold fashion, which means deploying a new service and restarting the service runtime environment. Therefore, the hot service deployment [6], which does not need to restart the runtime environment while deploying services, has become increasingly important. To achieve this feature, an archive format called a GAR file (grid archive) is proposed to encapsulate all the necessary files and configurations for a grid service. The GAR file can be moved to a target service container through SOAP/HTTP protocols. The target service container receives the GAR file and uncompresses it to update the container information without stopping the container.

Security issues are guaranteed through trust negotiation using the ATN technique. ATN is a new approach to access control in an open environment, which, in particular, successfully protects sensitive information while negotiating a trust relationship. With ATN, any individual can be fully autonomous. Two individuals, who are in the same security domain, may try to set up a trust relationship by exchanging credentials according to respective policies. With the availability of remote and hot service deployment, many applications will benefit, such as load balancing, job migration, and so on.

1.2.4.3 JSDL-Based Job Submission and BES-Based Job Scheduling

JSDL (Job Submission Description Language) and BES (Basic Execution Service) are adopted in the CROWN scheduler, with extensions to Web service-based job submission. Jobs can be submitted to the CROWN scheduler via any JSDL-compatible client, such as GridSAM and gLite, using a BES interface. Interoperability demonstrations are proposed in AHM (All Hands Meeting) 2006 and SC (International Conference on Super Computing) 2006, organized by the HPCP (High Performance Computing Profile) Working Group in OGF (Open Grid Forum).

1.2.4.4 Security Architecture Supporting Domain Interoperability

CROWN uses a federate construction to form the virtual organization. We use the term "region" to denote the area with a homogeneous security infrastructure such as PKI or Kerberos, and the term "domain" to denote the area of autonomous organization. When grid services are deployed in different domains, each domain may have its own security concerns about the services. CROWN provides a fine-grained and extensible architecture that maximizes the separation of service administrators and service developers. Besides this, CROWN enables the same imple-mented service to be deployed into the PKI and Kerberos domains without having to modify the source code of the service. Furthermore, CROWN-ST also supports users from domains with heterogeneous security infrastructures to access the resources from other domains.

1.3 Resource Management in CROWN

1.3.1 Overview

CROWN employs a three-layered structure of resource organization and management [5], as illustrated in Figure 1.4, based on the characteristics of e-Science applications and the resource subordination relationship. The three layers are the node server, RLDS (Resource Locating and Descrip-tion Service), and S-Club and RCT (Resource Category Tree).

   In CROWN, before a computer can become a node server (NS), it must have the CROWN middleware installed. The service container is the core component in the CROWN middleware and provides a runtime environ-ment for various services. Each NS usually belongs to a security domain. Every domain has at least one RLDS to provide information services, and RLDS maintains the dynamic information of the available services. S-Club and RCT are used for more efficient resource organization and service discovery.

1.3.2 Node Server

The node server contains one component: the node. All kinds of heteroge-neous resources are encapsulated into CROWN nodes, and services are deployed on these nodes to provide a homogeneous view for the upper middleware to access the resources.

   The CROWN node server is implemented on the basis of GT4 Java WSRF core, and Figure 1.5 shows its system architecture. GT4 pro-vides a stable implementation of the WSRF specification family and a

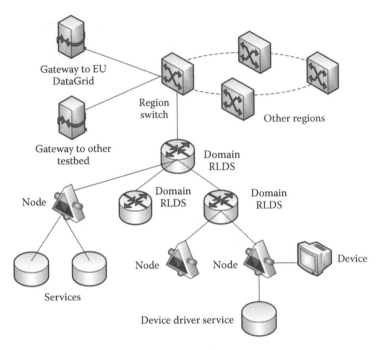

FIGURE 1.4   CROWN resource organization and management.

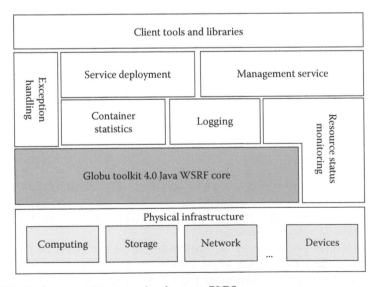

FIGURE 1.5   System architecture of node server RLDS.

lightweight embedded runtime. However, these basic functions are not enough to satisfy the requirements for a service container in a real grid environment.

RLDS contains three components:

1. Domains: Nodes are organized into different domains according to their subordination or resource catalogs. An RLDS is deployed in each domain as the center of grid information management. Domains may contain several subdomains such that all RLDSs come up with a tree-like topology.
2. Regions: For flexibility, the concept of a "region" is introduced to coordinate different domain trees. A region switch is deployed in each region such that flexible topologies and information sharing mechanisms can be applied among regions.
3. Gateways: To connect with other grid systems, we deploy several gateways on the edge of the CROWN system.

1.3.3 S-Club and RCT

CROWN employs a service club mechanism, called S-Club, for efficient resource organization and service discovery. S-Club is used to build an efficient overlay over the existing GIS (grid information service) mesh network [7–9]. In such an overlay, GISs provide the same types of services organized into a service club. An example of such a club overlay is shown in Figure 1.6, where nodes C, D, E, and G form a club. A search request could be forwarded to the corresponding club first such that searching the

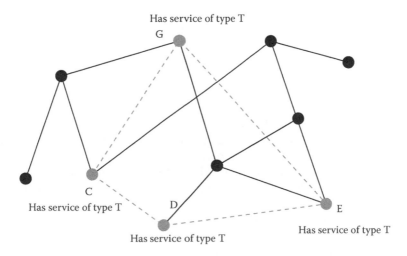

FIGURE 1.6   An example of service club.

response time and overhead can also be reduced if the desired result is available in the club.

Intuitively, to set up a club requires information exchange, and clubs need to be maintained dynamically because new GISs may join and some existing GISs may leave. Also, it is possible that some types of services become less popular after the club is built. Therefore, the system has to be careful on the trade-off between the potential benefit and the cost incurred. In general, the usage of services is not uniformly distributed. Some types of services can be very popular and others may not. When/how clubs are constructed/destroyed will be key issues in the S-Club scheme.

Assuming any search request is first sent to a GIS close to the user, on receiving a search request for a specific service type, the GIS checks locally whether there has been a club for this type. If so, the GIS forwards the request to the club, which will be flooded within the club only. If there is no club for this type, however, the GIS floods the request throughout the mesh network.

When a new GIS joins the GIS network, it has no idea what clubs are there. But since it has at least one neighbor in the underlying mesh network, it can ask one of its neighbors for the information of existing clubs. Namely, it simply copies the information of clubs from its neighbor. For a more detailed discussion, refer to [7].

Besides S-Club, there is an RCT (resource category tree) for the third layer's resource management. Computational resources are usually described by a set of attribute–value pairs. Among all attributes of a computational resource, one or several attributes are chosen to characterize the resource capacity of meeting application resource requirements as primary attributes (PA). An overlay called RCT (resource category tree) is used to organize computational resources based on PAs.

Grid applications can be characterized by their requirements for computational resources, for example, computing intensive and data-intensive applications and, in turn, categorizing computational resources based on certain resource characteristics that can meet application resource requirements. By doing so, resource discovery is performed on specific resource categories efficiently. For example, resources with huge storage can better serve a data-intensive application, thus they can be organized together based on an overlay structure.

Furthermore, according to the observation, the values of most resource attributes are numerical, for example, values of disk size. Also, attributes whose values are not numerical can be converted to numerical values through certain mathematical methods. Based on this consideration, RCT adopts an AVL tree (or balanced binary search tree) overlay structure to organize resources with similar characteristics. The attribute that can best describe the characteristic of resources organized by an RCT is named a primary attribute or PA. Figure 1.7 is an example of an RCT. The chosen PA is available memory size, and the value domain of available memory ranges from 0 MB to 1000 MB.

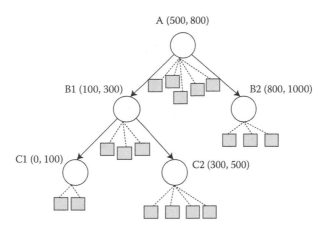

FIGURE 1.7 An example of an RCT.

Compared with traditional AVLs, each node of the RCT manages a range of values, instead of a single value. Each node only needs to maintain its connection with direct child nodes and a parent, and operations like registration, updating, and querying can start from any node. Unlike the traditional AVL structure, higher-level nodes of RCT are not required to maintain more information or bear more load than those in lower levels, which provide the basis for RCT to scale easily.

Suppose D is the value domain of the PA of an RCT. Each node n of an RCT is responsible for a subrange of D, or D_n. All resources with PA values belonging to D_n register themselves to node n. We name each RCT node an HR (head of a subrange). The terms "HR n" and "node n" will be used interchangeably in the rest of this chapter. In Figure 1.7, the circles denote HRs, while the squares below an HR denote computational resources registered with an HR.

Suppose N is the total number of HRs in an RCT, and lc(n) and rc(n) are the left and right child nodes of HR n, respectively. Since an RCT is a binary search tree, the following observations can be found:

$$D_i \cap D_j = \varphi, \quad \forall_{i,j} \in [1, N] \tag{1.1}$$

$$D_{\mathrm{lc}(i)} < D_i < D_{\mathrm{rc}(i)}, \quad \forall_i \in [1, N] \tag{1.2}$$

$D_i < D_j$ if the upper bound of D_i is less than the lower bound of D_j; for example, $[1, 2] < [3, 4]$.

If the ranges of node i and node j, that is, D_i and D_j, are adjacent, node i is referred to as a neighbor of node j, and vice versa. If node i is a neighbor of node j and $D_i < D_j$, node i is called the left neighbor of node j (denoted by L-neighbor(j)) and node j is called the right neighbor of node i (denoted by R-neighbor(i)). Note that there are two exceptions: the leftmost HR and

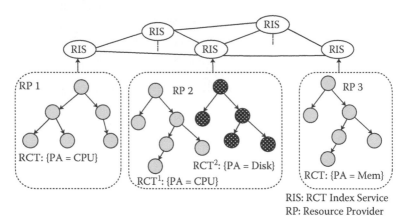

RIS: RCT Index Service
RP: Resource Provider

FIGURE 1.8   Resource organization with RCT.

the rightmost HR; the former has no left neighbor and the latter has no right neighbor. As shown in Figure 1.7, C2 and B2 are neighbors of A, while C2 is the L-neighbor of A and B2 is the R-neighbor of A. Note that C1 does not have a left neighbor and B2 has no right neighbor.

As resources are owned and managed by different resource providers, the providers may define different PAs for their resources, which results in constructing multiple RCTs. In Figure 1.8, we present a two-layer architecture for organizing resources across resource providers by using RCT. In the lower layer, each resource provider defines a set of PAs that can best describe their resources. Based on PAs, resources are organized through a certain number of RCTs. To enable wide-area resource discovery across different providers, an RCT index service (RIS) is deployed by each service provider in the upper layer. An RIS is a basic service that stores information about the PAs of a provider and the entry points of RCTs. RISs can be implemented, for example, as Web services or grid services, and find each other using services like UDDI.

In practice, a resource may have many attributes, but only a few of them are chosen as the primary attributes, so there will not be too many RCTs. When a query request cannot be satisfied by a resource provider, the RIS will contact other RISs to recommend another resource provider for further discovery operations.

1.4 Security Architecture in CROWN

1.4.1 Overview of CROWN Security

CROWN provides a hierarchical security solution to secure virtual organizations established via the CROWN middleware system. There are three

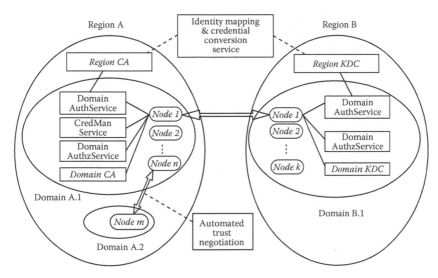

FIGURE 1.9 Architecture of CROWN security.

levels of security mechanisms in the CROWN security architecture: node-level, domain-level, and region-level security mechanisms. CROWN uses a federated construction to form the virtual organization and the architecture of security is designed accordingly, as shown in Figure 1.9. The term "region" is used to denote the area with a homogeneous security infrastructure such as PKI or Kerberos, and the term "domain" is used to denote the area of autonomous organization.

In order to wrap, share, and protect the raw resources in autonomous domains, the CROWN node should be deployed in the domain. It is the responsibility of the CROWN node to accept or intercept resource requests from grid users and do the security control. The raw resources to be protected are located in what we call protected areas, which may be a physical area or a conceptual area. When deploying CROWN security one should insist that all access to the resources in the protected area is mediated by a CROWN node.

1.4.1.1 Node-Level Security

In a CROWN node, CROWN Security implements communication security; fine-grained access control; basic message-level security, such as encryption/decryption, signing/verification, and authentication; and authorization mechanisms. Moreover, other new functionalities can easily be extended in this architecture due to its flexibility. CROWN Security is highly flexible through configuration, which makes it easier for administrators to specify fine-gained security policies for each service. For example, it is feasible to apply various security processing modes, such as a signature-only mode, an encryption-with-signature mode, and so on, to different services and even different methods or method parameter values in a service, in the same node.

Currently, CROWN Security supports two kinds of security infrastructure: Kerberos and PKI. Therefore the X.509 certificate [10] and Kerberos ticket are both supported in the authentication module. Both Kerberos and PKI authentication are implemented as a WSSecureConversation [11] service that conforms to the GSS-API standard [12]. For instance, a service deployed in the Kerberos region can use CROWN Security to authenticate and authorize users according to their Kerberos credentials, and in the meantime the same service can also be deployed in the PKI region with only slight configuration adjustments made by the administrator. This feature is the essential infrastructure for supporting further credential federation among regions.

In particular, during the dynamic trust establishment between two unknown nodes located in different security domains, the sensitive credentials or access control policies may be disclosed. In CROWN Security, a dedicated ATNService, namely, the Automated Trust Negotiation Service, which complies with the WS-Trust standard, is provided to preserve privacy for the nodes. If the service requestor has a trust ticket issued by a target service, then the trust can be established without negotiation. Otherwise, trust negotiation will be triggered, where the negotiation strategy enforcer in the ATNService will determine where and which credentials should be disclosed. In particular, an advanced trust chain construction component, which holds by trust management with various delegation credentials, is supported in ATNService.

1.4.1.2 Domain-Level Security

Although some security functions such as authentication and authorization are implemented as a node-level security mechanism in CROWN Security, sometimes it is a huge burden for administrators to maintain authentication and authorization policies on an enormous number of CROWN nodes in each domain. Therefore, several fundamental security services are provided by CROWN Security with the intention of easing security administration and reducing administration burden, including the authentication service (AuthService) and the authorization service (AuthzService). For example, a centralized authorization service can be deployed in a domain, and this authorization service will serve for all grid services residing in the CROWN nodes in this domain to make authorization decisions.

Furthermore, CROWN Security provides a credential management service (CredManService) as a MyProxy [13] replacement in CROWN middleware. CredManService allows users to access their credentials anywhere, anytime, even when they are on a system without a grid infrastructure or without secure access to their long-term credentials, as

MyProxy does. However, CredManService is implemented as a grid service and is decoupled with underlying security mechanisms. This actually benefits the administrators with immeasurable flexibility to tailor different security configurations for different service deployments. On the other hand, MyProxy is heavily coupled with SSL as a session security mechanism and a built-in access control model, which is hard coded and inflexible to extend. The domain administrator can deploy these services selectively. These services are all implemented as an extension to the WS-Trust standard [14], which has a policy-based design; therefore they are highly adaptable and easy to configure.

1.4.1.3 Region-Level Security

The region-level security mechanism in CROWN Security is realized by a credential federation service (CredFedService). In a multiparty collaboration, users in one region may have fundamental problems in accessing services provided by other regions because they have different authentication methods as well as different formats for user credentials, such as the X.509 certificate and the KerberosV5 [15] ticket. A credential conversion mechanism is an essential enabling mechanism for establishing profound collaboration among multiple parties. For example, CredFedService can be employed as a bridge between the PKI region and the Kerberos region. Therefore, users from one region can access the resources across different security infrastructures via the policy-based identity mapping and credential conversion feature provided by CredFedService. CredFedService is also implemented as a grid service, which is decoupled with underlying security mechanisms. Administrators can adapt different security configurations as well as identity mapping policies to their own requirements.

1.4.2 Design and Implementation

As discussed above, CROWN Security presents an extensible framework and implements basic communication security components inside the CROWN node. CROWN Security also provided four other components based on the framework, including credential management, policy-based authorization, trust management and negotiation, and credential federation. Implementation of CROWN Security is tightly integrated with the CROWN NodeServer, which is the core component of the CROWN middleware system. The basic function of CROWN Security comes together with CROWN NodeServer, and several fundamental security services are available as grid service archives, which can be remotely deployed into a CROWN NodeServer through the ROST service [16].

Before diving into design and implementation details, the security structure of the CROWN node, which provides the flexible and adaptable features of CROWN Security, will be discussed in the following subsection.

1.4.2.1 Security Structure for the CROWN Node

Figure 1.10 depicts the internal security structure for the CROWN node. The design of this structure is much inspired by Axis, although the purpose of CROWN Security is message-level security processing rather than SOAP message processing. The security processing depends on configurable security chains, and generally two chains, which deal with the request and response messages of service respectively, are configured for every grid service. For the sake of simplicity, there is only one processing chain shown in Figure 1.10. Each handler in the processing chain is in charge of some specific security functions. The grid services developers, deployers, and administrators can customize grid services protection by merely configuring these chains. The configuration is stored in security descriptors.

When the message interceptor embedded in the CROWN node intercepts a request or response message for a grid service, it will call the engine with information related to this message. The engine will then generate an appropriate chain by means of a configuration engine according to the security descriptor, and invoke the chain to process the message; that is, it will invoke each handler in the chain in a sequential order. After invoking a handler, the engine will choose to continue the process or terminate it according to the current result. The configuration engine can also be used to cache the instantiated processing chains and handlers in order to achieve better performance.

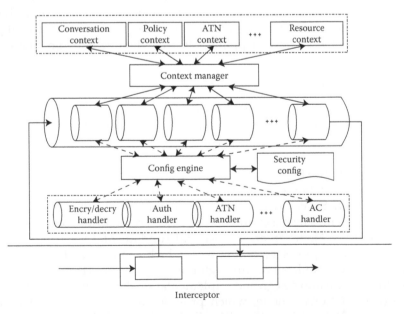

FIGURE 1.10 Security architecture of the CROWN node.

Besides the information related to the request or response message, some other information such as session keys, states of automated trust negotiation, and properties of resources is also needed by the handlers to finish their processing. In CROWN Security, this information is termed "security contexts" and is classified and managed by a context manager.

It should be noted that most handlers currently implemented in CROWN Security follow the policy-based design. For some handlers, such as the authentication handler and the authorization handler, there are two editions available with different modes, namely, the callout mode and the standalone mode. For instance, a standalone authentication handler will follow the authentication policy specified in a node-local security descriptor with all the policy decisions made locally, whereas a callout authentication handler will merely read the locations of access points of a centralized authentication service and consult the service for policy decisions.

As discussed above, to wrap, share, and protect the raw resources in autonomous domains, CROWN nodes should be deployed into the hosts of the domain. The software to be installed is called CROWN NodeServer, which is the core component of the CROWN middleware system. The NodeServer is implemented based on a GT WS-Core container with various new features and extensions, such as remote and hot service deployment, monitoring and management service, and so on.

The security structure for the CROWN node is tightly integrated with the CROWN NodeServer. Some functions of CROWN-ST come together with CROWN NodeServer as security handlers, which can be configured and customized by administrators in security processing chains.

1.4.2.2 Communication Security

The communication security module consists of both security handlers and security services, which can be used to secure corresponding messages between nodes, including encryption, decryption, signatures, authentication handlers, authentication services, and secure-conversation services. All handlers provided by CROWN Security conform to the WS-Security standard in terms of SOAP message encryption and signature. Moreover, the WS-Policy [17] language is used to express different policies for message processing, which makes CROWN Security highly flexible.

CROWN Security currently supports three modes of message-level security: username token mode, secure-message mode, and secure-conversation mode. The first two modes are similar to those implemented in GT4, which complies with WS-Security. Furthermore, our secure-conversation mode supports using both an X.509 certificate and a KerberosV5 ticket as a user's credentials for authentication and encryption, which conforms to WS-SecureConversation [11], WS-Trust [18], and IETF GSS-API standards [12].

1.4.2.3 Policy-Based Authorization

The policy-based authorization module in CROWN Security implements policy decision points in both the handler and the service. We adopt XACML (eXtensible Access Control Markup Language) [19] to express fine-grained access control policy in AuthzService. By using SAML assertions, the AuthzService can make authorization decisions based on user attributes rather than identity.

In Figure 1.11, the authorization module intercepts each request sent to the target service, and then collects attribute certificates signed by the attribute authority for both the user and the service to form a request context, which is conducted by a policy decision point to make an authorization decision for the request. As mentioned previously, the authorization policy is coded with XACML language and managed by the domain administrator.

1.4.2.4 Credential Management

Grid portals are increasingly used to provide user interfaces for grids. Through these interfaces, users can access a grid conveniently. When security is taken into account, a user requires access to his credentials in a secure and convenient way, anytime and anywhere. The credential management module in CROWN security, which consists of a CredManService and corresponding client tools, known as CredMan, is designed to meet this

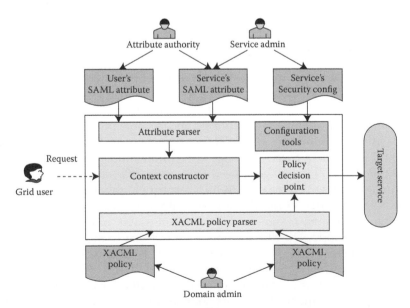

FIGURE 1.11   Structure of authorization module.

requirement. At first, a user would use a CredMan client command, named *credman-init*, to visit the CredManService and delegate a set of proxy credentials that are signed by the user's permanent credential to the service repository. At a later time when the user's credential is needed, the user, or the service acting on behalf of the user, gets a proxy credential delegated from the proxy credential stored in the repository.

In CROWN, client tools can be integrated with the portal; that is, a user can access his credential through the portal. By using the tools, a user can easily delegate to and retrieve credentials from the repository. Moreover, some client tools are provided for the user to manage the credentials stored in the repository. In order to protect the credentials in the service repository, CredManService provides a protected mechanism in which a user can specify authentication information and retrieval restrictions to protect his credentials in the repository.

1.4.2.5 Trust Management and Negotiation

A dedicated ATNService can be deployed with the target service to support the trust negotiation with the service requestor. As illustrated in Figure 1.12, a series of procedures are involved in the trust negotiation. When the client requests the target service, which is protected by the trust negotiation service, it will firstly initialize an *ATNEngine* through a local *RedirectHandler*. Upon receiving the negotiation request from a client, the service provider will create an *ATNEngine*, too. The state of negotiation will be stored in *ATNContext*. Then, the two participants may disclose their credentials according to the provider's policy or policies for sensitive credentials. This process will be interacted until a final decision ("success" or "failure") is reached. If the negotiation succeeds, ATNService will return a success status, and the context will be updated accordingly. The requestor can insert

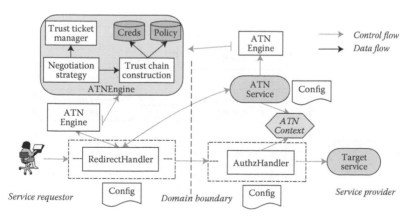

FIGURE 1.12 ATNService and ATNEngine.

the session ID into a SOAP header and sign it before sending it to the target service. The target service will verify the authenticity of the session ID through its AuthzHandler, and allow access if the verification succeeds.

1.4.2.6 Credential Federation

A credential federation component is provided as a grid service called CredFedService. The function of CredFedService is to convert an X.509 certificate to a KerberosV5 ticket according to a specified identity mapping policy, and vice versa. Figure 1.13 shows the relationships and data flow among the modules inside CredFedService implementation.

The input of CredFedService is the user's credential, and the output is a new credential in a format different from the input credential. Figure 1.13 demonstrates the procedure of mapping an X.509 credential to a Kerberos credential. First, the input credential is processed by the authentication module, which is realized by a secure-conversation mode offered by an underlying communication security component of CROWN-ST, to verify whether the user is the real owner of this credential. If so, the credential is then forwarded to the identity mapping module, which will map the identity of the user to another domain based on mapping policy. Then the new identity will be processed by the credential conversion module to generate a new credential for the user. Finally, this credential is returned to the user by CredFedService. As shown in Figure 1.13, each module has its corresponding policy that can be customized by the CredFedService administrator.

1.4.3 CROWN Security Summary

CROWN Security provides a fine-grained and extensible framework enabling trust federation and trust negotiation for resource sharing and collaboration in an open grid environment. We have also demonstrated

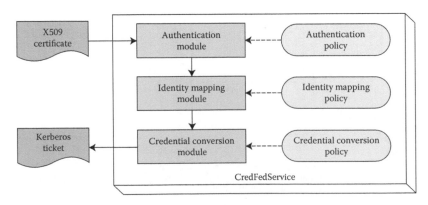

FIGURE 1.13 CredFedService.

the performance of our implementation through comprehensive experimental studies. CROWN Security aims to satisfy some security requirements of dynamic distributed resources sharing and integration, but much work remains to be done.

1.5 Testbed and Applications of CROWN

CROWN is now becoming one of the important e-science infrastructures in China. We have developed and deployed a series of applications from different disciplines, which include Advanced Regional Eta-Coordinate Numerical Prediction Model (AREM), Massive Multimedia Data Processing Platform (MDP), gViz for visualizing the temperature field of blood flow, Scientific Data Grid (SDG), and Digital Sky Survey Retrieval (DSSR) for virtual observation. These applications have been used as test cases to verify the technologies in CROWN.

AREM uses a grid as a tool to study and refine the numerical prediction models of weather and climate. Several numerical models are worked out by meteorologists during their research and prediction work. Typically these models use the raw weather data from a national meteorology authority as inputs and simulate the weather transformation according to the laws of atmospheric physics and fluid dynamics. The output can be used as a prediction result of future weather. The simulations are all based on complex numerical calculations and need large quantities of computing power and storage capacities. By using the resource organization and job scheduling technologies provided by CROWN, we successfully developed the AREM research system. We encapsulated the Fortran complier, visualization tools (GrADS), and the simulation framework of AREM as services, and a unified raw weather data center is also deployed. Meteorologists can submit simulation jobs to the system and refine their numerical models according to the results. Since the jobs are executed using the resources provided by the CROWN testbed, the execution procedure can be parallel, the execution time can be greatly reduced, and the efficiency of weather system research and prediction model refinement can be improved.

Large amounts of storage capability and computing power are needed when performing multimedia data processing, such as content recognition of voice or video. Traditionally a central processing model is applied and pieces of data are collected and processed in a single point. When the input data are increased, this method provides little scalability, especially for real-time applications. We combined the service grid technologies with massive data processing and implemented the MDP platform for multimedia data processing. MDP has been deployed into CROWN and has provided service since 2005. We encapsulated the related algorithms into services and deployed them on many grid nodes. Users can provide many ways of submitting multimedia data and jobs to the grid scheduling system. After analyzing the workload of the grid nodes, available resources can be found automatically and data can be processed by invoking the corresponding services. Since the platform is deployed in a wide-area environment, we also introduced the trust management and negotiation mechanisms. These technologies protect the user data and make the processing trustworthy. By using MDP, the resources that can be used to process multimedia data increase, and the throughput and dependability of processing can be much improved.

CROWN interoperates with other grid middleware through specifications. The testbed also links to some famous grid testbeds. For example, the gViz application is deployed in both CROWN and the White Rose Grid (WRG), which is a part of the UK National Grid Service (NGS). We demonstrated the application at the UK e-Science All Hands Meeting 2005 to show the interoperability of heterogeneous and autonomic grid systems.

Our experience with system development and deployment shows that CROWN provides the capabilities of resource management, distributed access control, and trust management and negotiation. It can be used to support applications that are computation intensive and/or data intensive. Eleven applications were deployed into CROWN by April 2006 and more than 25,000 requests have been processed.

Acknowledgments

Part of this work is supported by grants from the China Natural Science Foundation (No. 90412011, 60703056), China 863 High-tech Programme (Project No. 2006AA01A106), China 973 Fundamental R&D Program (No. 2005CB321803), and the National Natural Science Funds for Distinguished Young Scholars (No. 60525209). We would also like to thank Jianxin Li, Tianyu Wo, and Hailong Sun and other members of the CROWN team at Beihang University for their contributions to the design and implementations of CROWN middleware, testbed, applications, and related resource management technologies mentioned in this chapter.

References

1. CROWN website. Available at: http://www.crown.org.cn.
2. I. Foster, C. Kesselman, J. Nick, and S. Tuecke, "The physiology of the grid: An open grid services architecture for distributed systems integration," Open Grid Service Infrastructure WG, Global Grid Forum, 2002, USA. Available at: http://www.globus.org/research/papers/ogsa.pdf.

3. K. Brodlie, D. Duce, J. Gallop, M. Sagar, J. Walton, and J. Wood, "Visualization in grid computing environments," in *Proceedings of IEEE Visualization*, Austin, Texas, October 10–15, 2004, pp. 155–162.
4. H. Sun, W. Liu, T. Wo, and C. Hu, "CROWN Node Server: An enhanced grid service container based on GT4 WSRF core," in *Proceedings of the Fifth International Conference on Grid and Cooperative Computing Workshops*, Changsha, China, October 21–23, 2006, pp. 510–517.
5. J. Huai, T. Wo, and Y. Liu, "Resource management and organization in CROWN grid," in *Proceedings of the First International Conference on Scalable Information Systems*, 2006, p. 10.
6. J. Huai, H. Sun, C. Hu, Y. Zhu, Y. Liu, and J. Li, "ROST: Remote and hot service deployment with trustworthiness in CROWN grid," *Future Generation Computer Systems*, 23 (6), 825–835, 2007.
7. J. Frey and T. Tannenbaum, "Condor-G: A computation management agent for multi-institutional grids," *Journal of Cluster Computing*, 5, 237, 2002.
8. W. Hong, M. Lim, E. Kim, J. Lee, and H. Park, "GAIS: Grid advanced information service based on P2P mechanism," in *Proceedings of the 13th IEEE International Symposium on High Performance Distributed Computing (HPDC-13)*, Honolulu, Hawaii, June 4–6, 2004, pp. 276–277.
9. A. Iamnitchi, I. Foster, and D.C. Nurmi, "A peer-to-peer approach to resource location in grid environments," in *Proceedings of the 11th IEEE International Symposium on High Performance Distributed Computing (HPDC-11)*, 2002.
10. R. Housley, W. Ford, T. Polk, and D. Solo, "Internet X.509 public key infrastructure certificate and CRL profile," IEEE Request for Comments (IEEE RFC 2459), 1999.
11. S. Anderson, J. Bohren, and T. Boubez, "Web services secure conversation language," XMLSOAP.org Specifications, specs.xmlsoap.org/ws/2004/04/sc/ws-secureconversation.pdf, 2005.
12. J. Linn, "Generic security service application program interface, version 2," IEEE Request for Comments (IEEE RFC 2078), 1997.
13. J. Basney, M. Humphrey, and V. Welch, "The MyProxy online credential repository," *Software: Practice and Experience*, 35, 801–816, 2005.
14. S. Anderson, J. Bohren, and T. Boubez, "Web services trust language," XMLSOAP.org Specifications, specs.xmlsoap.org/ws/2005/02/trust/WS-Trust.pdf, 2005.
15. C. Neuman, T. Yu, S. Hartman, and K. Raeburn, "The Kerberos network authentication service (V5)," IEEE Request for Comments (IEEE RFC 4120), 2005. Available at: http://www.ietf.org/rfc/rfc4120.txt.
16. H. Sun, Y. Zhu, C. Hu, J. Huai, Y. Liu, and J. Li, "Early experience of remote and hot service deployment with trustworthiness in CROWN grid," in *Proceedings of Sixth International Workshop on Advanced Parallel Processing Technologies*, (APPT 2005): 301–302, Hong Kong, China, October 27–28, 2005.
17. S. Bajaj, D. Box, and D. Chappell, "Web services policy framework," World Wide Web Consortium (W3C) Specification. Available at: http://www.w3.org/Submission/WS-Policy/, 2005.
18. A. Nadalin, M. Goodner. et al., WS-Trust, version 1.3. OASIS standard. http://docs.oasis-open.org/ws-sx/ws-trust/200512/ws-trust-1.3-os.html.
19. T.M. Simon Godik, "OASIS eXtensible Access Control Markup Language (XACML)," 2003.

2

Cyberinfrastructure in New York State

Russ Miller, Jonathan J. Bednasz, Kenneth Chiu,
Steven M. Gallo, Madhu Govindaraju, Michael Lewis,
Catherine L. Ruby, and Charles M. Weeks

CONTENTS

2.1 Introduction

Cyberinfrastructure sits at the core of modern simulation and modeling, providing entirely new methods of investigation that allow scholars to address previously unsolvable problems. Specifically, the development of software, algorithms, portals, and interfaces that will enable research and scholarship by freeing end-users from dealing with the complexity of various computing environments is critical to extending the reach of high-end computing, storage, networking, visualization, and sophisticated instrumentation to the general user community.

The grid currently serves as a critical infrastructure for most activities in cyberinfrastructure. The grid is a rapidly emerging and expanding technology that allows geographically distributed and independently operated resources to be linked together in a transparent fashion (www.gridcomputing.com; www.globus.org; Berman et al., 2003; Foster and Kesselmann, 1999). These resources include CPU cycles, data storage systems, sensors, visualization devices, and a wide variety of Internet-ready instruments. The power of both computational grids (i.e., seamlessly connecting computer systems and their local storage) and data grids (i.e., seamlessly connecting large storage systems) lie not only in the aggregate computing power, data storage, and network bandwidth that can readily be brought to bear on a particular problem, but also on its ease of use.

Numerous reports state that grid computing is a key to twenty-first century discovery by providing seamless access to the high-end computational infrastructure that is required for revolutionary advances in contemporary science and engineering. Numerous grid projects have been initiated, including GriPhyN,[*] PPDG,[†] EGEE,[‡] EU DataGrid,[§] NASA's Information Power Grid (IPG),[¶] TeraGrid,[**] and Open Science Grid[††] to

[*] www.griphyn.org/.
[†] www.ppdg.net/.
[‡] www.eu-egee.org/.
[§] eu-datagrid.web.cern.ch/eu-datagrid/.
[¶] www.gloriad.org/gloriad/projects/project000053.html.
[**] www.teragrid.org/.
[††] www.opensciencegrid.org/.

name a few. However, the construction of a production-quality, heterogeneous, general-purpose ubiquitous grid is in its infancy. Such a grid will be transparent to the users and, as such, requires coordinated resource sharing and problem solving in a dynamic, multi-institutional scenario using standard, open, general-purpose protocols and interfaces that deliver a high quality of service. The immediate focus of grid deployment continues to be on the difficult issue of developing high-quality middleware* to deal with issues including interoperability, security, performance, management, and privacy.

2.2 Cyberinfrastructure in Buffalo

In the twenty-first century, leading academic institutions will embrace our digital data-driven society and empower students to compete in this knowledge-based economy. In order to support research, scholarship, education, and community outreach, Miller's Cyberinfrastructure Laboratory[†] is dedicated to the integration of research in disciplinary domains (e.g., science, engineering, nontraditional areas of high-performance computing), with research in enabling technologies and interfaces. The goal is to allow students and scientists to transparently collect, manage, organize, analyze, and visualize data without having to worry about details such as where the data are stored, where the data are processed, where the data are rendered, and so forth. This ease of use and high availability of data and information processing tools will allow for revolutionary advances in science, engineering, and beyond.

2.2.1 New York State Grid

The design, development, and deployment of the New York State Grid (NYS Grid) was led by Miller's Cyberinfrastructure Laboratory. The NYS Grid includes resources from institutions throughout New York State[‡] and is available in a simple and seamless fashion to users worldwide.[§] The NYS Grid contains a heterogeneous set of resources and utilizes general-purpose IP networks (Green and Miller, 2003, 2004a–c). A major feature of the NYS Grid is that it integrates a computational grid (compute clusters

---

* www.nsf-middleware.org/.
[†] www.cse.buffalo.edu/faculty/miller/CI/.
[‡] Binghamton University, Columbia University, Cornell University, Geneseo State College, the Hauptman-Woodward Medical Research Institute, Marist College, New York University, Niagara University, Rochester Institute of Technology, Stony Brook University (SUNY), Syracuse University, University at Albany, University at Buffalo, and the University of Rochester.
[§] The NYS Grid has been used by users who are members of virtual organizations affiliated with the Open Science Grid and directly with the NYS Grid.

that have the ability to cooperate in serving the user) with a data grid (storage devices that are similarly available to the user) so that the user may deploy computationally intensive applications that read or write large data files in a very simple fashion. In particular, the NYS Grid was designed so that the user does not need to know where the data files are physically stored or where an application is physically deployed, while providing the user with easy access to their files in terms of uploading, downloading, editing, viewing, and so on. Of course, a user who wishes to more closely manage where the data are stored and where the applications are running also has the option to retain such a control.

The core infrastructure for the NYS Grid includes the installation of standard grid middleware, the use of an active Web portal for deploying applications, dynamic resource allocation so that clusters and networks of workstations can be scheduled to provide resources on demand, a scalable and dynamic scheduling system, and a dynamic firewall, to name a few.

Several key packages were used in the implementation of the NYS Grid, and other packages have been identified in order to allow for the anticipated expansion of the system. The Globus Toolkit* provides APIs and tools using Java SDK to simplify the development of OGSI-compliant services and clients. It supplies database services and Monitoring and Discovery System† index services implemented in Java, GRAM‡ service implemented in C with a Java wrapper, GridFTP§ services implemented in C, and a full set of Globus Toolkit components. The recently proposed Web Service-Resource Framework¶ provides the concepts and interfaces developed by the OGSI specification exploiting the Web services architecture.

The NYS Grid is the current incarnation of a Cyberinfrastructure Laboratory-led grid that progressed from a Buffalo-based grid (ACDC-Grid) to a persistent, hardened, and heterogeneous** Western New York Grid (WNY Grid) before being enhanced, expanded, and deployed throughout New York State. This series of grids was largely funded by the National Science Foundation through a series of ITR, MRI, and CRI grants. The NYS Grid currently supports a variety of applications and users from NYS Grid institutions and the Open Science Grid. In addition, a grassroots New York State Cyberinfrastructure Initiative†† (NYS-CI) has been granted access to the NYS Grid due to its promise of users with the need for significant resources. Unfortunately, to date, such users have not been identified by the NYS-CI.

---

*   www.globus.org/toolkit/.
†   www.globus.org/toolkit/mds/.
‡   www.globus.org/toolkit/docs/2.4/gram/.
§   www.globus.org/grid_software/data/gridftp.php.
¶   www.globus.org/wsrf.
** Equipment on the Western New York Grid includes one or more significant Sun Microsystems clusters (Geneseo State College), Apple Clusters (Hauptman-Woodward Institute), and Dell Clusters (Niagara University, SUNY-Buffalo), in addition to a variety of storage systems.
†† www.nysgrid.org/.

2.2.2 Middleware Efforts

In the early 2000s, Miller's Cyberinfrastructure Laboratory had a vision to design an integrated computational and data grid that would provide disciplinary scientists with an easy-to-use extension of their desktops that would enable breakthrough science and engineering. In order to provide such a utility, the Cyberinfrastructure Laboratory partnered with the Center for Computational Research* and other organizations in the Buffalo area in order to create a prototype grid. This prototype, the Advanced Computational Data Center Grid (ACDC-Grid), provided a platform for the Cyberinfrastructure Laboratory to experiment with critical packages, such as Globus and Condor, and to begin to work with critical worldwide organizations, including the Open Science Grid.

However, the most important aspect of this prototype grid was that it provided a platform upon which members of the Cyberinfrastructure Laboratory could develop middleware that was deemed critical to the deployment of a transparent and integrated computational-, data-, and applications-oriented grid.

2.2.2.1 *Grid Portal and Grid-Enabling Application Templates*

A key project† was the development of a grid portal (cf. Roure, 2003, for a discussion of grid portals). While it is true that Globus and other packages provide a variety of avenues for command-line submission to a grid, most require that the user be logged into a system upon which the appropriate package (e.g., Globus or Condor) has been installed. The approach of the Cyberinfrastructure Laboratory was to provide critical middleware that was accessible to users worldwide through a Web browser.

The New York State Portal, as shown in Figure 2.1 for a package in molecular structure determination (Miller et al., 1993, 1994; Rappleye et al., 2002; Weeks and Miller, 1999; Weeks et al., 1994, 2002), can be found at http://grid.ccr.buffalo.edu. This portal currently provides access to a dozen or so compute-intensive software packages, large data storage devices, and the ability to submit applications to a variety of grids consisting of tens of thousands of processors. Our grid portal integrates several software packages and toolkits in order to produce a robust system that can be used to host a wide variety of scientific and engineering applications. Specifically, our portal is constructed using the Apache HTTP server, HTML, Java and PHP scripting, PHPMyAdmin, MDS/GRIS/GIIS from the Globus Toolkit, OpenLDAP, WSDL, and related open-source software that interfaces with a MySQL database.

---

* Miller was director of the Center for Computational Research during this period.
† More details of this section can be found in Green and Miller (2003, 2004c).

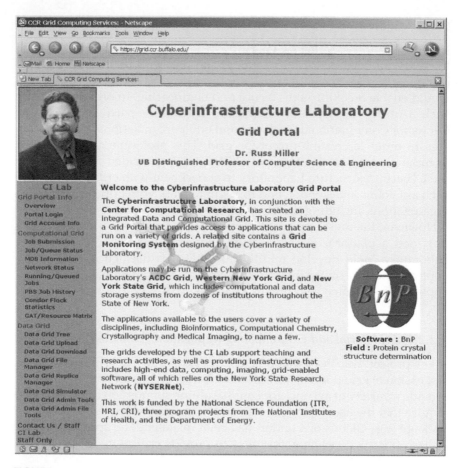

FIGURE 2.1   (a) Dr. Miller's Cyberinfrastructure Laboratory. Entering the New York State Grid Portal. (*Continued*)

Our grid portal provides a single point of access to the NYS Grid for those users who want to concentrate on their disciplinary efforts and prefer to avoid low-level commands. Applications are typically ported to the grid portal through our grid-enabling application template, which provides developers with a template for porting a fairly traditional science or engineering application to our grid-based Web portal. This approach provides the developer with access to various databases, APIs, PHP scripts, HTML files, shell scripts, and so on in order to provide a common platform to port applications and for users to efficiently utilize such applications. The generic template for developing an application provides a well-defined standard-scientific application workflow for a grid application. This workflow includes a variety of functions that include data grid interactions, intermediate processing, job specification, job submission,

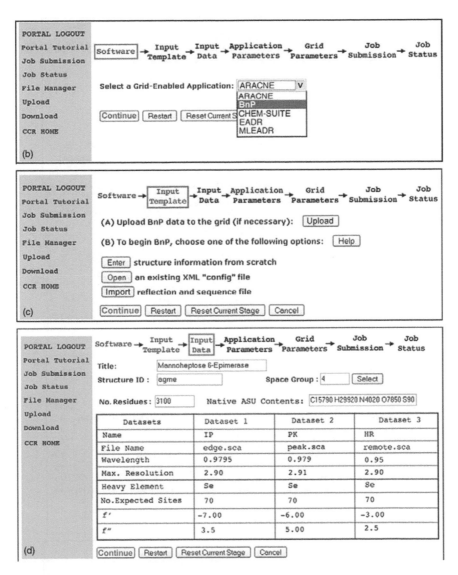

FIGURE 2.1 (*Continued*) (b) Select the software application to be used (*BnP*, a molecular structure determination package). (c) Select execution options and upload files from the local computer. (d) Input additional information about the molecular structure. (*Continued*)

collection of results, runtime status, and so forth. The template provides a flexible methodology that promotes efficient porting and utilization of scientific routines. It also provides a systematic approach for allowing users to take advantage of sophisticated applications by storing critical application and user information in a MySQL database. Most applications have been ported to our grid portal within two weeks.

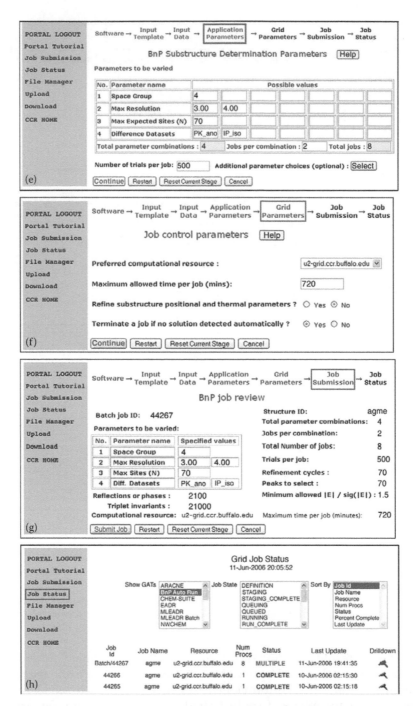

FIGURE 2.1 (*Continued*) (e) Assign values to program parameters that will be varied during execution (additional parameters are accessible from a pop-up window). (f) Supply information controlling job execution for a grid. (g) Review all parameters and start the execution. (h) Check whether the jobs are still running. (*Continued*)

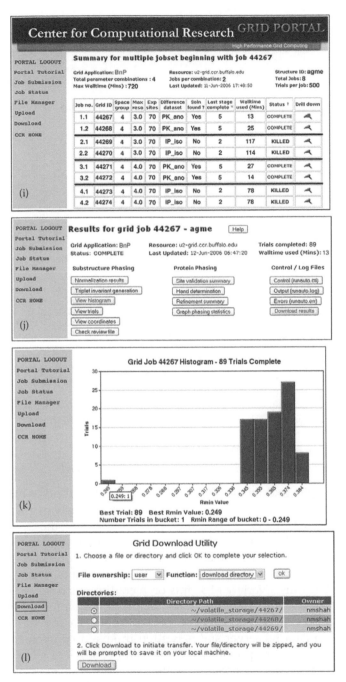

FIGURE 2.1 (*Continued*) (i) Drilldown to check the execution status of individual components of a multiple job set. (j) Drilldown to inspect the results of an individual job. (k) Figure-of-merit histogram for the selected job. (l) Download the output files to the local computer.

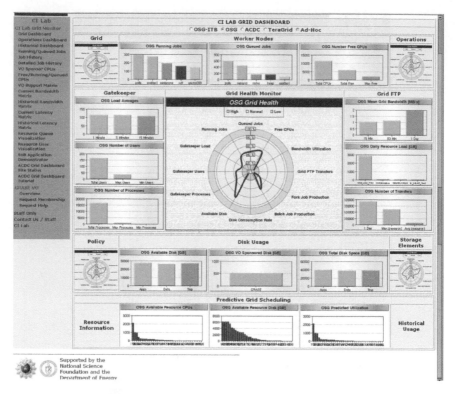

FIGURE 2.2   The dashboard used to display the home page of the grid monitor. Note that the user can click on any of the boxes to drill down for additional information.

2.2.2.2 *Grid Monitoring and the Operations Dashboard*

Our lightweight grid monitoring tool, which can be found at http://osg. ccr.buffalo.edu, is used to monitor resources from a variety of grids, including the NYS Grid, the WNY Grid, the Open Science Grid, the Open Science Grid Testbed, and the TeraGrid to name a few (see Figure 2.2).* With production grids still in their infancy, the ability to efficiently and effectively monitor a grid is important for users and administrators. Our grid monitoring system runs a variety of scripts on a continuous basis, stores information in a MySQL database, and displays the information in an easy-to-digest and easy-to-navigate grid dashboard. The dashboard is served by an Apache server and is written in Java accompanied by PHP scripts. The dashboard provides a display that consists of a radial plot in the center of the main page that presents an overview of an available grid, surrounded by histograms and other visual cues that present critical statistics. By clicking on any of these individual components, the user can drill

---

* More details on information presented in this section may be found in Ruby et al. (2006).

down to examine details on the information in question. These drilldown presentations include dynamic and interactive representations of current and historical information. For example, a user or administrator can easily determine the number of jobs running or queued on every system of any available grid; the amount of data being added or removed from nodes on a grid; as well as a wealth of current and historical information pertaining to the individual nodes, grids, or virtual organizations on an available grid. Our work contributes to the widespread monitoring initiative in the distributed computing community that includes NetLogger (Tierney et al., 1998), GridRM (Baker and Smith, 2002), Ganglia (Massie et al., 2004), and Network Weather Service (Wolski et al., 1999) to name a few.

Our grid operations dashboard, which can be found at http://osg.ccr. buffalo.edu/operations-dashboard.php, was designed to provide discovery, diagnosis, and the opportunity for rapid publication and repair of critical issues to grid administrators (see Figure 2.3). The operational status of a given resource is determined by its ability to support a wide variety of grid services, which are typically referred to as "site functional tests" (Prescott, 2005). Tests are performed regularly and sequentially in order to verify an ever more complex set of services on a node. These results are reported in our operations dashboard in an easy-to-read chart.

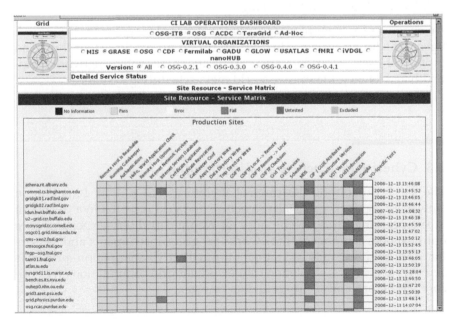

FIGURE 2.3 The operations dashboard shows the status of nodes on a selected grid or for a given virtual organization.

2.2.2.3 An Integrated Data and Computational Grid

The development of data storage solutions for the grid and the integration of such solutions into grid portals is critical to the success of heterogeneous production-level grids that incorporate high-end computing, storage, visualization, sensors, and instruments. Data grids typically house and serve data to grid users by providing virtualization services to effectively manage data in the storage network. The Storage Resource Broker* is an example of such a system. Our Intelligent Migrator is integrated into our grid portal in an effort to provide a scalable and robust data service to NYS Grid users. The Intelligent Migrator examines and models user utilization patterns in an effort to make efficient use of limited storage so that the performance of our physical data grid and the services provided to our computational grid are significantly enhanced. Our integrated data grid provides the NYS Grid users with seamless access to their files, which may be distributed across multiple storage devices. Our system implements data virtualization and a simple storage element installation procedure that provides a scalable and robust system to users. In addition, our system provides a set of online tools for the users so that they may maintain and utilize their data while not having to be burdened with details of physical storage location (see Ruby and Miller, in press, for additional information).

2.2.2.4 Intelligent Grid Scheduling

An ongoing project with very positive, yet preliminary, results is our intelligent scheduling system. This system uses optimization algorithms and profiles of users, their data, their applications, as well as network bandwidth and latency, to improve a grid meta-scheduling system.

2.3 Cyberinfrastructure in Binghamton

The Grid Computing Research Laboratory[†] (GCRL) at Binghamton University conducts cyberinfrastructure and grid research in a variety of areas, including various techniques for Web and grid service performance optimization, component frameworks for grids, instruments and sensors for grid environments, adaptive information dissemination protocols across grid overlays, emulation framework for grid computation on multi-core processors, and secure grid data transfer.

---

* www.sdsc.edu/srb/index.php/Main_Page
† www.grid.cs.binghamton.edu.

2.3.1 Optimizing Web and Grid Services

Very large scientific datasets are increasingly becoming available in XML formats. However, the current XML implementation landscape does not adequately meet the performance requirements of large-scale applications. To address this limitation, we have developed a high-performance XML processor that is designed to meet the needs of emerging grid applications that need to deal with high volume of XML data:

1. We have designed pre-fetching and piped implementation techniques that are ideally suited for processing very large XML documents on multicore nodes (Head and Govindaraju, 2007).
2. We have developed pre-parsing techniques along with static and dynamic partitioning techniques for load balancing (Lu et al., 2006a).
3. We have studied the effect of operating system-level caching on the parsing process of large documents that may be read more than one time.
4. We have conducted experiments to contrast the effect of our techniques on SMP, CMP, and classic uniprocessor CPU architectures.

2.3.2 Optimizing SOAP

SOAP is the standard communication protocol for Web services, which currently has a significant role in grid standards. SOAP's extensibility, platform and language independence, interoperability, and robustness can be offset by poor performance, an important limitation for many grid applications. In particular, converting in-memory binary data to and from ASCII format for transmission represents a significant bottleneck. To address this problem, we have built a SOAP implementation (bSOAP), which includes two effective optimization techniques: (1) differential serialization and (2) differential deserialization.

Differential serialization (Ghazaleh, 2004) can avoid serializing a SOAP message that is similar to previously sent messages by saving a copy of an outgoing SOAP message and then in subsequent deliveries only serializing those fields where values have changed. This approach requires no changes to the SOAP protocol, and the on-the-wire representation is fully compliant with the SOAP specification.

Differential deserialization (Ghazaleh, 2005) uses a different technique, on the server side, to address the same problem. Differential deserialization periodically checkpoints the state of the SOAP deserializer, and computes checksums of SOAP message portions. bSOAP compares checksums against those of the corresponding message portion in previous messages, to identify, with high probability, that the message portions match one another exactly. When message portions match, the deserializer can avoid duplicating the work of parsing and converting the SOAP message contents in that region.

2.3.3 Binary XML

XML provides flexible, extensible data models and type systems for structured data, and has found wide acceptance in many domains. XML processing can be slow, however, especially for scientific data, which has led to the widely held belief that XML is not appropriate for such data. Instead, data are stored in specialized binary formats and are transmitted via workarounds such as attachments and base64 encoding. Although these workarounds can be useful, they nonetheless relegate scientific data to second-class status within the Web services framework, and they generally require yet another API, data model, and type system. An alternative solution is to use more efficient encodings of XML, often known as "binary XML." Using XML uniformly throughout an application simplifies and unifies design and development. We have developed a binary XML format and implementation for scientific data called Binary XML for Scientific Applications (BXSA) (Chiu, 2005; Lu et al., 2006b; Devadithya et al., 2007). Our tests demonstrate that performance is comparable to that of commonly used scientific data formats such as netCDF. These results challenge the prevailing practice of handling control and data separately in scientific applications, with Web services for control and specialized binary formats for data.

2.3.4 Uniform Dynamic Service Deployment

Building, deploying, and evolving Web and grid services are difficult and inconvenient because different containers require that the services they host be written in specific languages to target particular internal interfaces (e.g., for state management). Therefore, services must be built and deployed differently for each hosting environment. Our uniform dynamic service code deployment solution works with three Web services containers (Tomcat, ASP.NET, and a gSOAP based C++ container), and two grid service containers (GT4 and WSRF.NET). Containers receive service code in an XML-based standard intermediate form, and then generate container-specific native code in different languages, without exposing these details to applications and grid services programmers. The dynamically deployed code can access states managed by the hosting container, utilize functionalities exposed by statically deployed services, and communicate with other dynamically deployed modules, running either in the same container or in different containers. Mobile code can run nearly as efficiently as it would if it had been deployed statically, through container-specific mechanisms.

2.3.5 Benchmarking SOAP and XML

Web services-based specifications have emerged as the underlying architecture for core grid services and standards, such as WSRF. XML and SOAP, and are inextricably intertwined with Web services-based specifications, and as a result the design and implementation of SOAP and XML processing tools play a significant role in grid applications. These applications use XML in a wide variety of ways, including workflow specifications, WS-Security-based documents, service descriptions in WSDL, and on-the-wire format in SOAP-based communication. The application characteristics also vary widely in the use of XML messages in their performance, memory, size, and processing requirements. Numerous XML processing tools exist today, each of which is optimized for specific features. To make the right decisions, grid application and middleware developers must, therefore, understand the complex dependencies between XML features and the applications. We have developed a standard benchmark suite for quantifying, comparing, and contrasting the performance of XML and SOAP processors under a wide range of representative use cases (Head et al., 2005, 2006). The benchmarks are defined by a set of XML schemas and conforming documents. To demonstrate the utility of the benchmarks and to provide a snapshot of the XML and SOAP implementation landscape, we have conducted a performance study of many different XML implementations on the benchmarks.

2.3.6 Component-Based Grid Computing

An important challenge in building and deploying high-performance scientific applications is providing a software development model that abstracts the complexity of the runtime environment and simplifies the task of scientists, allowing them to focus on the details of their particular application. A consortium of university and national laboratory researchers launched the CCA Forum to develop a common component architecture (CCA) specification for large-scale scientific computation. The CCA specification defines the roles and functionality of entities necessary for high-performance component-based application development. The specification is designed from the perspective of the required behavior of software components. However, the design and implementation of the framework, choice of communication protocol, and component containment mechanisms have not been formally specified. This has facilitated different research groups to design, develop, and evaluate the use of the same CCA specification to support a wide variety of applications. We have designed and implemented a C++-based CCA framework, XCAT-C++, which can efficiently marshal and unmarshal large datasets, as well as provide the necessary modules and hooks in the framework to meet the requirements of distributed scientific grid applications (Erdil et al., 2005; Govindaraju et al., 2005, 2007).

2.3.7 Grid-Enabled Instruments and Sensor Networks

Technological advances over the last several decades have greatly increased our ability to acquire massive quantities of data via various scientific instruments and sensors. At the same time, however, models, paradigms, and middleware for actually utilizing these data have not kept pace. Research in data curation, provenance, and metadata are crucial for scientists to exploit these large new sources of data. As the source of data, instruments and sensors are a crucial part of the data management process. Previously, sensors and instruments could be treated as off-grid entities, because the volume and rate of data were low. Such systems are no longer adequate, however.

Along with collaborators at Indiana University, the University of Wisconsin, and the San Diego Supercomputer Center, we are developing cyberinfrastructure to more fully integrate instruments and sensors into the grid. We are developing systems for remote X-ray crystallography to allow valuable instruments to be better utilized (Bramley et al., 2006; Devadithya et al., 2005; McMullen et al., 2005). We are also working on systems to better capture data provenance and metadata from sensor systems (Pan et al., 2006; Skovronski and Chiu, 2006).

2.3.8 Grid Emulation Framework for Multicore Processors

The microprocessor industry is rapidly moving toward chip multiprocessors (CMPs), commonly referred to as multicore processors, where multiple cores can independently execute different threads. This change in computer architecture requires corresponding design modifications to programming paradigms, including grid middleware tools, in order to harness the opportunities presented by multicore processors. Naive implementations of grid middleware on multicore systems can severely impact performance due to limitations of shared bus bandwidth, cache size and coherency, and communication between threads. The goal of developing an optimized multithreaded grid middleware for emerging multicore processors will be realized only if researchers and developers have access to an in-depth analysis of the impact of several low-level microarchitectural parameters on performance. None of the current grid simulators and emulators provide feedback at the microarchitectural level, which is essential for such an analysis. We have designed and developed a prototype emulation framework, the Multi-core Grid (McGrid), to analyze and provide insightful feedback on the performance limitations, bottlenecks, and optimization opportunities for grid middleware on multicore systems. McGrid is designed as a highly configurable framework with the ability to provide feedback at various levels of granularity, and for different hardware types and configurations used in heterogeneous grid environments. Using the McGrid framework, we can study the effect of various design options on the efficiency of key microarchitectural parameters for the processing of some representative XML-based grid documents on multicore nodes (Bhowmik et al., 2007).

2.3.9 Protected Grid Data Transfer

Widely available and utilized grid servers are vulnerable to a variety of threats from denial-of-service attacks, overloading caused by flash crowds, and compromised client machines. The focus of our work is the design, implementation, and evaluation of a set of admission control policies that permit the server to maintain sustained throughput to legitimate clients even in the face of such overloads and attacks. We have developed several schemes to effectively, and importantly in an automated fashion, deal with these attacks and overloads. We have implemented these schemes efficiently on an active network adapter-based gateway that controls access to a pool of backend data servers. Performance tests conducted on a system based on a dual-ported active NIC demonstrate that efficient optimization schemes can be implemented on such a gateway to minimize the grid service response time and to improve server throughputs under heavy loads and denial-of-service attacks. Our results, using the GridFTP server available with Globus Toolkit 4.0.1, demonstrate that even in adverse load conditions, the response times can be maintained at a level similar to normal, low-load conditions (Demir et al., 2005, 2007).

2.4 New York State Cyberinfrastructure Initiative

In this section, we present the timeline and chain of events that led to the establishment of a grassroots New York State Cyberinfrastructure initiative. We also discuss the relationship between this grassroots initiative, the NYS Grid, and Miller's Cyberinfrastructure Laboratory.

2.4.1 Timeline

In the early 1990s, Miller's research group began to use commercial parallel computers in an effort to solve the phase problem of crystallography, given atomic resolution X-ray diffraction data. We used a variety of such machines, including a Thinking Machines Inc. CM-2 and CM-5, an Intel iPSC/2, an Encore Multimax, and various other early commercial parallel machines. Simultaneously, Miller's group used a laboratory of Sun workstations and ran their Shake-and-Bake method of structure determination using the RPC (remote procedure call) protocol to employ a master/worker solution of Shake-and-Bake. This was similar to a CONDOR-based approach of stealing background cycles to perform computationally intensive tasks that can be broken up into discrete and loosely coupled tasks.

In 1998, with funding from the Department of Energy, Miller's group worked with Ian Foster's group at Argonne National Laboratory and employed Globus within the Shake-and-Bake framework to continue to refine the Shake-and-Bake method and solve ever-larger molecular structures. In 1999, with funding from the National Science Foundation (NSF), Miller's group initiated a Buffalo-based grid research project that included the Hauptman-Woodward Medical Research Institute, SUNY-Buffalo, and several Buffalo-area colleges. These early efforts represent the genesis of a WNY Grid.

In 2001, an NSF MRI grant funded a significant storage system that was shared by Miller and four other investigators at SUNY-Buffalo. The storage system was quickly incorporated into the grid that we had developed and served as a data repository for Shake-and-Bake results. All of this early work led to an NSF ITR grant that was funded in 2002 and focused on the deployment and efficient implementation of Shake-and-Bake on clusters and grids. In fact, this ITR grant also funded the design, development, deployment, and hardening of the aforementioned Buffalo-based grid (ACDC-Grid), the WNY Grid, and the NYS Grid.

In particular, these funds and the success of the ACDC-Grid and WNY Grid led to the establishment of the NYS Grid in 2004. The number of sites and variety of resources have grown substantially since 2004 and now include a heterogeneous set of compute and storage systems throughout New York State. The institutions include academic and non-profit organizations, although the NYS Grid is not restricted to such institutions. The NYS Grid has been used extensively by the Shake-and-Bake team as well as by numerous other users at SUNY-Buffalo and from the Open Science Grid.

A virtual organization, called GRASE (Grid Resource for Advanced Science and Engineering), was established by Miller's Cyberinfrastructure Laboratory to support general science and engineering users on the Open Science Grid and the NYS Grid. (The Cyberinfrastructure Laboratory applied for and continues to maintain control and oversight of GRASE.)

Given the success of the Buffalo-based ACDC-Grid, the WNY Grid, and the establishment of the NYS Grid, the NSF provided CRI funds in order to provide significant resources to the core sites in Western New York (SUNY-Buffalo, Niagara University, Hauptman-Woodward, and SUNY-Geneseo). More details, including publications, presentations, and the current status of these grids and their associated projects, as described earlier in this chapter, are available at www.cse.buffalo.edu/faculty/miller/CI/.

2.4.2 NYS Grid, Miller's Cyberinfrastructure Laboratory, and the Grassroots NYS Cyberinfrastructure Initiative

In July 2006, a group of interested parties gathered to discuss the possibility of initiating a state-wide effort in cyberinfrastructure at a meeting in Ithaca, New York. The meeting was entitled the "New York State Workshop on Data-Driven Science and Cyberinfrastructure." At this meeting, Miller presented the Cyberinstitute of the State of New York (CSNY), an effort that was being established within SUNY-Buffalo's Center of Excellence in Bioinformatics and Life Sciences (BCOEBLS). Critically, it was disclosed, with permission from the director of BCOEBLS and following the announcement of CSNY that was approved by SUNY-Buffalo's senior vice provost, that CSNY would include: (1) the Center for Computational Research; (2) faculty working in computational science and engineering; (3) faculty working on fundamental problems in cyberinfrastructure, and well as (4) enabling staff, including programmers, GUI designers, and personnel focused on integrating middleware with applications.

Miller also presented an overview of the Center for Computational Research and ongoing efforts in his Cyberinfrastructure Laboratory. At the end of the meeting, the attendees decided to move forward with another meeting to continue to discuss the possibility of a cyberinfrastructure initiative in New York State. In addition, the membership asked Miller during the open discussion at the end of the meeting whether the NYS Grid could serve as the underlying grid architecture for this potential state-wide cyberinfrastructure initiative so that the potential cyberinfrastructure initiative could focus on higher-level issues than a fundamental grid, avoid redundancy, and avoid a duplication of effort. Miller agreed and over the next several months his team brought a number of additional sites online and his group worked closely with other groups around the state to educate system administrators on how to deploy and maintain a node on a grid, how to obtain a certificate (using the GRASE certificate that was maintained by NYS Grid), and other related information.

In September 2006, Miller gave a presentation at the second NYS Cyberinfrastructure meeting. In this talk, he discussed the status of the NYS Grid; gave an overview of grids and cyberinfrastructure in general; and discussed related grids, details of installing a grid node, funding opportunities, and related information (please refer to the appropriate talks on www.cse.buffalo.edu/faculty/miller/talks.shtml). At the end of the September meeting, an inaugural board was voted on by the membership. This board would have a term of one year and would be required at the end of the year to (1) propose a set of by-laws to be voted on by the membership, (2) present a clear mission statement and vision for the cyberinfrastructure initiative, and (3) provide a status report of activities.

Each initial board member was in charge of one of seven working groups: (1) resource provider group; (2) user group; (3) technical working group; (4) communications group; (5) education, outreach, and training group; (6) funding group; and (7) infrastructure group.

Subsequent to this meeting, the group established a Website and chose the name NYSGrid as the name of this organization, causing significant confusion nationally between this grassroots NYS Cyberinfrastructure Initiative and the New York State Grid (NYS Grid), established years earlier by Miller's Cyberinfrastructure Laboratory.

In January 2007, at the third New York State Workshop on Data-Driven Science and Cyberinfrastructure, Miller presented* the status of the NYS Grid to the membership of this grassroots NYS Cyberinfrastructure initiative, along with details and demonstrations. Significant progress had been made in the technical working group and the communications group. Little or no progress was made in establishing or identifying issues with the other working groups. Critical results of this meeting included the recognition that the NYS Grid was stable and serving numerous users from outside of New York State; that high-end users from New York State required assistance in order to move them onto a grid; and that education and outreach to faculty, students, and staff throughout New York State were required. It was also noted that after identifying users within New York State, they should work with the middleware developers and the technical group affiliated with this initiative. However, the head of the user group and the head of one of the campus-based high-end compute centers stated clearly that they did not see the benefit of cyberinfrastructure or grid computing for users. This lack of cohesion in terms of existing opportunities and vision for the future served to shape the identity of this group as it moved forward. Subsequent to this meeting, the middleware group continued to meet and the user group and the user support group merged.

The board of the grassroots NYS Cyberinfrastructure Initiative (NYSGrid.org) then took several controversial steps:

1. They decided to ask the executive director of the group (who was also a member of the board) to resign.
2. They stated that the NYS Grid was now part of NYSGrid.org, the aforementioned grassroots NYS Cyberinfrastructure initiative that grew out of the series of three workshops.
3. They stated that NYSGrid.org was now in complete control of the NYS Grid.

Next, NYSGrid.org board took the following steps:

1. The remaining members of the inaugural board decided to establish their own virtual organization, called NYSGRID, which was sanctioned by the Open Science Grid even though the Open Science Grid was aware of the confusion and problems in New York State, as they had worked with the NYS Grid since its inception.
2. The creation of the NYSGRID virtual organization (VO) necessarily created additional confusion throughout the state and nationally in terms of understanding the distinction between this NYS Cyberinfrastructure Initiative and the NYS Grid, which was established years earlier by Miller's Cyberinfrastructure Laboratory.

---

* www.cse.buffalo.edu/faculty/miller/talks.shtml.

3. One of the members of the NYSGrid.org technical working group sent out an e-mail notice to many of the members of NYSGrid.org stating that the NYSGRID Virtual Organization has been established and that GRASE users were being moved without any discussion or input from GRASE to NYSGRID in terms of their primary VO. This was interesting for many reasons, not least of which was the fact that no new users had yet to emerge owing to the creation of the New York State Cyberinfrastructure Initiative and NYSGrid.org had never approached NYS Grid about this possibility.

4. The chair of the NYSGrid.org board sent out an e-mail to all members notifying them that (a) there were new board members added, (b) that this initiative was now part of the New York State Network for Research and Education (NYSERNet), and (c) that the board was now a steering committee to NYSERNet.

As of the time of writing, the approved one-year term of the initial board has passed, and the board has not proposed by-laws to the membership, has not asked for a vote or input from the membership on critical issues such as its attempt at taking over the NYS Grid, creating a new virtual organization (NYSGRID), adding new members to the board, and placing this organization within a nonprofit organization dedicated to networking within New York State (NYSERNet).

It is also interesting to note that none of the members of the board work in the area of cyberinfrastructure, many are opposed to cyberinfrastructure and grid computing, and many do not think that grid computing currently makes sense in New York State. In fact, most of these members are center directors, CIOs, and the like. However, this group has sent a proposal to the governor of New York requesting funds for the operation of the facilities and the establishment of a center(s) for education, outreach, and training (i.e., they have requested funds for personnel to run machines and do user support).

It will be very interesting to watch the progress of the NYS Cyberinfrastructure Initiative given its very awkward beginning.

2.5 Final Remarks

Cyberinfrastructure is critical to advances in simulation and modeling. In fact, one might claim that cyberinfrastructure is the cornerstone for modern science and engineering, which requires the "creation, dissemination, preservation, and application of knowledge."* In this chapter, we presented

---

* National Science Foundation.

overviews from cyberinfrastructure initiatives at two New York State institutions. We also discussed the efforts of the Cyberinfrastructure Laboratory, which designed and deployed an extensive, heterogeneous New York State Grid, which is well utilized by a worldwide community.

Acknowledgments

The authors would like to thank Mark Green, Jason Rappleye, Tony Kew, Sam Guercio, Adam Koniak, Martins Innus, Dori Macchioni, and Cynthia Cornelius for their contributions to the efforts described in this chapter.

Referencecs

Abu-Ghazaleh, N. and Lewis, M., "Differential deserialization for optimized SOAP performance," in *SC'05 (Supercomputing): International Conference for High Performance Computing, Networking, and Storage*, Seattle, WA, 2005, pp. 21–31.

Abu-Ghazaleh, N., Lewis, M., and Govindaraju, M., "Differential serialization for optimized SOAP performance," in *HPDC-13: IEEE International Symposium on High Performance Distributed Computing*, Honolulu, Hawaii, 2004, pp. 55–64.

Baker, M. and Smith, G., "GridRM: A resource monitoring architecture for the grid," in *Proceedings of Third International Workshop of Grid Computing*, Baltimore, MD, November 18, 2002, pp. 268–273, Lecture Notes in Computer Science 2536 Springer 2002, ISBN 3-540-00133-6.

Berman, F., Hey, A.J.G., and Fox, G.C., Eds., *Grid Computing: Making the Global Infrastructure a Reality*. New York: John Wiley, 2003.

Bhowmik, R., Gupta, C., Govindaraju, M., and Aggarwal, A., presented at the Workshop on Service-Oriented Computing Performance: Aspects, Issues, and Approaches, 2007.

Bramley, R., Chiu, K., Devadithya, T., Gupta, N., Hart, C., Huffman, J., Huffman, K., Ma, Y., and Donald, F., "McMullen: Instrument monitoring, data sharing, and archiving using common instrument middleware architecture (CIMA)," *Journal of Chemical Information and Modeling*, 46 (3), 1017–1025, 2006.

Chiu, K., Devadithya, T., Lu, W., and Slominski, A., presented at the IEEE International Conference on e-Science and Grid Computing, 2005.

Demir, O., Head, M., Ghose, K., and Govindaraju, M., *Grid 2005—Sixth IEEE/ACM International Workshop on Grid Computing*, 2005, pp. 9–16.

Demir, O., Head, M., Ghose, K., and Govindaraju, M., presented at the IEEE International Workshop on Parallel and Distributed Scientific and Engineering Computing, 2007.

Devadithya, T., Chiu, K., Huffman, K., and McMullen, D., presented at the IEEE Workshop on Instruments and Sensors on the Grid, 2005.

Devadithya, T., Liu, Z., Abu-Ghazaleh, N., Lu, W., Chiu, K., and Ethier, S., in *Proceedings of High Performance Computing Symposium,* 2007.

Erdil, D., Chiu, K., Govindaraju, M., and Lewis, M., presented at the Workshop on Component Models and Frameworks in High Performance Computing, 2005.

Erdil, D., Lewis, M., and Abu-Ghazaleh, N., presented at e-Science 2006: The 2nd IEEE International Conference on e-Science and Grid Computing, 2006.

Erdil, D., Lewis, M., and Abu-Ghazaleh, N., presented at PCGrid '07: Workshop on Large-Scale and Volatile Desktop Grids, 2007.

Foster, I. and Kesselmann, C. Eds., *The Grid: Blueprint for a New Computing Intrastructure.* San Francisco: Morgan Kaufmann, 1999.

Govindaraju, M., Head, M., and Chiu, K., in *Proceedings of the 12th Annual IEEE International Conference on High Performance Computing (HiPC),* 2005, pp. 270–279.

Govindaraju, M., Lewis, M., and K. Chiu, K., "Design and implementation issues for distributed CCA framework interoperability," *Concurrency and Computation: Practice and Experience,* 19 (5), 651–666, 2007.

Green, M.L. and Miller, R., "Grid computing in Buffalo," *Annals of the European Academy of Sciences,* New York, 191–218, 2003.

Green, M.L. and Miller, R., "Molecular structure determination on a computational and data grid," *Parallel Computing,* 30, 1001–1017, 2004a.

Green, M.L. and Miller, R., "Evolutionary molecular structure determination using grid-enabled data mining," *Parallel Computing,* 30, 1057–1071, 2004b.

Green, M.L. and Miller, R., "A client-server prototype for grid-enabling application template design," *Parallel Processing Letters,* 14 (2), 241–253, 2004c.

Head, M. and Govindaraju M., presented at the Workshop on Service-Oriented Computing Performance: Aspects, Issues, and Approaches, 2007.

Head, M., Govindaraju, M., Slominski, A., Liu, P., Abu-Ghazaleh, N., van Engelen, R. et al., presented at the SC'05 (Supercomputing): International Conference for High Performance Computing, Networking, and Storage, 2005.

Head, M., Govindaraju, M., van Engelen, R., and Zhang, W., presented at the SC'06 (Supercomputing): International Conference for High Performance Computing, Networking, and Storage, 2006.

Liu, P. and Lewis, M., in *Proceedings of the IEEE/ACM International Conference on Web Services,* 2005, pp. 167–174.

Liu, P. and Lewis, M., presented at ICWS 2007: The 2007 IEEE International Conference on Web Services, 2007.

Lu, W., Chiu, K., and Pan, Y., presented at Grid 2006: The 7th IEEE/ACM International Conference on Grid Computing, 2006a.

Lu, W., Chiu, K., and Gannon, D., presented at the 15th IEEE International Symposium on High Performance Distributed Computing, 2006b.

McMullen, D., Devadithya, T., and Chiu, K., in *Proceedings of the Third APAC Conference on Advanced Computing, Grid Applications and e-Research,* 2005.

Miller, R., DeTitta, G.T., Jones, R., Langs, D.A., Weeks, C.M., and Hauptman, H.A., *Science,* 259, 1430–1433, 1993.

Miller, R., Gallo, S.M., Khalak, H.G., and Weeks, C.M., "SnB: Crystal structure determination via Shake-and-Bake," *Journal of Applied Crystallography,* 27, 613–621, 1994.

Pan, Y., Kimura, N., Zhang, Y., Wu, C., and Chiu, K., in *Proceedings of the Seventh International Conference on Hydroscience and Engineering*, 2006.

Pan, Y., Lu, W., Zhang, Y., and Chiu, K., in *Proceedings of the Seventh IEEE International Symposium on Cluster Computing and the Grid (CCGrid 2007)*.

Prescott, C., Available at: http://osg-docdb.opensciencegrid.org/cgi-bin/ShowDocument?docid=83, 2005.

Rappleye, J., Innus, M., Weeks, C.M., and Miller, R., "SnB v2.2: An example of crystallographic multiprocessing," *Journal of Applied Crystallography*, 35, 374–376, 2002.

Roure, D.D., in *Proceedings of Grid Computing: Making the Global Infrastructure a Reality*, 2003, pp. 65–100.

Ruby, C.L., Green, M.L., and Miller, R., *Parallel Processing Letters*, 16, 2006, 485–500.

Ruby, C.L. and Miller, R., "Effectively managing data on a grid," in *Handbook of Parallel Computing: Models, Algorithms, and Applications*, S. Rajasekaran and J. Reif (eds.), CRC Press, 2007, pp. 46-1–46-36.

Skovronski, J. and Chiu, K., presented at the Eighth International Conference on Information Integration and Web-based Applications and Services 2006.

Tierney, B. et al., presented at the IEEE High Performance Distributed Computing Conference, 1998.

Weeks, C.M., Blessing, R.H., Miller, R., Mungee, R., Potter, S.A., Rappleye, J., Smith, G.D., Xu, H., and Furey, W., "Toward automated protein structure determination: BnP, the SnB-PHASES interface," *Zeitschrift für Kristallographie*, 217, 686–693, 2002.

Weeks, C.M., DeTitta, G.T., Hauptman, H.A., Thuman, P., and Miller, R., "Structure solution by minimal function phase refinement and Fourier filtering. II. Implementation and applications," *Acta Crystallographica*, A50, 210–220, 1994.

Weeks, C.M. and Miller, R., "The design and implementation of SnB v2.0," *Journal of Applied Crystallography*, 32, 120–124, 1999.

Wolski, R., Spring, N., and Hayes, J., "The network weather service: A distributed resource performance forecasting service for metacomputing," *Journal of Future Generation Computing Systems*, 15, 757–768, 1999.

3

Enabling Grids for e-Science: The EGEE Project

Erwin Laure and Bob Jones

CONTENTS

3.1 Introduction

Modern science is increasingly dependent on information and communication technologies, analyzing huge amounts of data (in the terabyte and petabyte range), running large-scale simulations requiring thousands of CPUs, and sharing results between different research groups. This collaborative way of doing science has led to the creation of virtual organizations (VOs) that combine researches and resources (instruments, computing, data) across traditional administrative and organizational domains [1]. Advances in networking and distributed computing techniques have enabled the establishment of such virtual organizations and more and more scientific disciplines are using this concept, which is also referred to as "grid" computing [2–4].

The past years have shown the benefit of basing grid computing on a well-managed infrastructure federating the network, storage, and compute resources across different institutions and making them available to different scientific communities via well-defined protocols and interfaces exposed by a software layer (grid middleware). This kind of federated infrastructure is referred to as "e-Infrastructure" and Europe, through ambitious national research and infrastructure programs [5] as well as dedicated programs of the European Commission, is playing a leading role in building multinational, multidisciplinary e-Infrastructures and has devised a roadmap for a pan-European e-Infrastructure [6].

Initially, these efforts were driven by academic proof-of-concept and testbed projects, such as the European Data Grid project [7], but have since developed into large-scale, production e-Infrastructures supporting numerous scientific communities. Leading these efforts is a small number of large-scale flagship projects, mostly co-funded by the European Commission, which take the collected results of predecessor projects forward into new areas. Among these flagship projects, the EGEE (Enabling Grids for E-sciencE) project unites thematic, national, and regional grid initiatives in order to provide an e-Infrastructure available to all scientific research in Europe and beyond in support of the European Research Area [8].

The project is a multiphase program started in 2004 and expected to end in 2010. EGEE has built a pan-European e-Infrastructure that is being increasingly used by a variety of scientific disciplines. EGEE has also expanded to the Americas and Asia Pacific working toward a worldwide e-Infrastructure. EGEE currently federates some 250 resource centers from 48 countries providing over 50,000 CPUs and several petabytes of storage. The infrastructure is routinely being used by over 5000 users forming some 200 VOs and running over 140,000 jobs per day. EGEE users come from disciplines as diverse as archaeology, astronomy, astrophysics, computational chemistry, earth science, finance, fusion, geophysics, high-energy physics, life sciences, material sciences, and many more.

In this chapter we discuss the challenges and successes EGEE met and provide lessons learned in building the EGEE infrastructure. The remainder of this chapter is organized as follows: The next section discusses the EGEE infrastructure, operation processes, and organizations. This is followed by an overview on the middleware distribution deployed by EGEE. Representative examples of applications using EGEE are then given followed by a discussion of the global grid landscape. Directions of future work with a particular focus on sustainability follow, and we end the chapter with concluding remarks.

3.2 The EGEE Infrastructure

The EGEE infrastructure consists of a set of services and testbeds, and the management processes and organizations that support them. Table 3.1 gives an overview of these entities and we will discuss them in the remainder of this section.

3.2.1 The EGEE Production Service

The computing and storage resources that EGEE integrates are provided by a large and growing number of resource centers (RCs), mostly in Europe

TABLE 3.1

Components of the EGEE Infrastructure

	EGEE Infrastructure
Testbeds and services	Certification testbeds
	Preproduction service
	Production service
Support structures	Operations coordination
	Regional operations centers
	Global grid user support
	Operational security coordination
	Certification and testing
	Middleware distribution release management
	Grid Security Vulnerability Group
	EGEE Network Operations Centre
	Training activities
Policy groups	Joint Security Policy Group
	EUGridPMA (and IGTF),
	Local CA management
	Resource Access Group

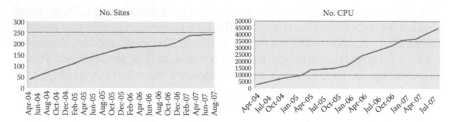

FIGURE 3.1   Development of the EGEE infrastructure.

but also in the Americas and Asia Pacific. Figure 3.1 shows the development of the EGEE infrastructure in number of sites and CPUs from the start of the project in April 2004 up to summer 2007.

EGEE has decided early on that a central coordination of these resource centers would not be viable, mainly because of scaling issues. Hence, EGEE has developed a distributed operations model where regional operations centers (ROCs) take over the responsibility of managing the resource centers in their region. Regions are defined geographically and include up to eight countries. This setup allows adjusting the operational procedures to local peculiarities like legal constraints, languages, best practices, and so on. Currently, EGEE has 11 ROCs managing between 12 and 38 sites as shown in Table 3.2. CERN acts as a special "catch all" ROC taking over responsibilities for sites in regions not covered by an existing ROC. Once the number of sites in that region has reached a significant number (typically over ten), the creation of a new ROC is envisaged.

TABLE 3.2

EGEE Resource Operation Centers

Region	Number of Countries	Number of RCs
UK/Ireland	2	26
France	1	12
Italy	1	36
Northern Europe	8	29
Germany/Switzerland	2	15
Southwest Europe	2	18
Southeast Europe	8	38
Central Europe	7	23
Russia	1	15
Asia-Pacific	8	21
CERN	6	14

EGEE's ROC model has proven very successful, allowing the four-fold growth of the infrastructure (in terms of sites) to happen without any impact on the service.

ROCs are coordinated by the overall Operations Coordination located at CERN. Coordination mainly happens through weekly operations meetings where issues from the different ROCs are discussed and resolved. A so-called "grid operator on duty" constantly monitors the health of the EGEE infrastructure using a variety of monitoring tools and initiates actions where services or sites are not in a good state. EGEE requires personnel at resource centers that manage the grid services offered and take corrective measures in case of problems. Service-level agreements (SLAs) are being set up between EGEE and the resource centers to fully define the required level of commitment, which may differ between centers. The Operations Coordination is also responsible for releasing the middleware distribution for deployment; individual sites are supported by their ROC in installing and configuring the middleware. The EGEE middleware distribution is discussed on p. 000.

Security is a cornerstone of EGEE's operation and the Operational Security Coordination group coordinates the security-related operations at all EGEE sites. In particular, the EGEE security officer interacts with the site's security contacts and ensures that security-related problems are properly handled, security policies are adhered to, and the general awareness of security-related issues is raised. EGEE's security policies are defined by the Joint Security Policy Group, a group jointly operated by EGEE, the U.S. Open Science Grid project (OSG, http://www.opensciencegrid.org), and the Worldwide LHC Computing Grid project (LCG, http://cern.ch/lcg). Further security-related policies are set by the EUGridPMA and IGTF (http://www.gridpma.org/), the bodies approving certificate authorities and thus establishing trust between different actors on the EGEE infrastructure. A dedicated Grid Security Vulnerability Group is proactively analyzing the EGEE infrastructure and its services to detect potential security problems early on and initiate their remedy: as discussed on p. 000 the EGEE security infrastructure is based on X.509 proxy certificates, which allow are to implement the security policies both at site level and on grid-wide services.

The EGEE infrastructure is federating resources and making them easily accessible but does not own the resources itself. Instead, the resources accessible belong to independent resource centers that procure their resources and allow access to them based on their particular funding schemes and policies. Federating the resources through EGEE allows the resource centers to offer seamless, homogeneous access mechanisms to their users as well as to support a variety of application domains through the EGEE VOs. Hence, EGEE on its own cannot take any decision on how to assign resources to VOs and applications. EGEE merely provides a marketplace where resource providers and potential users negotiate the terms of usage on their own. EGEE facilitates this negotiation in terms of the Resource Access Group that brings together application representatives and the ROCs. Overall, some 20% of the EGEE resources are provided by non-partner organizations and while the majority of the EGEE resource centers are academic institutions, several industrial resource centers participate on the EGEE infrastructure to gain operational experience and to offer their resources to selected research groups.

All EGEE resource centers and particularly the ROCs are actively supporting their users. As it is sometimes difficult for a user to understand which part of the EGEE infrastructure is responsible for a particular support action, EGEE operates the Global Grid User Support (GGUS). This support system is used throughout the project as a central entry point for managing problem reports and tickets, for operations, as well as for user, VO, and application support. The system is interfaced to a variety of other ticketing systems in use in the regions/ROCs in order that tickets reported locally can be passed to GGUS or other areas, and that operational problem tickets can be pushed down into local support infrastructures. Overall, GGUS deals on average with over 1000 tickets per month, half of them due to the ticket exchange with network operators as explained below.

All usage of the EGEE infrastructure is accounted and all sites collect usage records based on the Open-Grid Forum (OGF) usage record recommendations. These records are both stored at the site and pushed into a global database that allows the retrieval of statistics on the usage per VO, regions, countries, and sites. Usage records are anonymized when pushed into the global database, which allows retrieving them via a Web interface (http://www3.egee.cesga.es/gridsite/accounting/CESGA/egee_view.php). Figure 3.2 shows an example of the reports generated by the accounting portal. The accounting data are also gradually used to monitor SLAs that VOs have set up with sites.

EGEE also provides an extensive training program to enable users to efficiently exploit the infrastructure. A description of this training program is out of the scope of this chapter and the interested reader is referred to the EGEE training Website (http://www.egee.nesc.ac.uk/) for further information.

An e-Infrastructure is highly dependent on underlying network provisioning. EGEE relies on the European Research Network operated by DANTE and the National Research and Educational Networks. The EGEE Network Operations Center (ENOC) links EGEE to the network operations and ensures that both EGEE and the network operations are aware of requirements, problems, and new developments. In particular, the network operators push their trouble tickets into GGUS, thus integrating them into the standard EGEE support structures.

Further information on EGEE's operation including all the operational documentation can be found at http://egee-sa1.web.cern.ch/egee-sa1/.

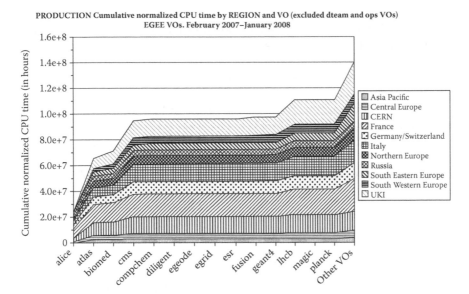

PRODUCTION Cumulative normalized CPU time by REGION and VO (excluded dteam and ops VOs)
EGEE VOs. February 2007–January 2008

FIGURE 3.2 Accounting example.

3.2.2 The Preproduction Service

Before new middleware distributions are deployed on the production service, they are thoroughly Tested on a small-scale system, encompassing some 20 sites, called the preproduction service. This testing is vital for such a large grid, as any errors or faults in deployed software would not only expose many sites to security risks, but would also be time-consuming to fix owing to the diverse properties and geographical spread of the grid sites connected to the EGEE infrastructure. Also, the preproduction sites are in most cases identical to ROCs and thus the preproduction service allows them to gain experiences with new distributions they can then pass on to their resource centers.

The preproduction service works in a truly distributed manner that as far as possible mimics the properties of the production service and allows the project to test new software in the most rigorous way possible prior to deployment across the whole production service. A set of predefined tests is typically run and users can also test new middleware versions against their applications.

3.2.3 Certification and Testing Testbeds

Prior to their deployment on the preproduction service, the middleware components need to be individually tested and certified. EGEE operates a dedicated, distributed infrastructure for these tasks. On this infrastructure, typical deployment scenarios and multiple versions of services are tested. These tests ensure the deployability, basic functionality, and robustness of the middleware components. This infrastructure is not intended for long-term or scalability tests that are performed on the pre-production service.

The certification and testing testbeds need to be quite flexible in their configuration, allowing the quick installation and removal of different versions of middleware services. The usage of virtualization techniques, in particular XEN [9], facilitates this flexibility and helps in keeping the resources used for the testbeds relatively small (in the order of 200 nodes).

3.3 The gLite Middleware

EGEE deploys the gLite middleware [10], a middleware distribution that combines components developed in various related projects, in particular Condor [11], the Globus Toolkit (GT) [12], LCG [13], and VDT [14], complemented by EGEE developed services. This middleware provides the user with high-level services for scheduling and running computational jobs, accessing and moving data, and obtaining information on the grid infrastructure as well as grid applications, all embedded into a consistent security framework.

The gLite middleware is released with a business-friendly open-source license (the Apache 2 license), allowing both academic and commercial users to download, install, and even modify the code for their own purposes. The gLite grid services follow a service-oriented architecture [15], which facilitates interoperability among grid services and allows the services provided by gLite with application-specific services, such as metadata catalogs and meta-schedulers, to be complemented easily, which is a common usage pattern on EGEE. Figure 3.3 gives a high-level overview of the different services, which can thematically be grouped into five service group:

Security services encompass authentication, authorization, and auditing services, which enable the identification of entities (users, systems, and services), allow or deny access to services and resources, and provide information for postmortem analysis of security-related events. The gLite security infrastructure is based on X.509 certificates, involves sophisticated tools for VO management and local authorization, and is currently being integrated with upcoming organization membership services based on Shibboleth [16].

Information and monitoring services provide mechanisms to publish and consume information on the existence of services, their status, as well

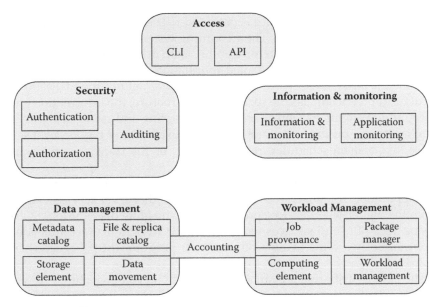

FIGURE 3.3 gLite components.

as user job-related information. The information system is based on hierarchical LDAP servers.

Job management services allow the execution and management of jobs, in particular the computing element provides an abstraction of the physical computing resources and the workload management services is a metascheduler dispatching jobs to appropriate computing elements. A basic package manager allows VOs to install VO-specific software at different sites. The job provenance prototype allows the long-term storage of job-related information and thus the replay of full workflows. Access to compute resources is provided via Condor and GT2 GRAM (which is currently being updated to GT4 GRAM), and work on an HPC-Profile/BES [17] -compliant interface is ongoing [18].

Data services allow the manipulation of and access to data and files. The storage element provides an abstraction from the underlying physical storage location, the file and replica catalogue allows managing the logical namespace and locating files on the infrastructure, and the data management components allow managed transfers of data and provide data management workflows like creating and deleting replicas of files. EGEE has standardized on the SRM [19] interface to storage resources, for which a number of production-level implementations are available. GridFTP is used as underlying protocol for data transfers.

Finally, an accounting service is provided to track the usage of the computing and storage resources. The accounting system uses the OGF-defined usage records facilitating interoperability with other infrastructures.

The focus of gLite is on so-called foundation services. These services provide the basic middleware infrastructure that defines the EGEE infrastructure and encompass the security infrastructure, the information and accounting systems, as well as access to computing and storage resources; that is, the computing and storage element. Examples of higher-level services, like a meta-scheduler, replica catalog, or file transfer service are included in the gLite distribution as well.

As depicted in Figure 3.4, foundation services are used either directly by the application frameworks or indirectly via higher level services. Apart from the higher-level services provided with the gLite distribution, EGEE keeps track of useful services via the so-called RESPECT (Recommended External Software for EGEE Communities) program (http://egeena4.lal. in2p3.fr/).

The gLite distribution is available from the gLite Webpage: http://www. glite.org. The current version 3.1 was gradually released during 2007 and some services of the 3.0 release are still in production. More details on gLite are available in [10] and via the gLite documentation available from the gLite Webpage.

EGEE has taken a conservative approach in defining the gLite composition, avoiding frequently changing cutting-edge software while tracking emerging standards. For a production infrastructure, reliability and scalability are of higher value than the exploitation of the very latest advances

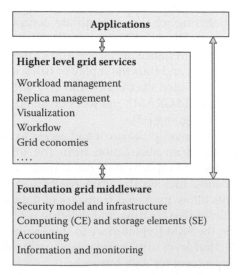

FIGURE 3.4   gLite middleware layers.

that still need time to mature. EGEE is moving toward Web services adhering to WS-Interoperability recommendations wherever possible.

In order to achieve the required reliability and scalability to support the size and workloads of EGEE, a strict software process, involving several stages of testing has been applied. This process, which has been modeled after industry-strength processes, includes software configuration management, version control, defect tracking, and an automatic build system. The experiences gained with the gLite developments are made available to other software projects via the ETICS project [20], which offers the software development tools to a wider community: as described on p. 000, a series of testing infrastructures are used to bring the gLite components to production readiness.

3.4 EGEE Applications

EGEE actively engages with application communities, beginning with high, energy physics (HEP) and biomedicine at the start of EGEE but expanding to support many other domains. These include astronomy, computational chemistry, earth sciences, financial simulations, fusion science, and geophysics, with many other applications currently evaluating the infrastructure. EGEE also works with industrial users and industrial service providers to ensure technology transfer to business. Further details on EGEE's work with business can be found in [21].

The applications running on the EGEE grid are rapidly moving from testing to routine usage, with some communities already running large numbers of jobs on a daily basis. Overall, the infrastructure serves some 140,000 jobs per day. Figure 3.5 shows the development of the usage of the infrastructure, normalized to kilo SpecInt200. There has been a steady increase in usage and although the high-energy physics domain is still the dominant user responsible for some two-thirds of the resource consumption, the usage by other domains is steadily increasing and now equivalent to the total HEP usage a year ago. EGEE allows user groups to federate their distributed resources into a seamlessly accessible infrastructure, thus optimizing their usage; at the same time, resource centers supporting multiple disciplines can offer their services to all the supported disciplines via a common interface provided by EGEE. An overview on applications using EGEE is provided in [3].

The following sections highlight three of the application areas supported by EGEE: high-energy physics and biomedicine, which are the pilot applications of EGEE, and a digital image analysis application that thanks to the service-oriented architecture of EGEE is building an application-specific infrastructure on top of EGEE.

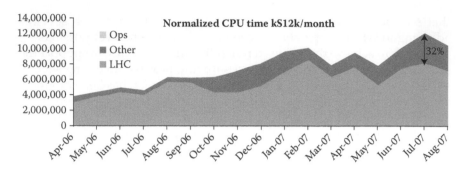

FIGURE 3.5   Usage of the EGEE resources.

3.4.1 Large Hadron Collider

Support for processing data coming from the forthcoming Large Hadron Collider (LHC), a next-generation particle accelerator completing commissioning at CERN, Switzerland, was a major driver in the development of the EGEE program. The LHC experiments are predicted to produce on the order of 15 petabytes of data per year, implying a scale of data processing that made grid technology a natural choice. EGEE works closely with the Worldwide LHC Computing Grid (WLCG) collaboration set up to distribute and process this data. Through the EGEE infrastructure, members of the global physics community will be able to collaborate on the analysis of these data, which it is hoped will find the Higgs boson, an important step in the confirmation of the so-called "standard model" in particle physics.

These LHC experiments also help to stress-test the EGEE infrastructure due to their large-scale requirements. An example of this is the network bandwidth needed for the distribution of the LHC data, where the experiments have each demonstrated capacity of 1 petabyte per month transfer, and aggregate speeds for the LHC experiments have reached 1 gigabyte per second with real workloads (see Figure 3.6).

3.4.2 The WISDOM Initiative

WISDOM (Wide In Silico Docking On Malaria) was launched in 2005 to use emerging information technologies to search for drugs for Malaria and other so-called "neglected" diseases. WISDOM works closely with EGEE, and has made use of the EGEE infrastructure to run a number of large-scale "data challenges." These are tests that screen large-scale databases of molecules for potential treatments for disease. The first of these was carried out in summer 2005, when 42 million compounds were screened for efficacy against a malarial protein in just 6 weeks. This was followed by a data challenge looking for potential treatments for the H5N1

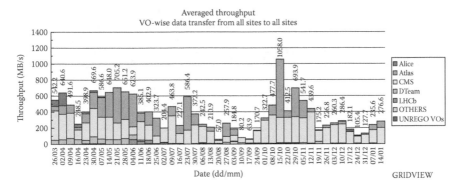

FIGURE 3.6 Average throughput for high-energy physics experiments.

avian influenza, and a second challenge on malaria in winter 2006/2007. This challenge processed up to 80,000 compounds per hour, totaling 140 million compounds tested in the 4 months of the challenge. Furthermore, the grid-enabled testing is much more cost effective than testing in research laboratories. WISDOM estimates the total cost of computing resources, power and time of running such a study on the grid as €3.6 million* on the grid rather than some €350 million in a lab.

3.4.3 Digital Image Analysis

The DILIGENT project (http://www.diligentproject.org/), co-funded by the European Commission, has the objective creating an advanced test-bed that allows virtual e-Science communities to share knowledge and collaborate in a secure, coordinated, dynamic and cost-effective way. DILIGENT integrates grid and digital library technologies, which will lay the foundations for a next-generation e-Science knowledge infrastructure with many different research and industrial applications. Exploiting the service-oriented architecture of EGEE, the DILIGENT services rely on EGEE to provide the computing and (partially) storage resources required for this goal.

In autumn 2007 the DILIGENT team used the EGEE infrastructure to analyze 37 million images from the Flickr online database (http://www.flickr.com) at a rate of over 300,000 images per day. This analysis created some 1120 million text and image objects (nearly 5 terabytes of data) containing more than 150 million extracted features. This collection will be used by the SAPIR project (http://www.sapir.eu) to develop new large-scale

---

* EGEE is not paid for this service. The costs include an estimation of the equipment the tests are run on (which are contributed to the project by the resource centers), and the time of members of EGEE and WISDOM that is funded by the EC or contributed by the WISDOM members.

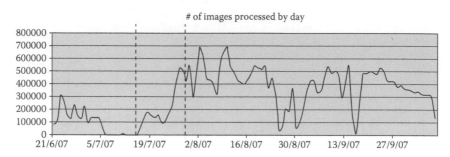

FIGURE 3.7 DILIGENT data challenge.

content-based data retrieval and automatic data classification techniques that combine both text and image content, expanding the limits of conventional search engines, which can only search text associated with images and audio-visual content.

Figure 3.7 shows the images processes per day. Three phases can be identified, separated by vertical lines in the figure: the first phase was used to test the overall setup, particularly interactions with the Flickr database and the EGEE infrastructure; in the second phase, EGEE resource centers have been gradually added until all resource centers supporting this work (10) were able to contribute, which is when the phase third started. The negative dips in the figure correspond to times where major upgrades in the software have been performed.

3.5 The Worldwide Grid Landscape

The vision of grid computing that launched EGEE implies a federation of resources across institutional, disciplinary, and national boundaries, in contrast to "enterprise grids" that often exist within individual companies. The broader vision pursued by EGEE and others requires members of individual grids to be aware of and cooperate with other related grid efforts to work toward interoperability at both European and global levels. Hence EGEE is working closely with other grid projects in Europe and worldwide to implement the vision of a seamless worldwide e-Infrastructure. These projects include the US-based Open Science Grid (OSG, http://www.opensciencegrid.org) and TeraGrid (http://www.teragrid.org) projects, the Japanese NAREGI (http://www.naregi.org) project, and the European DEISA (http://www.deisa.org) project, a number of projects that extend the EGEE infrastructure to new regions (see Table 3.3) as well as (multi)national efforts like the UK National Science Grid (NGS, http://www.ngs.ac.uk), the

TABLE 3.3

Infrastructure Projects Connected to EGEE and Deploying gLite

Project	Webpage	Countries Involved
BalticGrid	http://www.balticgrid.org	Estonia, Latvia, Lithuania, Poland, Sweden, and Switzerland
EELA	http://www.eu-eela.org	Argentina, Brazil, Chile, Cuba, Italy, Mexico, Peru, Portugal, Spain, Switzerland, and Venezuela
EUChinaGrid	http://www.euchinagrid.eu	China, Greece, Italy, Poland, Switzerland, and Taiwan
EUIndiaGrid	http://www.eumedgrid.eu	India, Italy, and United Kingdom
EUMedGrid	http://www.euindiagrid.eu	Algeria, Cyprus, Egypt, Greece, Jordan, Israel, Italy, Malta, Morocco, Palestine, Spain, Syria, Switzerland, Tunisia, Turkey, and United Kingdom
SEE-GRID	http://www.see-grid.eu	Albania, Bulgaria, Bosnia and Herzegovina, Croatia, Greece, Hungary, FYR of Macedonia, Moldova, Montenegro, Romania, Serbia, Switzerland, and Turkey

Northern Data Grid Facility (NDGF, http://www.ndgf.org) in Northern Europe, and the German D-Grid (http://www.d-grid.org).

Together, these projects cover large parts of the world (see Figure 3.8) and a wide variety of hardware systems. In particular, infrastructures like EGEE and OSG primarily federate centers with clusters of commodity PCs whereas DEISA and TeraGrid federate supercomputing centers. However, the technologies deployed by the different projects vary widely exposing different interfaces and protocols and offer different service characteristics.

The goal of the Open Grid Forum (OGF, http://www.ogf.org) is to harmonize these different interfaces and protocols and develop standards that will allow interoperable, seamlessly accessible services. One notable effort in this organization where all of the above-mentioned projects work together is the Grid Interoperation Now (GIN) [22] group. Through this framework, the mentioned infrastructures work to make their systems interoperate with one another through best practices and common rules. GIN does not intend to define standards but instead to provide the groundwork for future standardization performed in other groups of OGF. Already this has led to seamless interoperation between OSG and EGEE, allowing jobs to freely migrate between the infrastructures as well as seamless data access using a single sign-on. Similar efforts are currently ongoing with NAREGI, DEISA, and NDGF. As part of the OGF GIN effort, a common service discovery index of nine major grid infrastructures worldwide has been created, allowing users to discover the different services available

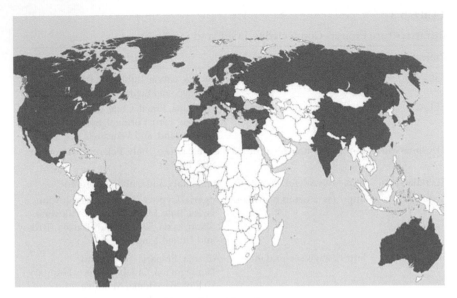

FIGURE 3.8 Countries covered by major grid projects.

from a single portal [23]. A similar effort aiming at harmonizing the policies for gaining access to these infrastructures is underway.

3.6 Future Work

EGEE is currently preparing for its next phase running from 2008 to 2010, during which it will maintain, enhance, and simplify the use of the infrastructure for an increasing range of researchers in diverse scientific fields. Particular focus will be laid on simplifying the operation of the infrastructure through improved automation (such as automatic installation and configuration, monitoring, error recovery, etc.) as well as through improving the stability and scalability of the gLite services. In this context the scalability of the information system, a common authorization system, and the scalability of the job and data management systems will be focused on. In addition, the usability of the system will be increased through collaborations with application groups, the provision of high-level access mechanisms (portals), and the increased usage of established standards as well as increased user support.

The usage and enforcement of SLAs between VOs and sites will increase thanks to the sophisticated EGEE accounting infrastructure. In addition, SLAs will be set up between EGEE and sites as well, clearly defining the level of service a site is contributing to EGEE; this includes, for instance, the definition of support hours, services run by sites, up-time of services, mean intervention times, and so on. The EGEE monitoring tools will be used to enforce these SLAs.

While EGEE's success has been clearly demonstrated, its status as a series of short-term projects raises the question of the future of the infrastructure it has built. Already today, many scientific applications depend on the EGEE production grid infrastructure, new scientific collaborations have been formed thanks to the advancements of grids, and business and industry are rapidly adopting grid and virtualization technologies. However, a long-term plan for the availability of the grid service model is needed to protect the investments of user communities.

As a consequence, EGEE has performed a series of activities together with national European grid efforts to establish the production grid infrastructure as a sustained, permanent research tool for science and engineering. The initial outcome of these discussions is an operational model based on National Grid Infrastructures (NGIs) coordinated by a body tentatively named the European Grid Initiative (EGI). Driven by the needs and requirements of the research community, it is expected that this setup will enable the next leap in research infrastructures.

The details of this structure are currently being defined by the EGI Design Study (EGI_DS) project that started in September 2007. Initial considerations have been made in [24] building on the experiences of EGEE and other national and international grid efforts. Particularly, the NGIs are expected to provide the building blocks of the infrastructure, much like the current ROCs in EGEE. However, from an organizational point of view NGIs are not yet fully operational today. In fact, while most countries have established NGIs to enable the construction and operation of a national grid, the diversity between these countries is very high. They range from single institutes that act as a point of contact for the national grid community to operational national grid infrastructures; a wide variety of levels of maturity can be observed.

It is therefore essential that EGI supports the development and progress of NGIs as much as possible, that EGI defines the workload distribution between the individual NGIs and the EGI together with the appointed NGI representatives, and that the NGI representatives drive the establishment of the EGI organization. Another contributor to the EGI is the e-Infrastructure Reflection Group (eIRG), which has provided the political roadmap for the establishment of EGI [25,26]. A first blueprint of the EGI structure is expected in summer 2008 and will be available via the EGI_DS Website (http://web.eu-egi.org).

One of the major technical challenges will be the provision of a continuous grid infrastructure during the transition from operations funded by projects like EGEE and others to the NGI/EGI structure. EGEE is preparing for that through its decentralized model, which will be further refined in the coming two years, and we expect a significant overlap time of both structures to ensure a smooth transition.

3.7 Conclusions

Since the launch of the EGEE program in 2004, the project has made strides forward in both delivering a world-class production grid infrastructure and strengthening the grid field in general. Through provision of a reliable, scalable production infrastructure, high-quality middleware, and well-developed support services, EGEE has attracted a wide range of users from the research and business communities, which are now using the infrastructure on a daily basis and are critically dependent on the existence of a production grid infrastructure. The EGEE infrastructure will continue to expand and optimize the services it provides to its users while contributing to efforts turning the currently short-term financed infrastructure operation into a sustainable model, based on national grid infrastructures coordinated by a European body.

Acknowledgments

The authors would like to thank all the members of EGEE for their enthusiasm and excellent work that made the EGEE vision reality. This work was co-funded by the European Commission through the EGEE-II project, contract number INFSO-RI-031688.

References

1. I. Foster, C. Kesselman, and S. Tuecke, "The anatomy of the grid," *International Journal of High Performance Computing Applications*, 15, 200–222, 2001.
2. I. Foster and C. Kesselman (Eds.), *The Grid: Blueprint for a New Computing Infrastructure*, San Francisco, CA, Morgan Kaufmann, 2003.
3. V. Floros, F. Harris, M. Kereksizova, A-L. Goisset, and F. Harris (Eds.), "EGEE user forum—book of abstracts," Technical Report EGEE-TR-2007-002, November, 2007. Available at: http://cdsweb.cern.ch/record/1068842.

4. M. Lamanna and E. Laure (Eds.), "The EGEE user forum—experiences in using grid infrastructures," *Journal of Grid Computing*, 6 (1), 1–123, 2006.
5. GridCoord "Deliverable 3.1.1—survey of funding bodies and industry," EU Project Number IST-2003-511618, March, 2005. Available at: http://www.gridcoord.org/grid/portal/information/public/D.3.1.1_V.2.2_221105.doc.
6. e-Infrastructure Reflection Group "e-Infrastructures roadmap (2005–2006)," December, 2005. Available at: http://www.e-irg.org/roadmap/eIRG-roadmap.pdf.
7. F. Gagliardi, B. Jones, and E. Laure, "The EU DataGrid project: Building and operating a large scale grid infrastructure," in B. Di Martino, J. Dongarra, A. Hoisie et al. (Eds.) *Engineering the Grid: Status and Perspective.* American Scientific Publishers, January 2006.
8. For details see http://www.eu-egee.org.
9. P. Barham et al., "XEN and the art of virtualization," in *Proceedings of the Nineteenth ACM Symposium on Operating Systems Principles*, Bolton Landing, NY, 2003.
10. E. Laure, S. Fisher, A. Frohner, et al., "Programming the grid with gLite," *Computational Methods in Science and Technology*, 12 (1), 33–45, 2006.
11. D. Thain, T. Tannenbaum, and M. Livny, "Distributed computing in practice: The condor experience," *Concurrency and Computation: Practice and Experience*, 17 (2–4), 323–356, 2005.
12. I. Foster, "Globus toolkit version 4: Software for service-oriented systems," in *IFIP International Conference on Network and Parallel Computing*, LNCS 3779, 2–13, 2005.
13. For details see http://lcg.web.cern.ch/LCG/.
14. For details see http://vdt.cs.wisc.edu/.
15. D. Sprott and L. Wilkes, "Understanding service-oriented architecture." Available at: http://msdn.microsoft.com/library/default.asp?url=/library/enus/dnmaj/html/aj1soa.asp.
16. For details see http://shibboleth.internet2.edu/.
17. For details see https://forge.gridforum.org/sf/projects/ogsa-bes-wg.
18. For details see http://grid.pd.infn.it/cream/.
19. A. Shoshani, A. Sim, and J. Gu, "Storage resource managers—essential components for the grid," in J. Nabrzyski, J. Schopf, and J. Weglarz (Eds.), *Grid Resource Management State of the Art and Future Trends,* Kluwer, Dordrecht, The Netherlands, pp. 321–340, 2003.
20. M. Begin, G.D-A., Sancho, A. Di Meglio, E. Ferro, E. Ronchieri, M. Selmi, and M. Żurek, "Build, configuration, integration and testing tools for large software projects: ETICS," in *Proceedings of RISE 2006, International Workshop on Rapid Integration of Software Engineering Techniques*, September 13–15, 2006, University of Geneva, Switzerland, LNCS 4401, pp. 81–97, 2007.
21. S. Holsinger, "Annual report of the industry forum," EGEE-II Deliverable DNA2.5.1, April, 2007. Available at: https://edms.cern.ch/document/819325/.
22. For details see https://forge.gridforum.org/sf/projects/gin.
23. For details see http://forge.gridforum.org/sf/wiki/do/viewPage/projects.gin/wiki/GINInfoWiki.
24. EGI Prep Team, "The future European grid infrastructure—toward a common sustainable e-infrastructure," paper prepared for the EGI Workshop,

Munich, Germany, February, 2007. Available at: http://www.euegi.org/public/EGI Vision (v1.1).pdf.

25. M. Read, "A European vision for a universal e-infrastructure for research," e-IRG (e-Infrastructure Reflection Group, London, December, 2005. Available at: http://www.e-irg.org/meetings/2005-UK/A_European_vision_for_a_Universal_e-Infrastructure_for_Research.pdf.

26. e-IRG "Report of the e-infrastructure reflection group task force on sustainable e-infrastructures," Vienna, June, 2006. Available at: http://www.e-irg.org/publ/2006-Report_e-IRG_TF-SEI.pdf.

4

ChinaGrid and Related Dependability Research

Xuanhua Shi and Hai Jin

CONTENTS

4.1 Introduction

In recent years, grid computing has become very popular for its potential of aggregating high-performance computational and large-scale storage resources that distribute over the Internet. According to [1,2], grid computing is "resource sharing and coordinated problem solving in dynamic, multi-institutional virtual organizations." The purpose of grid computing is to eliminate the resource islands in the application level, and to make computing and services ubiquitous.

The prerequisites for grid computing lie in three aspects: network infrastructure, wide-area distribution of computational resources, and continuous increasing requirement for resource sharing. Nearly all the existing grid computing projects are based on existing network infrastructure, such as the UK e-Science Programme [3], Information Power Grid (IPG) [4], and TeraGrid [5]. In TeraGrid, the five key grid computing sites are interconnected via 30 or 40 Tb/s fast network connections. The ChinaGrid project [6], which will be discussed in detail in this chapter, is also based on a long running network infrastructure, which is called China Education and Research Network (CERNET) [7].

Grid computing provides new potentials for distributed computing, as well as many challenges. Because of the specific features of grids (such as dynamic, large-scale, wide-area distribution), there are many challenges in constructing dependable grid services; for example: (1) power failure leading to power loss of one part of the distributed system; (2) physical damage to the grid computing fabric as a result of natural events or human acts; and (3) failure of system or an application software leading to loss of service. Owing to the diverse failures and error conditions in the grid environments, developing, deploying, and executing applications over the grid is a challenge. Dependability is a key factor for grid computing, and dependability has great influence on the practice of grid platform. To target the practical vision of the ChinaGrid, dependability research in the ChinaGrid is very active, covering the platform through to applications.

The remainder of this chapter is organized as follows: the next section gives an illustration of the vision and mission of the ChinaGrid project; this is followed by an illustration of the ChinaGrid Support Platform; the application platforms in the ChinaGrid are presented next; and dependability research is discussed in the final section, covering grid monitoring, hardware and middleware fault tolerance, grid fault detection, and application fault tolerance.

4.2 ChinaGrid Project: Vision and Mission

In 2002, China's Ministry of Education (MoE) launched the largest grid computing project in China, called the ChinaGrid project, aiming to provide a nationwide grid computing platform and services for research and education purpose among 100 key universities in China. The vision for the ChinaGrid project is to deploy the largest, most advanced and most practical grid computing project in China or even worldwide.

The underlying infrastructure for the ChinaGrid project is the CERNET, which began to run from 1994, covering 800 more universities, colleges, and institutes in China. Currently, it is the second largest nationwide network in China. The bandwidth of the CERNET backbone is 2.5 Gbps, and it is connected by seven cities called local network centers. The bandwidth of the CERNET local backbone is 155 Mbps.

The ChinaGrid project is a long-term project with three different stages. The first stage period was from 2002 to 2005, covering 12 top universities in China. They are: Huazhong University of Science and Technology (HUST), Tsinghua University (THU), Peking University (PKU), Beihang University (BUAA), South China University of Technology (SCUT), Shanghai Jiao Tong University (SJTU), Southeast University (SEU), Xi'an Jiaotong University (XJTU), National University of Defense Technology (NUDT), Northeastern University (NEU), Shandong University (SDU), and Sun Yat-Sen University (ZSU). The focus for the first stage of the ChinaGrid project was on platforms and applications in the computation grid (e-Science). These applications covered all scientific disciplines, from life sciences to computational physics. The second stage of the ChinaGrid project was from 2005 to 2007, covering 20–30 key universities in China. The focus extended from computational grid applications to information service grid (e-Info), including applications for distance learning grid,

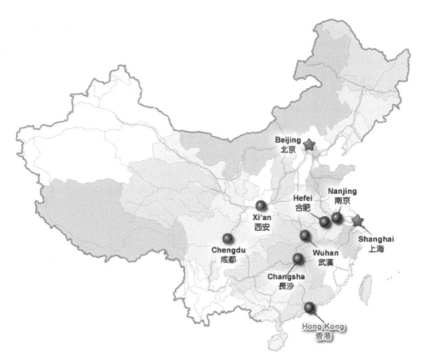

FIGURE 4.1   ChinaGrid member universities.

digital Olympic grid, and so on. The third stage will, from 2007 to 2010, extend the coverage of the ChinaGrid project to all 100 key universities. The focus of third-stage grid applications will be even more diverse, and include instrument sharing (e-Instrument). Figure 4.1 shows the current distribution of the ChinaGrid members.

4.3 ChinaGrid Support Platform

The underlying common grid middleware for the ChinaGrid project is called the ChinaGrid Support Platform (CGSP) [8], supporting all grid applications of the three different stages, which are e-Science, e-Info, and e-Instrument. CGSP integrates all kinds of resources in education and research areas, makes the heterogeneous and dynamic nature of resource transparent to the users, and provides high-performance, highly reliable, secure, convenient, and transparent grid services for the scientific research. CGSP provides the ChinaGrid service portal and a set of development environment for deploying various grid applications. The current version,

CGSP 2.0 released in April 2006, is based on the core of Globus Toolkit 4.0.1. CGSP is WSRF [9] and OGSA [10] compatible.

4.3.1 Functions

The main functions provided by CGSP2 are presented in this section.

1. Job execution: CGSP2 aims to provide a powerful platform to execute various jobs, including a legacy program [11], Web service, WS-Resource, and composite services that are actually defined workflows. For the legacy program, CGSP provides an efficient runtime environment which supports both common command-line programs and MPI programs. And these programs can be invoked not only with a simple Web service interface (WSDL) [12] but also with a JSDL description [13]. Moreover, CGSP also provides a uniform way to invoke Web services and WS-Resource. Besides the ability to execute these atomic jobs, CGSP provides a workflow balancer server to execute composite jobs defined by BPEL [14] effectively and efficiently, through which Web services and WS-Resource can be orchestrated to finish a complex job.

2. Data space: CGSP2 also aims to provide a virtual storage space and a reliable data transfer mechanism with high performance. The purpose of the storage virtualization is to integrate the heterogeneous storage resources distributed in the grid environment. From the perspective of a grid user, the data management component provides a virtual file system. One can execute operations on the virtual file system as he/she can do on local file systems. The requirements of the data transfer in CGSP are efficiency, manageability, and reliability. We plan to construct a mechanism implemented by WSRF service so that each process of data transfer can be managed as a WSRF resource. The creator of the resource is able to start, suspend, restart, stop the transfer process, and get its up-to-date status.

3. Heterogeneous database: CGSP2 aims to enable users in the grid environment to access the services provided by various heterogeneous database more conveniently. It provides unified accessing methods, which makes data integration possible and brings heterogeneity transparency, naming transparency, and distribution transparency. Hence, CGSP is capable of being a supporting platform for accessing the very large and basic heterogeneous databases.

4. User interface: CGSP2 aims to provide a user-friendly interface to utilize CGSP functions. CGSP Portal provides Webpage interface for the end user. From the Web portal, users can browse services and resources in the grid, view users' data space, upload and

download files with http or GridFtp [15], deploy their applications to the ChinaGrid that is managed by CGSP, and submit jobs to applications and services.

5. Programming interface: CGSP2 aims to provide a powerful programming interface to develop applications in a grid environment. CGSP2 provides an MPI-like programming interface and a bunch of grid services, such as service publishing and querying services, resource monitoring services, and data transferring services, to help developers to deliver distributed, high-performance, large-scale applications.

4.3.2 Building Blocks

The detail software building blocks for CGSP are shown in Figure 4.2. There are five building blocks in CGSP 2.0 as follows:

1. Grid portal: The grid portal is the entrance for the end user to use grid services. By using the grid portal, users can submit their jobs, monitor the running of jobs, manage and transfer data, query the grid resource information. The grid portal also provides other

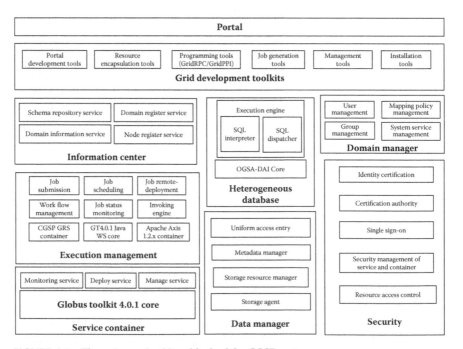

FIGURE 4.2   The software building block of the CGSP system.

facilities such as user management and accounting of grid resource usage.

2. Grid development toolkits: These toolkits provide a toolkit to pack the resource to grid services, the deployment and management toolkit for grid, and programming model to deploy complex grid application in grid environment.

3. Information center: This is responsible for the management of various resources within the grid environment, provides a global resource view and grid information services, and updates grid resource information in a real-time manner. The main purpose is to provide real-time information of various grid resources for end users and other modules in the grid environment.

4. Grid management: This provides basic support for various jobs in the grid environment. It consists of the following five parts:

   • Service container: Provides a grid service installation, deployment, running, and monitoring environment on each node in grid environment. It also provides necessary support to monitor the real-time resources status of each grid node.

   • Data manager: Responsible for the management of various storage resources and data files in the grid environment. It provides a global file view, so that users can access various data files transparent.

   • Heterogeneous database: Based on OGSA-DAI [16,17], CGSPHDB (CGSP Heterogeneous Database) provides uniform access to heterogeneous databases, which enables CGSP users to obtain more data services provided by different kinds of data sources.

   • Execution management: The most complicated and most kernel module of CGSP. Based on the information center, data management, and CGSPHDB, it provides support for job management, scheduling, and monitoring for end users' computational task, so that data and resources can be accessed transparently within grid and cooperative working among distributed resources.

   • Domain manager: The ChinaGrid is organized into a domain. A domain refers to an independent grid system to provide services to the others. A domain can be a specialized grid or a regional grid. The domain manager is mainly responsible for user management, logging, accounting within domain, and interacting with other domains. It allows the domain administrator to easily manage the users, services, and resources within the domain, and interactive policies among domains.

5. Grid security: Provides user authentication, resources and services authorization, encrypted transmission, and the mapping between users to resource authorization.

4.4 Grid Computing Application Platforms in the ChinaGrid

Besides the development of CGSP, another major task for the ChinaGrid project is the deployment of the grid computing application platform and some pilot grid applications, making the ChinaGrid very unique compared to other existing grid projects. In the first stage of the ChinaGrid project, five ongoing main grid computing application platforms were under development: the bioinformatics grid [18], image grid [19], computational fluid dynamics grid [20], course online grid [21], and massive data processing grid [22]. The first three grids are actually computationally oriented. The course online grid is a very unique grid, which is the first step to an information service grid. The last one, the massive data processing grid, is the data grid in nature [23]. These grid application platforms will now be discussed in detail.

4.4.1 Bioinformatics Grid

Bioinformatics merges biology, computer science, and information technology into one discipline. The ultimate goal of the field is to enable the discovery of new biological insights. It mainly focuses on grasping the biological meaning of plentiful biological data. Bioinformatics uses databases, data processing methods, and software to get results through mass computation. Each bioinformatics research institute holds their own computing facilities, software resources, and storage devices. They set up their own research platform independently. Although most of these resources are connected to the Internet, they are only used by their owners and the utilization rate is low. At the same time, many researchers in Chinese universities have no such research resources to use.

The bioinformatics grid integrates heterogeneous large-scale computing and storage facilities within the ChinaGrid to provide bioinformatics supercomputing services for bioinformatics researchers through a user-friendly Web interface. Only two steps are required for users to use bioinformatics grid through this portal: first inputting the computing requests according to the submission form from Web page, and second getting the computing results from bioinformatics grid. Where the requests are submitted, how the computing requests are allocated, and the monitoring and management for computing tasks are all completed by the bioinformatics grid itself.

When the bioinformatics grid server receives the computational requests from the client, it locates a suitable node in the grid to perform the mathematical computation according to the users' requirement and task allocation rule, or integrates a virtual supercomputer to perform the larger computational requests from users. There are three key modules in bioinformatics grid. They are as given below.

4.4.1.1 Management of Heterogeneous Resources

In the bioinformatics grid, resource management consists of both resource abstraction and resource organization. In order to integrate the diverse resources in a uniform and seamless way, all hardware and software resources are packaged into certain services. A set of attributes is abstracted to describe the services (such as OS information, computation power, software category, software name). Objectives of OMR (organization and management of resources) are as follows:

- Integrated heterogeneous computer
- Make bioinformatics computing tools grid-enabled
- Basic communication mechanisms: These mechanisms must permit the efficient implementation of a wide range of communication methods, including message passing, remote procedure call, and multicasting
- Process creation: This component is used to initiate computation on a computing grid node once it has been located and allocated. This task includes setting up executables, creating an execution environment, starting an executable, passing arguments, integrating the new process into the rest of the computation, and managing termination and process shutdown

4.4.1.2 Monitoring

The bioinformatics grid provides uniform protocol and interface specification for grid resources monitoring. Monitoring is used to identify system bottlenecks. The system can be adjusted and optimized by using the information gathered by monitoring. Objectives of grid resources monitoring (GRM) in the bioinformatics grid are as follows:

- Maintaining the latest copy of crucial data in the root node
- Task scheduling
- Monitoring bioinformatics grid
- Keeping log of operations, errors, and statistics

4.4.1.3 Integration of Heterogeneous Bioinformatics Computing Tools

At present, many bioinformatics computing tools have been designed for different application purposes. All of these tools are installed into the grid hardware environment and are used through the Web interface easily and widely. In the bioinformatics grid, objectives of integration of heterogeneous bioinformatics computing tools (IBCT) are as follows:

- Classification and installation of shared bioinformatics computing and research tools
- Integration of bioinformatics database
- Pipeline running control of computing flow
- Graphic user interface
- Billing system

4.4.2 Image Grid

The image grid is a grid image-processing platform based on the ChinaGrid infrastructure, which is an integrated environment hiding inner heterogeneous resources and dynamic information. It not only realizes cooperative characteristics, but also affords a secure and transparent image-processing environment with correlative services [24,25]. The remarkable characteristics of the image-processing platform are: application development, runtime environment, and remote visualization.

The image-processing grid application development and runtime environment is based on Web service and grid service technology, in which image processing toolkits or software are deployed as Web services or grid services components. These services are usually deployed redundantly for all kinds of reasons such as fault-tolerance, availability, and performance. The concept of virtual services, which are the abstract of several physical services, is introduced to implement common functions and interfaces. Therefore during the development of image processing grid application, users only care about the virtual services of image applications, which then convert into different physical services running on the ChinaGrid through the service-scheduling system.

Various image-processing functions are afforded and extended in the image grid environment with much higher performance and QoS. Currently, the following applications are integrated into the image grid as pilot applications: reconstruction of digital virtual human body, image processing of remote sensing, and medical image diagnoses.

The Virtual Human Project (VHP) is the creation of complete, anatomically detailed, three-dimensional representations of normal male and female human bodies. The long-term goal of VHP is to describe perfectly the conformations and functions of gene, protein, cell, and organ so as to achieve the whole accurate simulation of body information.

The basic realization of a digital virtual human is to identify and classify inner viscera, rebuild the edge profile data of outer body and three-dimensional profile data of inner viscera, construct a high-precision digital human grid, and use these data to achieve the visualization of a human body. In addition to the massive amount of body slices (a suit of raw data requires 20 GB storage space), the amount will increase greatly for mesh construction. Only a computer resource grid can afford this processing. Before we analyze and dispose these data, some optimized resource scheduling algorithms must be brought forward, such as the highly precise algorithm for contour extraction and highly precise parallel algorithm for reconstructing three-dimensional grids of the digital virtual human.

Remote sensing can be defined as the collection of data about an object from a distance. Geographers use the technique of remote sensing to monitor or measure phenomena found in the Earth's lithosphere, biosphere, hydrosphere, and atmosphere. Remote sensing imagery has many applications in mapping land use and cover, agriculture, soil mapping, forestry, city planning, archeological investigations, military observation, and geomorphological surveying, among other uses.

Medical imaging processes are dedicated to the storage, retrieval, distribution and presentation of images, which are handled by various modalities, such as ultrasonography, radiography, magnetic resonance imaging, positron emission tomography, and computed tomography. It consists of a central server that stores a database containing the images. A doctor and his patients can get more reliable and precise medical images (such as X-ray slices, CT, and MR) with more quick, exact, and simple modes, so as to increase the rate of inchoate morbid detection and diagnoses and living opportunity.

4.4.3 Computational Fluid Dynamics Grid

The computational fluid dynamics (CFD) grid provides the infrastructure to access different CFD applications across different physical domains and security firewalls. By defining a standard CFD workflow and general interfaces for exchanging mesh data, the platform facilitates interoperating between CFD applications. An application development package (ADP) is being developed and an application development specifications (ADS) is being defined, shipping with the platform to make migrating CFD applications feasible. The major function of ADP is to schedule jobs, recover from faults, and discover services effectively.

The CFD platform includes two levels: a high level for service-oriented applications (SOA) and a low level for parallel computing. The physical architecture of the CFD platform is shown in Figure 4.3.

In the high level, all services are registered in the main hub or subhubs. The function of the main hub is to provide job definition. Service requestor visits the main hub for the service interface definition and finds the

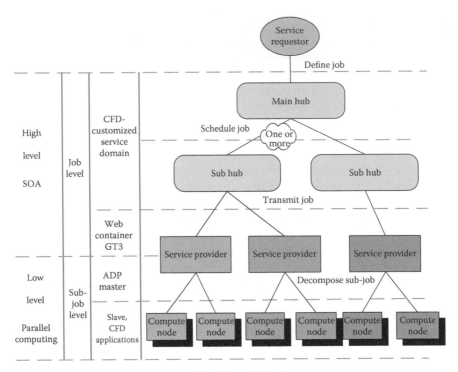

FIGURE 4.3   Physical architecture of the CFD grid platform.

invocation entrance. The main hub is also responsible for the task scheduling and resource allocation. It collects all the service information from subhubs and then find the best resource solution for a given service requestor. The subhubs also have those functions that the main hub has, but within a smaller area. Subhubs deliver the subtasks (jobs) to the specific service provider, and then collect the results. All the services are provided as a Web service. In this way (hierarchy architecture and general service interface), the SOA has a good scalability.

In the low level for parallel computing, all kinds of parallel environments could be employed. In this system, MPICH2 is used as the low-layer parallel environment. The service provider decomposes tasks that are distributed by subhubs and then sends them to computing nodes via an MPI mechanism. This feature enables us to generally utilize the distributed computing nodes to provide scalable and stable services.

4.4.4 Course Online Grid

Course Online Grid (*realcourse* in short) is a video stream service supported by a collection of physical servers distributed all over China [26]. Since its birth in the April of 2003, *realcourse* was in nonstop operation for more than 18 months, in spite of some of the servers being up-and-down and in-and-out from time to time. At the time of writing, *realcourse* is composed of 20 servers spanned over 10 major cities distributed on CERNET with 1500 hours quality course video, which has grown from four servers initially within the Peking University campus. As a result, we normally observe 20–100 online users at any time. In turn, these user activities help us to bring the service to a level of robustness.

The 1500 hours of video are from different universities within the ChinaGrid. They are uploaded to different servers with a permanent backup copy created in a server other than the original one, and flow from server to server on demand. If one server is down for some reason, its duty is automatically taken up by another server(s). Once it is up again, a maintenance session will occur automatically to bring it up to a consistent state, in case any transactions occurred during its down time (e.g., some more videos were uploaded to the system).

Realcourse is a successful application of distributed computing technologies in a geographically wide area. In contrast to some traditional distributed fault-tolerant services, *realcourse* emphasizes giving clients access to the service—with reasonable response times—for as much of the time as possible, even if some results do not conform to sequential consistency.

Realcourse possesses some distinguished characteristics. It is aimed at non-critical application of video on demand and file storage in which temporal inconsistency is acceptable only if consistency of the whole system is finally reached. By eliminating the "update" operation, *realcourse* greatly reduces the consistency requirement. The length of exchanged message among servers is kept constantly small and thus heavy communication traffic is observed with the growth of server numbers.

The servers in *realcourse* are not equal to each other. Only two servers are chosen to keep two permanent physical copies of files. No effort is made to keep track of those temporal copies of files but the existence of these copies greatly improves the performance of downloading operations by way of a wide-area network cache. In the new version of the consistency-check procedure, a pervasive search is started to find those "lost" copies in case both permanent copies of the file are corrupted. The loop topology of all servers makes it possible for each server to run a consistency-check procedure independently without any overhead. In fact, a more sophisticated consistency-check procedure is in development at the time of writing. Knowledge from the area of "autonomous computing" is of much help. By exploiting the reliable communication provided by asynchronous messaging middleware, *realcourse* hides the failures of the network, which are relatively common in our observation of a wide-area network from servers. The consistency of servers is eventually kept in a "delayed" fashion.

4.4.5 Massive Data Processing Grid

Massive Data Processing (MDP) Grid includes three data-intensive grid applications: high-energy physics computing, the alpha magnetic spectrometer (AMS) experiment, and the University Digital Museum Grid.

High-energy physics computing based on MDPGrid is a solution for the processing and analyzing of the massive amounts of data generated by CERN's ATLAS experiment, the Sino–Italy cosmic ray experiment at YangBaJing, Tibet, and the BEPC/BES project at Shandong University.

The AMS experiment project is a large-scale physics experiment on the International Space Station (ISS), the main purpose of which is to look for antimatter, search for the source of dark matter, and measure the source of cosmic rays. Space Detector Type Two (AMS-02) sent the space shuttle in 2005 to the ISS to carry on the experiment for three to five years. The data collected by AMS-02 will be eventually stored, indexed, and analyzed in the Science Operation Centers (SOCs). The Data Processing Center in Southeast University will be able to directly receive the data sent by NASA and carry on data analysis independently. Currently, the SOC in South East University has put up a grid platform SEUGrid for the Monte-Carlo simulation computing of AMS-02 MC.

The University Digital Museum Grid aims to integrate the enormous dispersed resources of various digital museums, to share the resources effectively and eliminate the information island, to filter and classify the collection information, and to provide an appropriate information service to users according to their knowledge levels and motivation through a unified grid portal. For the time being, the university digital museums involved include the Digital Museum of Aeronautics and Astronautics (BUAA) [27], the Archaeological Digital Museum (SDU) [28], the Geoscience Digital Museum (NJU) [29], and the Mineralogical Digital Museum (KUST) [30]. The services and resources in each university digital museum compose one site. The collection provider service gives access to digital museum collections as raw data.

4.5 Grid Dependable Research

To make the ChinaGrid practical, dependable grid computing research is very active, which covers grid infrastructure dependability through to application dependability. In the ChinaGrid, there are some systems related to dependable grid computing developed and released; for example, ChinaGrid SuperVison [31,32]. There are also some specific models presented; for example, DRIC provides an adaptive application fault-tolerant model [33]. The selected systems and models are presented in the following sections: ChinaGrid SuperVison, a national-wide grid monitoring system; ALTER, an adaptive grid fault detector [34]; FTGP, a fault-tolerant grid architecture [35]; and DRIC, an adaptive application fault-tolerance framework.

4.5.1 Grid Monitor

Monitoring resources is an important aspect of the overall efficient usage and control of any distributed system, and for improving the dependability of the whole system. The primary idea of grid monitoring is to integrate heterogeneous monitoring deployments or facilities of grid participants. In the ChinaGrid, a nationwide grid monitoring system has been developed and deployed, called ChinaGrid SuperVison (CGSV).

CGSV, sponsored by HP, is a program launched to handle monitoring-related issues for the ChinaGrid project. It was exhibited at the China National Computer Conference (CNCC) 2005. In accordance with the latest efforts in SOA and Grid communities, for example, WSRF and WSDM [36], CGSV seeks a set of resource measurement and control services for the operation of the ChinaGrid ecology. CGSV tries to provide system status information for CGSP components and application users. Moreover, CGSV tries to provide system trace data for analysis, to improve the system dependability and resource usability. The difference between CGSV and the CGSP Information Center is that CGSV puts more effort into dynamic information while the Information Center focuses more on static information.

4.5.1.1 Architecture

CGSV is designed based on the grid monitoring architecture [37] proposed by the Global Grid Forum. As shown in Figure 4.4, CGSV is built up by several components, which can be divided into three layers: the collection layer, service layer, and presentation layer. At the collection layer, sensors are deployed for data collection. Sensor-I is responsible for collecting hardware resource information. Other sources of information, including network condition and dynamic service information or data generated by other monitoring tools, are wrapped to unify the data format by an adaptor. Above the data interface, based on a transfer and control protocol, lies the service layer. In each domain of the ChinaGrid, this layer is presented as a logical domain monitor center, where the domain registry service and monitor services are deployed. A message gateway service is provided to alleviate the communication cost of monitor services since over-notification and over-subscription will disastrously decease service performance. The top layer is the presentation layer, where data analysis, visualization, and management can be performed. A detailed description of sensors, protocol and stream-integration design will be given in the next section.

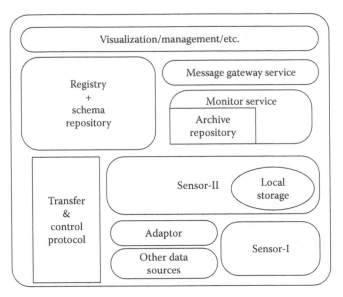

FIGURE 4.4 Architecture of CGSV.

4.5.1.1.1 Collection Layer

Unlike many existing monitor tools, the most significant characteristics of the CGSV sensor is that the sensor is runtime configurable, which means that the monitor metadata, such as each metric's switch, collection frequency, and granularity, is able to be changed over runtime on demand; for example, demanding to turn off all the resource monitoring metrics except CPU load and also lowering down the information collecting frequency to alleviate intrusiveness on some machines with heavy load. For most of the existing monitoring tools, configuration is preloaded at startup, therefore the required action needs users to login on the monitoring node, to shutdown the tool, to change the configuration file and start the tool again. While in CGSV, this action only needs users to send a command according to the protocol, the sensor will automatically change the configuration. The configuration file is also used in CGSV implementation, which is used for initialization and logging configuration when changes occur. In other words, this file is a real-time hard-disk backup of the sensor configuration, and is read only at startup.

There are two main types of sensors, called Sensor-I and Sensor-II, in CGSV. The difference between the two types is their function and deployment location. For corresponding components in GMA, Sensor-I is the producer and Sensor-II can be treated as the re-publisher. Sensor-I is deployed on computational resources, and is responsible for collecting resource information. Broadcast discovery and data transmission between Sensor-Is are performed via UDP packets with the specific format. Sensor-II is deployed on the front-end node of clusters, which is responsible for pulling resource information from Sensor-I-like data sources through UDP messages and processing incoming data request through TCP massages. Dynamic information of Web services are available through the service container, so the APIs can be treated as sensors and wrapped with an adaptor. In addition, Sensor-II is also responsible for processing control messages and adjusting behaviors of Sensor-I. All the messages are under the protocol that will be discussed in the next section.

4.5.1.1.2 *Transfer and Control Protocol*

A message protocol is designed for measurement data transmission and sensor control. Inspired by the classical file transfer protocol (FTP) [38] and Supermon project [39], the protocol used in CGSV is a client–server protocol, which is based on symbolic expressions (or s-expressions). The underlying transfer protocol varies from UDP to TCP. S-expressions originates from LISP as a recursively defined, simple format for data representation. This format of data has the following features:

- Extensibility: The protocol can be extended to new data types and new type of commands, which allows the system to be capable of evolving.
- Self-descriptive: Each measurement data is associated with its metric name and timestamp, so the packet can be independently interpreted without any other knowledge, which increases the system's interoperability.
- Compactness: Though the packets are self-descriptive, the format is very compact compared to XML. This feature saves network transmission bandwidth and memory cost for protocol interpretation, which decrease the intrusiveness to host systems.
- Architecture independence: This is achieved by plain textual representation, which facilitates system portability.

To improve the protocol parsing efficiency, command packets are encoded in s-expressions. This unified data transmission and control method simplifies the implementation of monitoring components and makes protocol interpretation and protocol execution logically separated.

This effort also makes CGSV components loose-coupled and easy to collaborate with other monitor systems. Table 4.1 lists the basic packet types.

4.5.1.1.3 *Service Layer*

- Registry: There is a logical domain monitor center in each domain of the ChinaGrid, where registry, archive module, and monitor

TABLE 4.1

Five Basic Packet Types Implemented in CGSV Protocol

Type	Packet Example	Purpose	Comments
QUERY	(QUERY (get 1.2.3.4*))	Issue a data query request for all measurement data of host 1.2.3.4	If this IP is a cluster, all back-end nodes' information are returned
DATA	(DATA(hostname 1.2.3.4)(timestamp 1115715626) (OSType 1 10) (CPUNumber 2 10))	Data packet indicating 2 monitor items (OS type and CPU number) with host IP and collect time	Metric information tuple is composed metric name, value and time difference with the complete timestamp.
SCHEMA	(SCHEMA (hostname 1.2.3.4) (OSType 1 1500) (CPULoad 0 15))	Schema packet indicating 2 metrics are supported on host 1.2.3.4 with their switches and monitor intervals	Metric schema tuple is composed the name switch flag and monitor interval in seconds
CTRL	(CTRL (close 1.2.3.4 4))	Issue a control request to switch MemFree metric off on host 1.2.3.4	"4" is predefined number for MemFree metric, full metric name is also accepted
RESULT	(RESULT (1 (get 1.2.3.4 *)))	Indicates execution result (1) of command "get 1.2.3.4 *"	"1" is predefined error code for success

services are deployed. Registry of CGSV performs two tasks. One is to provide a lookup service for consumers to locate producer/republisher registration; the other is to store producers' metric schema for system extension and modification. Since the adaptable implementation of sensors and protocol allows producers to accept control packets from any trusted source, the schema held by the registry needs to be synchronized periodically.

- Archive: Archiving is an optional component. It periodically acquires measurement data from all data sources and stores them as historical information in DBMS. This mechanism works similarly to registry schema synchronization. The difference is that the archive needs more resources, such as storage, so a distributed DBMS is used to share the loads when the domain grows larger.

- Services and message gateway: Domain monitor services are WSRF-compliant services that are responsible for providing monitoring information and management information interfaces. WSDM specification has been studied and applied to monitoring services for management issues. Each domain is treated as a WSDM manageability capability.

To alleviate the communication cost of monitor services, a message gateway service is used only for transmitting requests and responding between monitor services and grid users. As a result, data processing and data transmission are separated.

4.5.1.1.4 *Presentation Layer*

Visualization is implemented with an http server plus servlet and java applet, which has several forms (tables, histograms) of data presentation. Measurement data are retrieved from monitor services or a monitoring service gateway. Users can view the grid by simply browsing the Web pages. Geographical Information System (GIS) is introduced to locate resources from their geographical location on maps. An open-source toolkit JFreeChart [40] is used for diagram plotting support. Basic real-time visualization is implemented with dynamic resource information. Management actions can also be performed through the GUI client.

4.5.1.2 Features

Compared with existing distributed monitoring system, CGSV has the following features:

- Supporting different data sensors: CGSV provides an adaptor, which can interoperate with third-party monitoring tools. This feature makes CGSV capable of providing as much performance data as possible.
- History data collecting: CGSV collects the history data with the achieved repository. The stored data can be used as data analysis for later use; for example, performance forecasting based on the history data.
- In accordance with latest efforts in SOA and grid communities: CGSV supports WSRF and WSDM, which uses java RMI to implement WS-Notification, is compatible with the latest Apache Muse specification [41].

4.5.2 Fault-Tolerant Grid Platform

The fault-tolerant grid platform is one of the first attempts at dependable computing in the ChinaGrid project. The fault-tolerant grid platform (FTGP) is mainly trying to target hardware crashing and middleware crashing. FTGP is based on the Globus fault detection service [42]. FTGP addresses the fault tolerance with the following points:

- Multiple levels: The design of the FTGP provides system-level components fault tolerance and application-level fault tolerance.

- Accuracy and completeness: FTGP identifies faults accurately, with lower false positives and false negatives.
- Simplicity: FTGP provides toolkits to solve any type of faults in grid systems, and provides a virtual interface for grid users.

4.5.2.1 Fault-Tolerant Grid Topology

In grid systems, there are two types of hardware failure: node crashing and network appliance. It is quite difficult for application developers to determine when no remote response arises from node crashing or network failure. To cope with this, FTGP constructs a network with fault-tolerant topologies. A simple network topology with fault-tolerant features is shown in Figure 4.5. Any two single machines have two routes in a fault-tolerant grid system.

In order to prevent one network interface card (NIC) in a single machine from crashing a single point of failure, two NICs are required for one machine. Under this network architecture, two routes for any end-to-end information would not overlap with each other at the same network appliance. Any network appliance failure would not affect network communication of the grid. Based on this network topology, FTGP detects the network appliance status based on simple network management protocol (SNMP) [43]. Double links provided by such a topology reduce the unreachability of a machine. For a node crash, FTGP should not only recover applications and tasks from those failed nodes by transferring them to other correct grid nodes, but also dynamically extend the grid system to provide

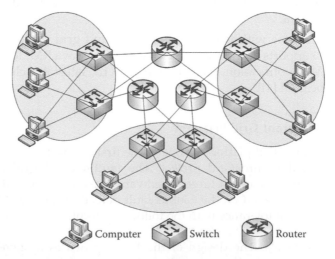

FIGURE 4.5   A simple network topology with fault tolerance.

enough resources. Once the workload of grid resources exceeds a threshold value, the grid system needs to scale up dynamically.

4.5.2.2 Fault Tolerance of Information Server

The information server fault-tolerant framework is shown in Figure 4.6. It has three levels: distributed information server (DIS), grid information index server (GIIS)- and resource provider. The Globus discovery service records the identity and essential characteristics of "services" available to domain members, and a GRIS registers its local resource information to a GIIS in the same domain. Most of the time grid users access their domain GIIS to get registered resources. They can also get local resource information directly from a GRIS. Based on the distributed architecture and softstate registration, FTGP makes the system highly fault-tolerant.

In Figure 4.6, there are two GIISs in one domain. Server B is a backup of server A, which records replicated categories. A grid notification framework is added to GIIS. Server A is implemented as a notification source, and server B is implemented as a notification sink. Server B subscribes to server A as a listener. Once a new resource registers to server A or a registered resource leaves, server A will quickly notify such a change to server B. Other information such as task results and the status of grid applications will also be propagated to server B with the help of the grid notification framework. In addition, server B detects server A to make sure whether server A is valid or not. Once server A fails, server B will

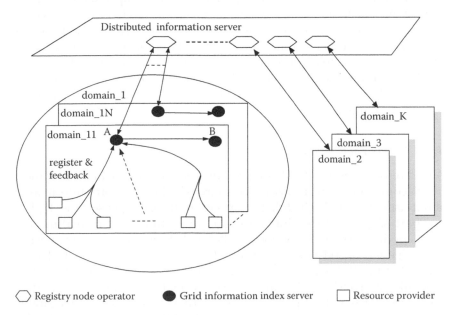

FIGURE 4.6 Information server fault-tolerant framework.

take over the work of server A. The registered information includes host information (such as static and dynamic information about CPU, memory, disk, and system software), task results, and applications that provide specific services for final users.

By introducing DIS to the Globus information server framework, FTGP provides a convenient means for grid users to search all the grid resources. DIS is a distributed system, which is internally constructed as a peer-to-peer [44,45] system. DIS consists of several registry node operators. Information registered to each registry node operator is regularly copied to the other node operator; that is, information is replicated among internal nodes. The architecture of the DIS works just like distributed UDDI. The GIIS interacts with the DIS with three API types: registered API, updated API, and terminated API. An active GIIS summarizes its registered resources and application services and reports them to the DIS. Grid users can get more resources and service information of other communities from the DIS. The DIS is also implemented as a notification sink, and accepts the resource changes from the active GIISs.

4.5.2.3 Fault Tolerance of the System-Level Component

There are complicated relationships among system-level components. Any component failure may lead to system crash. The local monitor is also a system-level component. When other components start, the processes register to the local monitor, and submit a handling script simultaneously. The handling script defines their dependent relationship and their startup order. Once the local monitor detects that any process fails, it will recover that process from failure according to the handling script. FTGP provides a self-maintenance mechanism for the local monitor. There are two local monitors in the system: the active and backup. There is a shared slot table used to record registered processes. When the active one fails, the backup one will take over the work according to the slot table, and restart another backup local monitor.

4.5.2.4 Task Fault Tolerance

A grid resource broker (GRB), such as Nimrod/G [46], provides a graphical interface to monitor and utilize grid resources [47]. GRB records information about user tasks and scheduled information. Information service [48] is a key part of grid software infrastructure, providing fundamental mechanisms of discovery and monitoring. Globus Information Services consists of a GIIS and a grid resource information server (GRIS). User tasks are submitted to GRB, which searches resource information from GIIS and schedules user tasks to the suitable grid node. Local components in grid nodes register to GIIS to publish their resource information and locations by lightweight directory access protocol (LDAP). GRIS is responsible for recording local resources and registers them to its related GIIS. In order to simplify task fault tolerance, we only focus on sequential programs; that is, a task is only submitted to one grid node at a time.

The workflow of task fault detection is depicted in Figure 4.7. A data collector is configured in the GRB. When many tasks are submitted to the GRB, the GRB looks up a GIIS for registered resources by LDAP, gets the required resource information, and specifies the suitable resource to any submitted task according to the requirement of the task. Tasks with a fault-tolerant requirement register to the data collector in the GRB and are scheduled to a selected grid node. At the same time, a callback mechanism is added to the submitted task. The grid resource management component verifies the corresponding scheduled task [49]. Once the task is legal, it is accepted by the grid resource management component. Otherwise, it is refused.

The work completed by user program codes is transferred to GFTP, instead of being implemented with Globus fault detection APIs in program codes. The resource management component analyzes the task's resource specification language (RSL) and determines a job-scheduling system, such as fork, LSF, and so on. The selected job-scheduling system starts the task in the local node and registers it with the local monitor. The grid resource management component provides the local monitor with the identity of the user task process to be monitored, the data collector to which the process heartbeat is to be sent, and a heartbeat interval. The local monitor is responsible for detecting the task status and reporting it to the data collector. If the process terminates, the component disconnects it

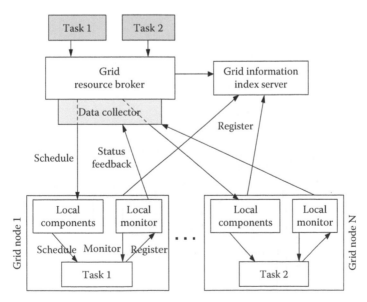

FIGURE 4.7 Task fault detection.

from the local monitor. The local monitor reports the status of each monitored process to the data collector between the time of process registration
and unregistration at a fixed specified interval.

The reasons why grid users could not get the results of submitted tasks
include task failure, grid node crash, and GRB crash. Fault recovery from
task failure is shown in Figure 4.8. The local monitor maintains a slot table
for all monitored processes. When the local monitor detects the failure of a
registered task, it will report the error status to a local resource management
component and disconnect the error process. The component also maintains
a table responsible for recording the scheduled task information. Task information in the table consists of the program name, parameter files, identification of the corresponding process, and other execution conditions. Once a
monitored process fails, the resource management component restarts the
same task according to registration information in the table, and registers the
process with a fault-tolerant requirement to the local monitor again.

Fault recovery from a grid node crash is shown in Figure 4.9. If the data
collector could not receive information from a local monitor in a valid
period, it will consider the grid node with the local monitor to have failed
and reports such information to the GRB. The GRB starts a thread to
search for the tasks located in that grid node from its database, and records
their error status. These failed tasks will enter the scheduling queue of
GRB again. The GRB allocates required resources to these tasks and
reschedules them to the selected resources.

Fault recovery from the GRB crash is shown in Figure 4.10. In order to
guarantee that the scheduled tasks continue to operate when the GRB

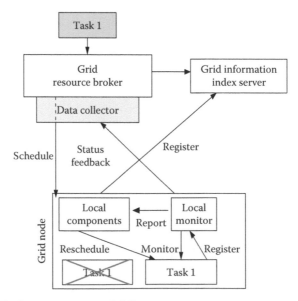

FIGURE 4.8   Fault recovery from task failure.

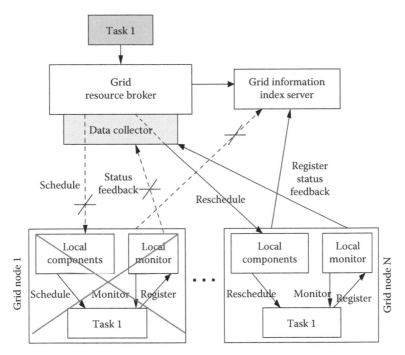

FIGURE 4.9 Fault recovery from a grid node crash.

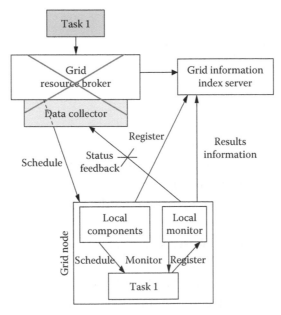

FIGURE 4.10 Fault recovery from a GRB crash.

crashes, the local monitors register result information of the scheduled tasks to a GIIS. When the GRB recovers from failure, it gets its information from the GIIS. In order to release network congestion, status information in the wide-area network is transferred by UDP, so the local monitor could not detect the status of data collector. The local monitor reports the task result by TCP. When the local monitor detects the failure of the GRB, it reports the result to GIIS. The local monitor need not report task status information to the GIIS even if the GRB crashes. From the above analysis, we can design a fault-tolerant task-scheduling algorithm as follows:

1 When GRB system starts or recovers:

   for all (task is running)

   task active time = current time;

   // Lookup back-end database to find the former running tasks

   task[ ] = Query(task_DB, "running");

   If (task[] not NULL)

       lookup GIIS and get tasks' result;

   // There are task status reports from local monitors even if the GRB crashes while (an incoming task status is normal and this task ∈ task[])

   task active time = current time;

2 task scheduling:

   get a task from back-end database;

   form the task RSL;

   job submit, get job identifier;

   // Callback function is completed by data collector

   add callback mechanism to the job;

   if (incoming task status is normal)

       Task active time = current time;

   else the task enters queue according to its requirement;

3 Analyze tasks status timely:

   if (current time – task active time > valid time)

   {

       lookup SNMP database, get network status;

       if (network is blocked)

       then wait until enough bandwidth is available;

       the task enters queue according to its requirement;}

4 Get task result.

4.5.3 Grid Fault Detection

Fault detection is well known as a fundamental building block for dependable systems [50–53]. In the ChinaGrid, an adaptive fault detection service named ALTER has been designed by the Cluster and Grid Computing Lab, Huazhong University of Science and Technology. ALTER addresses the scalability and flexibility of the grid system [54,55], which blends the unreliable failure detectors in distributed systems [56] and R-GMA (relational grid monitoring architecture). ALTER can fine tune system performance with different QoS requirements, and change its topology according to changes of grid environments, such as the addition of some resources, some crashing key components, or even crashing failure detectors.

4.5.3.1 Architecture

ALTER is organized in a hierarchical structure as shown in Figure 4.11. The system architecture is composed of two levels: local groups and global groups. In the local group, there is a unique group leader. Failure detectors in the local groups monitor the objects in the local space, and the monitored objects in one local space may be in one LAN or cross-LANs, but the network condition in one space should be good. Failure detectors in the global groups monitor the global objects by means of monitoring the detectors in local groups. Thus, in this architecture there are two different types of failure detectors: a local failure detector and a group leader. The monitored objects send "I'm alive" messages to the local failure detectors periodically, while the messages that a group leader sends is a list containing the monitored objects and their status, and the group leaders share failure detection messages with epidemic methods [57]. For management simplicity, there is an index service in the system implementation, and the index service works as a directory registry of the global group failure detectors and local group failure detectors. Also, the index service provides some decision-making capability for the organization of group leaders. There are three components or roles in ALTER system: consumers, producers, and an index service. A consumer who wants to detect some objects in a grid, first queries

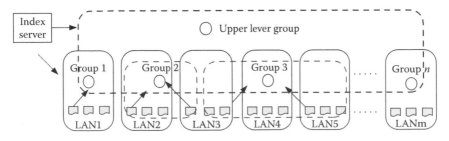

FIGURE 4.11   Architecture of ALTER.

the index service and gets the location of the producer; the failure detector is a producer that gives the status of the monitored objects.

4.5.3.2 Adaptive Model

A grid fault-detection service should stress two things: first, how to satisfy the QoS between two processes; and second, how to satisfy the grid dynamic nature. The QoS between the monitored process and the detector can be donated in a tuple $\left(T_D^U, T_{MR}^L, T_M^U\right)$, where T_D^U is an upper bound on the detection time, T_{MR}^L is a lower bound on the average mistake recurrence time, T_M^U is the upper bound on the average mistake duration. The QoS requirement can be expressed as in Equation 4.1:

$$T_D \leq T_D^U, \quad T_{MR} \geq T_{MR}^L, \quad T_M \leq T_M^U \tag{4.1}$$

It is clear that the heartbeat interval $\Delta_{interval}$ is a very important factor that contributes to the detection time:

$$T_D^U \geq \Delta_{interval} + \Delta_{tr} \tag{4.2}$$

where Δ_{tr} is a safety margin. A recursive method has been used to get $\Delta_{interval}$, as depicted in algorithm 1, which is adopted from [56,58]. After getting $\Delta_{interval}$, a sliding window algorithm is taken, and this sliding method records the message behaviors, thus the $\Delta_{interval}$ will change adaptively to the system conditions. Another problem for failure detection is when to suspect failure. The details are presented in Algorithm 4.1.

Algorithm 4.1

Assumption: The inter-arrival of the "I'm alive" message follows a Gaussian distribution. The parameters of the distribution are estimated from a sample window. The probability of a given message arriving more that t time unit is given as

$$p(t) = \frac{1}{\sigma\sqrt{2\pi}} \int_t^{+\infty} e^{-[(x-\mu)^2/2\sigma^2]} dx \tag{4.3}$$

Step 1: [find $\Delta_{intervalmax}$]
Compute

$$\gamma = \frac{(1 - P_L)(T_D^U)^2}{\Delta_{tr} + (T_D^U)^2} \tag{4.4}$$

$$\Delta_{intervalmax} = \max(\gamma T_M^U, T_D^U) \tag{4.5}$$

If $\Delta_{intervalmax} = 0$, the QoS cannot be achieved. P_L is the probability of message loss and P_L can be simply computed as in Equation 4.6:

$$P_L = \frac{C_{To} - C_{Re}}{C_{To}} \tag{4.6}$$

C_{To} is the count of total messages that are sent in a sample window, and C_{Re} is the count of the received message in the sample window.

Step 2: [get $\Delta_{interval}$]
Let

$$f(\Delta_{interval}) = \Delta_{interval} \prod_{j=1}^{T_D^U / \Delta_{interval}} \frac{\Delta_{tr} + (T_D^U - j\Delta_{interval})^2}{\Delta_{tr} + P_L(T_D^U - j\Delta_{interval})^2} \tag{4.7}$$

find the largest $\Delta_{interval}$ which is less than $\Delta_{intervalmax}$ and with $f(\Delta_{interval}) \geq T_{MR}^L$. Such a $\Delta_{interval}$ always exists.

Step 3: [estimate arrival time of next heartbeat message]
Compute μ and σ with Equation 4.3 by using the message behavior in the sliding window. Estimate the next message arrival time as

$$ET_{n+1} = W_s\mu + \sigma \tag{4.8}$$

Here, W_s is the window size of heartbeat messages.

Step 4: [get freshness point τ_{n+1}]
Compute the freshness point:

$$\tau_{n+1} = ET_{n+1} + T_D^U - \Delta_{interval} \tag{4.9}$$

If no fresh message is received in τ_{n+1} time, suspect the monitored process, that is,

$$\Delta_{timeout} = \tau_{n+1} - T_{now} \tag{4.10}$$

With Algorithm 4.1, the failure detector between two processes can be adaptive to the system conditions. Next the algorithm organizing the deployment of the failure detectors in grids is presented. The details are shown in Algorithm 4.2.

Algorithm 4.2

Step 1: [index service pulls information]
The index service pulls host environments information of all failure detectors with the round-robin algorithm, and system information includes static and dynamic information.

Step 2: [compute SL_i of all hosts]
Compute service level of host environments as

$$SL_i = \frac{\alpha FR_i + \beta Mem_i}{Load_i} \tag{4.11}$$

and take the host i which has Max(SL_i). FR_i is the frequency of CPU_i, Mem_i is the memory of host i and $Load_i$ is the load of host i; α and β are coefficients.

Step 3: [find a group leader on the specific host]
Get the list of failure detectors on host i, which is chosen in step 2. If one of them is a group leader, go to step 6.

Step 4: [deploy a new local group leader on the specific host]
Deploy a group leader failure detector on host i, and all the failure detectors in this group send detection messages to this failure detector.

Step 5: [global notification about the local leader]
The index service sends a trigger message to upper-level failure detector to tell the new local group leader, and the new group leader sends detection messages to the upper-level failure detector.

Step 6: [handle the new monitored objects]
If there are new monitored objects added in, first try to find the lightest-load failure detector in this group by getting the Min(N_{fdi}). If Min(N_{fdi}) is less than $F_s - 1$, then the new object sends "I'm alive" message to this failure detector; go to step 8.

Step 7: [deploy a new failure detector]
Create one more failure detector, and the new coming object sends "I'm alive" to this failure detector. If the number of failure detectors is larger than S_g, add a new group with a group leader failure detector and this new group leader registers itself to the upper-level failure detector.

Step 8: [change the messages sent to the upper-level failure detector]
Change the message list sent to the upper-level failure detector with the new object; at the same time send a message to index service to notify the global change.

Step 9: [handle failure of local detectors]
The group leader can detect the local detector's failure and, if it fails, just add on and let the monitored objects send "I'm alive" to it.

Step 10: [handle failure of group leader]
The top failure detector and the index service can detect the group leader's failure and, if one of them fails, just deploy a new one on the same host which is registered to the index service.

Step 11: [merge small groups]
When monitored objects decreased, compute number of groups needed as

$$n = \left\lfloor 2\log_{S_g} N_{total} \right\rfloor \tag{4.12}$$

Choose S_g as in Equation 4.13.

$$S_g = \left\lfloor \lambda \sqrt{N_{total}} \lg N_{total} \right\rfloor \tag{4.13}$$

The network conditions between two failure detectors is valued as in Equation 4.14:

$$\forall i, j, T_{ij} \leq \Phi \mid i, j \in LAN \rightarrow good \tag{4.14}$$

Try to annex the smallest group to the second smallest one if their network conditions are good enough, and loop this process, till the number of groups is less than n. With the algorithms presented above, the failure detector is able to be adaptive according to different grid environments.

4.5.4 Adaptive Application Fault Tolerance

There are different types of failures in grid systems and grid applications due to the diverse nature of the grid components and grid applications. The existing failure handling techniques in distributed systems, parallel systems, or even grid systems address failure handling with one scheme, which cannot handle failures in grids with different semantics. The failure handling method in grids should address the following requirements:

- Support diverse failure handling strategies. This is driven by the heterogeneous nature of the grid context, such as heterogeneous tasks and heterogeneous execution environments.
- Separation of failure handling policies from application codes. This is driven by the dynamic nature of the grid system. The separation of policies from the application codes provides a high-level

handling method without having any knowledge about the grid resources and the applications.

- Task-specific failure handling support. This is driven by the diversity of grid applications and grid resources. The user should be able to specify an appropriate failure handling method for performance and cost consideration.

In this section, an adaptive policy-based fault-tolerance approach is presented, which is called DRIC. First the overview of the approach is presented, with application-level fault-tolerance techniques reviewed later. Finally, the adaptive model of failure handling will be presented.

4.5.4.1 Overview of Failure Handling

As depicted in Figure 4.12, the fault tolerance in DRIC framework comprises two phases: failure detection and failure handling. Figure 4.12 presents the overview of the failure handling approach. The failure-handling approach uses a decision-making method to attain the QoS requirements described by users, in which almost all kinds of application-level failure-recovery techniques are integrated, such as checkpointing [59–62], replication [63–66], and workflow [67]. First, the user submits a task with QoS requirements. The policy engine analyzes the QoS requirements and the attributes of the application to constitute a failure-recovery policy. Based on the policy, the policy engine carries out the policy with appropriate techniques with the help of job management, data management, and information management illustrated in Figure 4.13.

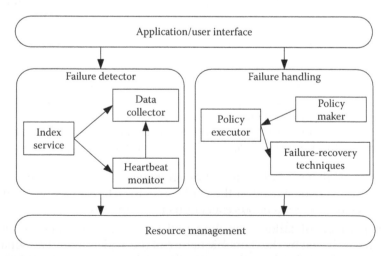

FIGURE 4.12   Overview of failure handling.

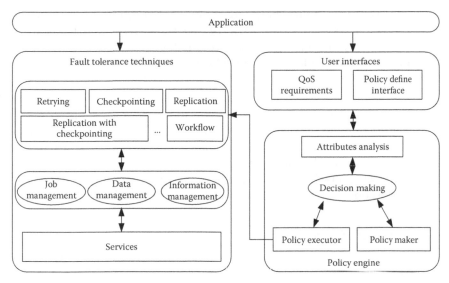

FIGURE 4.13 Overview of DRIC.

Also the user can specify the failure-handling policy with the policy definition interface, and the policy engine carries it out.

4.5.4.2 Application-Level Fault-Tolerance Techniques

In this section the task-level fault-tolerance techniques to handle task failures are reviewed.

- Retrying: This is a simple and obvious failure-handling technique, with which the system retries to run the task on the same resources when a failure is detected [68,69]. Generally, the user or the system specifies the maximum number of retries and the intervals between the retries. After these retries, if the failure still exists, the system prompts an error, and the user or the system should provide other failure-handling methods to finish the task.
- Replication: The basic idea of replication is to have replicas of tasks running on different resources, so that as long as not all replicated tasks crash, the execution of the associated activity would succeed. The failure detector detects these replicated tasks during the task execution and the policy executor kills other replicated tasks when one of them finishes with the appropriate result.
- Checkpointing: This has been widely studied in distributed systems. Traditionally, for a single system, checkpointing can be realized at three levels: kernel level, library level, and application

level [70]. Our failure-handling framework supports checkpoint-enabled tasks so that when task fails, the task can be restarted from the recently checkpointed state. Beside the three levels of check-pointing, we implement a workflow-level checkpointing, which defines the task as a workflow and saves states of the atom-task as checkpoints.

- Workflow: This is another technique to handle failures in grids. There are several workflow methods for failure handling, such as alternative task, workflow redundancy, and user-defined exception handling. The main idea of this method is to construct a new workflow for task execution. The workflow method can be regarded as the composition of other failure-handling techniques.

4.5.4.3 Model of Policy Making

In this section, some basic definitions about failure-recovery policies are presented, and some quality criteria for failure recovery are given. The policy-making model is illustrated later.

4.5.4.3.1 Basic Definitions

We present some basic definitions about the failure-recovery policies.

1. Execution time (Q_{et}): This is the execution duration of a task with some failure-recovery methods. The task can be a single running service or a composition of services. Execution time is important for a failure-recovery method, especially in some mission-critical applications. The execution time of the same task varies violently with different failure-recovery methods.

2. Failure-recovery policy (P): This is a policy specified by the user or the system to identify which failure-recovery techniques should be applied for a particular application. In this chapter, the failure-recovery policy specifies the techniques presented above or specifies the composition of them, such as replication with checkpointing and replication with retrying.

3. Execution cost (Q_{co}): This is the amount of money for a service request to pay for executing the specific operation for a service. The execution cost can be computed using Equation 4.15, where Q_{price} is the price of the specific resource, and R_{num} is the number of resources.

$$Q_{co} = Q_{price} \times Q_{et} \times R_{num} \qquad (4.15)$$

4. Task: This is a state chart with a sequence of states $[t_1, t_2, \ldots, t_n]$, such that t_1 is the initial state and t_n is the final state.

With the above definitions, we define the QoS of a service s as a tuple:

$$Q(s) = [Q_{et}(s), Q_{co}(s)] \tag{4.16}$$

4.5.4.3.2 Quality Criteria

In this section, the quality criteria for different failure-recovery policies are presented. For simplicity and feasibility, only five policies are discussed in DRIC, and they are retrying, replication, checkpointing, retrying with replication, and replication with checkpointing. The workflow method such as alternative task is specified by the user, and the system cannot control it but only execute it. Here are the definitions of the five failure-recovery policies.

1. Retrying: Given a task $[t_1, t_2, \dots, t_n]$, if the task fails at t_i, it will be re-executed from t_1.

2. Replication: Given a task $[t_1, t_2, \dots, t_n]$, there are m replications for this task, they are $[t_{11}, t_{12}, \dots, t_{1n}], [t_{21}, t_{22}, \dots, t_{2n}], \dots, [t_{m1}, t_{m2}, \dots, t_{mn}]$. If any task fails at t_{ij}, replica i is killed. If any replicated task successfully finishes at t_{in}, other replicas are killed.

3. Checkpointing: Give a task $[t_1, t_2, \dots, t_n]$, the running states are saved to a file. If the task fails at t_i, the task will be executed from t_i.

4. Replication with checkpointing: Given a task $[t_1, t_2, \dots, t_n]$, there are m replications for this task, they are $[t_{11}, t_{12}, \dots, t_{1n}], [t_{21}, t_{22}, \dots, t_{2n}], \dots, [t_{m1}, t_{m2}, \dots, t_{mn}]$. The running states are saved to files. If any task fails at t_{ij}, the task will be executed from t_{ij}. If any replicated task successfully finished at t_{in}, other replicas are killed.

5. Retrying with replication: Given a task $[t_1, t_2, \dots, t_n]$, there are m replications for this task, they are $[t_{11}, t_{12}, \dots, t_{1n}], [t_{21}, t_{22}, \dots, t_{2n}], \dots, [t_{m1}, t_{m2}, \dots, t_{mn}]$. If any task fails at t_{ij}, the task will be re-executed from t_{ij}. If any replicated task successfully finished at t_{in}, other replicas are killed.

After the definition of the failure-recovery policies, the QoS criteria about fault tolerance are presented. For expression simplicity and feasibility, some propositions about the failure behaviors in grid computing are given first.

Assumption 1

1. The time a task at state t_i takes is the duration that the task performs specific operation on a specific service.

2. The time switch from state t_i to state t_{i+1} is zero, that is, the task runs into state t_{i+1} immediately when state t_i finishes.

Assumption 2

1. The probability the task fails at state t_i is λ (or failure rate), and follows a Poisson distribution. Mean time to failure (MTTF) is mathematically defined by $1/\lambda$.
2. The task fails at state t_i, and the time when a failure occurs follows a Poisson distribution.
3. The failure execution time of the task is T_F.
4. The average down time of the system is T_D.

Assumption 3

1. The number of replicas of a task is N_R for replication policy or replication with checkpointing policy.
2. The overhead of checkpoint (T_C) is the same for different policies.
3. The recovery time of checkpointing (T_R) is the same for different policies.
4. The checkpoint interval is T_I.

With these definitions and assumptions, a brief explanation of QoS criteria for different failure-recovery policies is presented.

Retrying

The computation of the execution time with retrying was first specified by Duda [71].

$$Q_{et(ret)} = \frac{e^{\lambda T_D}(e^{\lambda T_F} - 1)}{\lambda} \qquad (4.17)$$

The execution cost for retrying is computed as

$$Q_{co} = Q_{price} \times Q_{et(ret)} \qquad (4.18)$$

Checkpointing

The execution time with checkpointing is computed from Equation 4.19 and adopted from [72].

$$Q_{et(ch)} = \left(\frac{T_F}{T_I}\right)\frac{e^{\lambda(T_D - T_C + T_R)}(e^{\lambda(T_I + T_C)} - 1)}{\lambda} \qquad (4.19)$$

The execution cost for checkpointing is as

$$Q_{co} = Q_{price} \times Q_{et(ch)} \qquad (4.20)$$

Replication with Checkpointing

The execution time for replication with checkpointing is as follows:

$$Q_{et(rep+ch)} = \left(\frac{T_F}{T_I}\right) \frac{e^{\lambda/N_R(T_D - T_C + T_R)}\left(e^{\lambda/N_R(T_I + T_C)} - 1\right)}{\lambda/N_R} \tag{4.21}$$

The execution cost for replication with checkpointing is computed as

$$Q_{co} = Q_{price} \times Q_{et(rep+ch)} \times N_R \tag{4.22}$$

Retrying with Replication

The execution time for retrying with replication is

$$Q_{et(ret+rep)} = \frac{e^{(\lambda/N_R)T_D}\left(e^{(\lambda/N_R)T_F} - 1\right)}{\lambda/N_R} \tag{4.23}$$

The execution cost for retrying with replication is computed as

$$Q_{co} = Q_{price} \times Q_{et(ret+rep)} \times N_R \tag{4.24}$$

Policy-Making Model

The basic idea of adaptive model in DRIC is to set the optimized failure-recovery method using the multiple attribute decision making (MADM) approach [73]. With the above definitions and the equations, the decision-making matrix M is obtained from Equation 4.25 as

$$M_{i,j} = \begin{pmatrix} Q_{et(ret)} & Q_{co(ret)} \\ Q_{et(ch)} & Q_{co(ch)} \\ Q_{et(rep+ch)} & Q_{co(rep+ch)} \\ Q_{et(ret+rep)} & Q_{co(ret+rep)} \end{pmatrix} = \begin{pmatrix} Q_{1,1} & Q_{1,2} \\ Q_{2,1} & Q_{2,2} \\ Q_{3,1} & Q_{3,2} \\ Q_{4,1} & Q_{4,2} \end{pmatrix} \tag{4.25}$$

As execution time and execution cost have different quantum, the matrix M need to be formalized with Equation 4.27.

$$V_{i,j} = \begin{cases} \dfrac{Q_i^{\max} - Q_{i,j}}{Q_i^{\max} - Q_i^{\min}} & \text{if } Q_i^{\max} - Q_i^{\min} \neq 0 \\ 1 & \text{if } Q_i^{\max} - Q_i^{\min} = 0 \end{cases}, \quad i = 1, 2 \qquad (4.26)$$

In Equation 4.26, $Q_i^{\max}$ and $Q_i^{\min}$ are the maximal value and the minimal value of a quality criterion in matrix M. With Equation 4.26, we obtain a new matrix M':

$$M' = (V_{i,j})2 \times 4 \qquad (4.27)$$

The following formula is used to compute the overall quality score. The policy that has the maximum score and with the execution time less than the user required will be chosen by DRIC.

$$Score(P_j) = \sum_{i=1}^{2} (V_{i,j} \times W_i) \qquad (4.28)$$

where W_j represents the weight of each criterion, and W_j is computed with the entropy method, as shown in the following equation:

$$\begin{cases} W_i \in [0,1] \\ \sum_{i=1}^{2} W = 1 \end{cases} \qquad (4.29)$$

$$F_T : \begin{cases} \max(Score(P_j)) \\ Q_{et}(j) \leq T_{re} \end{cases} \qquad (4.30)$$

where f_{ij} is depicted as in Equation 4.31.

$$f_{i,j} = \frac{V_{i,j}}{\sum_{j=1}^{4} V_{i,j}} \qquad (4.31)$$

$$H_i = -\frac{1}{\ln 4} \sum_{j=1}^{4} f_{i,j} \ln f_{i,j}, \quad i = 1, 2 \qquad (4.32)$$

With Equation 4.32, W_i is computed as

$$W_i = \frac{1 - H_i}{2 - \sum_{i=1}^{2} H_i} \tag{4.33}$$

In Equation 4.28, the end user can also specify the weight W_i for their preference on QoS. In DRIC, the entropy weight is the default choice, because entropy weight considers both the QoS requirement of the end user and the execution cost of the tasks. With the model presented above, the application can always choose a suitable method to handle faults.

4.6 Conclusion

This chapter has sought to give a brief introduction about the ChinaGrid and to related dependable computing research. While the research involved is plentiful, the chapter cannot cover everything; for example, an in-course online project, the IBM MQ, has been used to improve the reliability of the application platform. This chapter covered the vision and mission for the ChinaGrid project and five different applications and its application-supporting platform. To illustrate the practical issues of the ChinaGrid, the chapter covered the related dependability research in the ChinaGrid project. The dependability research covers grid monitoring, grid platform fault tolerance, efficient failure detectors, and failure handling of grid applications. The chapter gave an overview about the design and implementation about this research.

4.7 Acknowledgments

This work is supported by the National Science Foundation of China under grant 90412010, and 60603058, ChinaGrid project from the Ministry of Education of China. Special thanks to Jing Tie for providing materials for CGSV.

4.8 References

1. I. Foster and C. Kesselman, "Globus: A metacomputing infrastructure toolkit," *International Journal of Supercomputer Applications*, 11 (2), 115–128, 1997.

2. I. Foster, C. Kesselman, and S. Tuecke, "The anatomy of the grid: Enabling scalable virtual organizations," *International Journal of High Performance Computing Applications*, 15 (3), 200–222, 2001.
3. UK e-Science Programme. Available at: http://www.rcuk.ac.uk/escience.
4. W.E. Johnston, D. Gannon, and B. Nitzberg, "Grids as production computing environments: The engineering aspects of NASA's information power grid," in *Proceedings of 8th IEEE Symposium on High Performance Distributed Computing*, Redondo Beach, CA, 1999.
5. The TeraGrid Project. Available at: http://www.teragrid.org.
6. ChinaGrid. Available at: http://www.chinagrid.edu.cn.
7. China Education and Research Network. Available at: http://www.edu.cn.
8. Yongwei Wu, Song Wu, Huashan Yu, and Chunming Hu, "CGSP: An extensible and reconfigurable grid framework," in *Proceedings of 6th International Workshop Advanced Parallel Processing Technologies (APPT 2005)*, Hong Kong, 2005, pp. 292–300.
9. The Web Services Resource Framework. Available at: http://www.globus.org/wsrf.
10. Open Grid Services Architecture. Available at: http://www.ggf.org/Public_Comment_Docs/Documents/draft-ggf-ogsa-specv1.pdf.
11. Bin Wang, Zhuoqun Xu, Cheng Xu, Yanbin Yin, Wenkui Ding, and Huashan Yu, "A study of gridifying scientific computing legacy codes," in *Proceedings of Third International Conference Grid and Cooperative Computing—GCC*, 2004, Wuhan, China, pp. 404–412.
12. Web Services Description Language. Available at: http://www.w3.org/TR/wsdl.
13. Job Submission Description Language. Available at: http:// www.ogf.org/documents/GFD.56.pdf.
14. Business Process Execution Language for Web Services. Available at: http://download.boulder.ibm.com/ibmdl/pub/software/dw/specs/ws-bpel/ws-bpel.pdf.
15. W. Allcock (Ed.) *GridFTP Protocol Specification*. Global Grid Forum Recommendation GFD.20, March 2003.
16. OGSA-DAI. Available at: http://www.ogsadai.org.uk/about/ogsa-dai.
17. K. Karasavvas, M. Antonioletti, M.P. Atkinson, N.P. Chue Hong, T. Sugden, A.C. Hume, M. Jackson, A. Krause, and C. Palansuriya. "Introduction to OGSA-DAI services," in *Proceedings of Scientific Applications of Grid Computing*, LNCS 3458, 2005, Beijing, China, pp. 1–12.
18. ChinaGrid Bioinformatics Grid. Available at: http://166.111.68.168/bioinfo/tools/index.jsp.
19. ChinaGrid Image Processing Grid. Available at: http://grid.hust.edu.cn/ImageGrid.
20. ChinaGrid Computational Fluid Dynamics (CFD) Grid. Available at: http://grid.sjtu.edu.cn:7080/grid.
21. ChinaGrid Course Online Grid. Available at: http://realcourse.grids.cn.
22. ChinaGrid Mass Data Processing Grid. Available at: http://athena.vrlab.buaa.edu.cn/gcc.
23. DataGrid Project WP1, "Definition of architecture, technical plan and evaluation criteria for scheduling, resource management, security and job description," Datagrid document DataGrid-01-D1.2-0112-03, 14/09/2001.

Available at http://server11.infn.it/workload-grid/docs/DataGrid-01-TED-0102-1_0.pdf.

24. R. Zheng, H. Jin, Q. Zhang, Y. Li, and J. Chen, "IPGE: Image processing grid environment using components and workflow techniques," in *Proceedings of International Conference on Grid and Cooperative Computing*, LNCS 3251, Wuhan, China, 2004.

25. Hai Jin, Ran Zheng, Qin Zhang, and Ying Li, "Components and workflow based grid programming environment for integrated image-processing applications," *Concurrency and Computation: Practice and Experience*, 18 (14), 1857–1869, 2006.

26. J. Zhang and X. Li, "The model, architecture and mechanism behind realcourse," Technical Report, Beijing University, 2004.

27. Digital Museum of Aeronautics and Astronautics. Available at: http://digitalmuseum.buaa.edu.cn.

28. Archaeological Digital Museum. Available at: http://museum.sdu.edu.cn/index/index.asp.

29. Geoscience Digital Museum. Available at: http://202.119.49.29/museum/default.htm.

30. Mineralogical Digital Museum. Available at: http://www.kmust.edu.cn/dm/index.htm.

31. CGSV. Available at: http://www.chinagrid.edu.cn/cgsv.

32. Weimin Zheng, Lin Liu, Meizhi Hu, et al., "CGSV: An adaptable stream-integrated grid monitoring system," in *Proceedings of IFIP International Conference on Network and Parallel Computing (NPC'05)*, November, 2005. Beijing, pp. 612–623.

33. Hai Jin, Xuanhua Shi, Weizhong Qiang, and Deqing Zou, "DRIC: Dependable grid computing framework," *IEICE Transactions on Information and Systems*, E89-D (2), 612–623, February 2006.

34. Xuanhua Shi, Hai Jin, Zongfen Han, et al., "ALTER: Adaptive failure detection services for grids," in *Proceedings of SCC 2005*, Florida, pp. 355–358.

35. Hai Jin, Deqing Zou, Hanhua Chen, Janhua Sun, Song Wu, "Grid fault tolerance and practice," *Journal of Computer Science and Technology*, 18 (4), 423–433, July 2003.

36. OASIS Web Services Distributed Management (WSDM). Available at: http://www.oasis-open.org/committees/tc_home.php?wg_abbrev=wsdm.

37. Brian L. Tierney, Brain Crowley, Dan Gunter, Mason Holding, Janson Lee, Mary Thompson, B.L. Tierney, A grid monitoring architecture. Global Grid Forum, Toronto, 2002. Available at: http://www-didc.lbl.gov/GGF-PERF/GMA-WG/papers/GWD-GP-16-3.pdf.

38. Postel, J. and Reynolds, J. "File Transfer Protocol (FTP)." Available at: http://www.ietf.org /rfc/rfc959.txt.

39. Sottile, M.J. and Minnich, R.G. "Supermon: A high-speed cluster monitoring system," in *Proceedings of the IEEE International Conference on Cluster Computing*, Washington DC, 2002, p. 39.

40. JFreeChart Project. Available at: http://www.jfree.org/jfreechart/index.html.

41. Apache Muse. Available at: http://ws.apache.org/muse/.

42. P. Stelling, I. Foster, C. Kesselman, C. Lee, and G. von Laszewski, "A fault detection service for wide area distributed computations," in *Proceedings of the 7th IEEE Symposium on High Performance Distributed Computing*, Chicago, 1998, pp. 268–278.

43. W. Stallings, "SNMP and SNMPv2: The infrastructure for network management," *IEEE Communications Magazine*, 36 (3), 37–43, March 1998.
44. M. Ripeanu, "Peer-to-Peer architecture case study: Gnutella network," in *Proceedings of International Conference on Peer-to-peer Computing*, Sweden, 2001.
45. I. Stoica, R. Morris, D. Karger, M. Kaashoek, and H. Balakrishnan, "Chord: A scalable peer-to-peer lookup service for internet applications," in *Proceedings of ACM SIGCOMM*, San Diego, CA, 2001, pp. 149–160.
46. R. Buyya, D. Abramson, and J. Giddy, "Nimrod/G: An architecture for a resource management and scheduling system in a global computational grid," in *Proceedings of International Conference on High Performance Computing in Asia-Pacific Region (HPC Asia 2000)*, Beijing, 2000.
47. F. Berman, R. Wolski, S. Figueira, J. Schopf, and G. Shao, "Application-level scheduling on distributed heterogeneous networks," in *Proceedings of Supercomputing '96*, Pittsburgh, PA, 1996, pp. 39–39.
48. K. Czajkowski, S. Fitzgerald, I. Foster, and C. Kesselman, "Grid information services for distributed resource sharing," in *Proceedings of the 10th IEEE International Symposium on High-Performance Distributed Computing (HPDC-10)*, Washington, DC, 2001, pp. 181–194.
49. J. Howell and D. Kotz, "End-to-End authorization," in *Proceedings of the 2000 Symposium on Operating Systems Design and Implementation*, San Diego, CA, 2000, pp. 151–164.
50. P. Felber, X. D'efago, R. Guerraoui, and P. Oser, "Failure detector as first class objects," in *Proceedings of the 9th IEEE International Symposium on Distributed Objects and Applications (DOA'99)*, Vilamoura, Algarve, September 1999, pp. 132–141.
51. M. Bertier, O. Marin, and P. Sens, "Implementation and performance evaluation of an adaptable failure detector," in *Proceedings of the International Conference on Dependable Systems and Networks*, Washington DC, June 2002, pp. 354–363.
52. R. van Renesse, Y. Minsky, and M. Hayden, "A gossip-style failure detection service," in *Proceedings of Middleware '98*, The Lake District, England, 1998, pp. 55–70.
53. T.D. Chandra and S. Toueg, "Unreliable failure detectors for reliable distributed systems," *Journal of the ACM*, 43 (2), 225–267, 1996.
54. X. D'efago, N. Hayashibara, and T. Katayama, "On the design of a failure detection service for large scale distributed systems," in *Proceedings of the International Symposium toward Peta-Bit Ultra-Networks (PBit)*, Ishikawa, Japan, September 2003, pp. 88–95.
55. N. Hayashibara, A. Cherif, and T. Katayama, "Failure detectors for large-scale distributed systems," in *Proceedings of the 21st IEEE Symposium on Reliable Distributed Systems*, Suita, Japan, October 2002, pp. 404–409.
56. W. Chen, S. Toueg, and M.K. Aguilera, "On the quality of service of failure detectors," *IEEE Transactions on Computers*, 51 (2), 13–32, 2002.
57. A. Demers, D. Greene, C. Hauser, W. Irish, J. Larson, S. Shenker, H. Sturgis, D. Swinehart, and D. Terry, "Epidemic algorithms for replicated database maintenance," *Operating Systems Review*, 22 (1), 8–32, January 1988.
58. Marin Bertier, Olivier Marin, and Pierre Sens, "Performance analysis of a hierarchical failure detector," in *Proceedings of the International Conference on*

Dependable Systems and Networks (DSN 2003), San Francisco, 22–25 June 2003, pp. 635–644.

59. Asser N. Tantawi and Manfred Ruschitzka, Performance analysis of check-pointing strategies. *ACM Trans. Comput. Syst.* 2 (2), 123–144, 1984.

60. Y. Wang, Y. Huang, K. Vo, P. Chung, C. Kintala, "Checkpointing and its applications," *Proceedings of the 25th International Symposium on Fault-Tolerant Computing*, 1995.

61. A. Gianelle, R. Peluso, and M. Sgaravatto, "Job partitioning and checkpointing," Technical Report DataGrid-01-TED-0119-0 3, European DataGrid Project, 2001.

62. A. Nguyen-Tuong, "Integrating fault-tolerance techniques in grid applications," Ph.D. dissertation, University of Virginia. Available at: www.cs.virginia.edu/an7s/publications/thesis/thesis.pdf.

63. R. Guerraoui and A. Schiper, "Software-based replication for fault tolerance," *IEEE Computer*, 30 (4), 68–74, 1997.

64. P. Felber and P. Narasimhan, "Experience, strategies, and challenges in building fault-tolerant CORBA systems," *IEEE Transactions on Computers*, 53 (5), 497–511, 2004.

65. A.S. Grimshaw, A. Ferrari, and E.A. West, "Menta," in G.V. Wilson and P. Lu (Eds.) *Parallel Programming Using C++*, MIT Press, Cambridge, MA, 1996, pp. 382–427.

66. G.J. Popek, R.G. Guy, T.W. Page, Jr., and J.S. Heidemann, "Replication in ficus distributed file systems," *IEEE Computer Society Technical Committee on Operating Systems and Application Environments Newsletter*, 4, 24–29, 1990.

67. S. Hwang and C. Kesselman, "Grid workflow: A flexible failure handling framework for the grid," in *Proceedings of the 12th IEEE International Symposium on High Performance Distributed Computing*, Seattle, Washington, 2003, pp. 126–137.

68. H. Casanova, J. Dongarra, C. Johnson, and M. Miller, "Application-specific tools," in I. Foster and C. Kesselman (Eds.) *The grid: Blueprint for a New Computing Infrastructure*, Morgan Kaufmann, San Francisco, CA, 1998, pp. 159–180.

69. K. Seymour, A. YarKhan, S. Agrawal, and J. Dongarra, "NetSolve: Grid enabling scientific computing environments," in Grandinetti, L. (Ed.) *Grid Computing and New Frontiers of High Performance Processing*, Elsevier, Advances in Parallel Computing, North-Holland, 14, 2005.

70. K. Hwang and Z. Xu, *Scalable Parallel Computing, Technology, Architecture, Programming*, McGraw-Hill, New York, NY, 1997, pp. 468–472.

71. A. Duda, "The effects of checkpointing on program execution time," *Information Processing Letters*, 16, 221–229, 1983.

72. N.H. Vaidya, "Impact of checkpoint latency on overhead ratio of a check-pointing scheme," *IEEE Transactions on Computing*, 46 (8), 942–947, August 1997.

73. M. Kksalan and S. Zionts, *Multiple Criteria Decision Making in the New Millennium*, Springer-Verlag, Ankara, Turkey, 2001.

5

Gridbus Workflow Enactment Engine

Jia Yu and Rajkumar Buyya

CONTENTS

5.1 Introduction

With the advent of grid technologies, scientists and engineers are building complex and sophisticated applications to manage and process large datasets and execute scientific experiments on distributed grid resources [1]. Building complex workflows requires means for composing and executing distributed applications. A workflow expresses an automation of procedures wherein files and data are passed between procedures applications, according to a defined set of rules, to achieve an overall goal [2]. A workflow management system defines, manages, and executes workflows on computing resources. The use of the workflow paradigm for application composition on grids offers several advantages [3] such as:

- Ability to build dynamic applications and orchestrate the use of distributed resources
- Utilization of resources that are located in a suitable domain to increase throughput or reduce execution costs
- Execution of spanning multiple administrative domains to obtain specific processing capabilities
- Integration of multiple teams involved in managing different parts of the experiment workflow—thus promoting interorganizational collaborations

Executing a grid workflow application is a complex endeavor. Workflow tasks are expected to be executed on heterogeneous resources that may be geographically distributed. Different resources may be involved in the execution of one workflow. For example, in a scientific experiment, one needs to acquire data from an instrument, and analyze it on resources owned by other organizations, in sequence or in parallel with other tasks. Therefore, the discovery and selection of resources for executing workflow tasks could be quite complicated. In addition, a large number of tasks may be required to be executed and monitored in parallel and the location of intermediate data may be known only at runtime.

This chapter presents a workflow enactment engine developed as part of the Gridbus Project at the University of Melbourne, Australia [4]. It utilizes tuple spaces to provide an event-driven mechanism for workflow execution entities. The benefits of this design include the ease of deployment for various strategies of resource selection and allocation, and supporting complex control and data dependencies of tasks with scientific workflows.

5.2 Architecture

The primary components of the workflow enactment engine (WFEE) [5] and their relationship with other services in the grid infrastructure are shown in Figure 5.1. Workflow applications, such as scientific application portals, submit task definitions along with their dependencies, expressed in a workflow language, as well as associated QoS requirements to WFEE. WFEE schedules tasks through grid middleware on the grid resources.

The key components of WFEE are workflow submission, workflow language parser, resource discovery, dispatcher, data movement, and workflow scheduler.

- Workflow submission accepts workflow enactment requests from planner level applications.
- Workflow language parser converts workflow description from XML format into Java objects, *Task*, *Parameter*, and *DataConstraint* (workflow dependency), which can be accessed by workflow scheduler.
- Resource discovery is carried out by querying grid information services such as Globus MDS [6], directory services, and replica catalogs to locate suitable resources for the tasks.

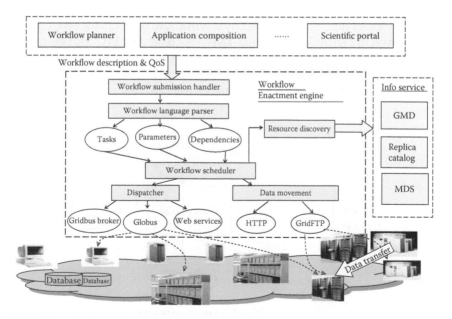

FIGURE 5.1 Architecture of WFEE.

- Dispatcher is used to access middleware. Resources may be grid-enabled by different middleware such as Globus [6] or Web services [7]. WFEE had been designed to support different middleware by creating dispatchers for each middleware to support interaction with resources.

- Data movement system enables data transfer between grid nodes by using HTTP and GridFTP [8] protocols.

- Workflow executor is the central component in WFEE. It interacts with *resource discovery* to find suitable grid resources at runtime; it locates a task on resources by using the dispatcher component; it controls input data transfer between task execution nodes through *data movement*.

5.3 Workflow Execution Management

The workflow execution is managed using a decentralized architecture. Instead of a central scheduler for handling whole workflow execution, a task manager is created for handling the processing of a task or a group of tasks, including resource discovery and allocation, task dispatcher, and failure processing. Different scheduling strategies can be deployed in different task managers (TMs) for resource selection, QoS negotiation, and data transmission optimization. The lifetimes of TMs, as well as the whole workflow execution, are controlled by a workflow coordinator (WCO).

As shown in Figure 5.2, dedicated TMs are created by WCO for each task group. Each TM has its own monitor, which is responsible for monitoring the health of the task execution on the remote node. Every TM maintains a resource group, which is a set of resources that provides services required

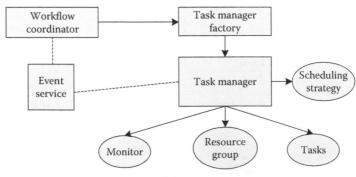

Workflow execution management

FIGURE 5.2   Execution management.

for the execution of an assigned task. TMs and WCO communicate through an event service server (ESS).

5.3.1 Communication Approach

A communication approach is needed for task managers. On the one hand, every task manager is an independent thread of execution and they can be run in parallel. On the other hand, the behavior of each task manager may depend on the processing status of other task managers according to the task dependencies. For example, a task manager should not execute the task on a remote node if the input generated by its parent tasks is not available for any reason.

In addition, in a workflow, a task may have more than one input that comes from different tasks. Furthermore, the output of these tasks may also be required by other task managers. Hence the communication model between the task managers is not just one-to-one or one-to-many, but it could be many-to-many depending on task dependencies of the workflow. Given this motivation, an event-driven mechanism with a subscription-notification model has been developed to control and manage execution activities. In the system, the behaviors of task managers and workflow coordinator are driven by events. A task manager is not required to handle communication with others and only generates events according to a task's processing status. At the same time, task managers take actions only depending on the events occurring without concern for details of other task managers. The benefit of this event-driven mechanism is that it provides loosely coupled control; hence the design and development of the system is very flexible and additional components can be easily plugged in.

The event notification is based on the subscription-notification model. WCO and TMs just subscribe to events of interest after activation, and then are informed immediately when a subscribed event occurs. There are three basic types of events, *status events*, *output events*, and *control events*. Status events are sent by the TMs to provide information on the status of task execution. Output events are sent by TMs to announce that the task output is ready along with the location of its storage. Control events are used to send control messages, such as to *pause* and *resume* the execution, to task managers.

As illustrated in Figure 5.3, TMs inform each other and communicate with the WCO through the ESS. For example, TMs put their task execution status (e.g., executing, done, failure) into the ESS, which notifies the WCO. If the output of a task is required by its child tasks, the task managers of the child tasks can subscribe to output events of the task. Once the task generates the required output, an output event is sent to the ESS, which immediately notifies the child TMs that have subscripted to the output event. A user can control and monitor the workflow execution by subscribing to status events and sending control events through a visual user interface.

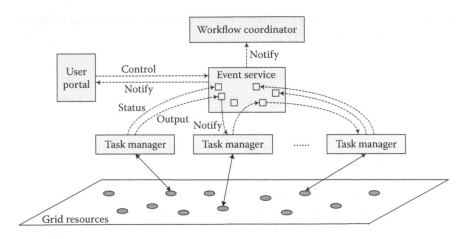

FIGURE 5.3 Event-driven mechanism.

5.3.2 State Transition

The state transition of a WCO is illustrated in Figure 5.4. WCO registers with the ESS and start TMs of first-level tasks and then monitors activated TMs. Upon receiving execution status from a TM, WCO starts the TMs of its child tasks. If WCO receives a status *done* event, it checks whether other TMs are still running. If so, WCO goes back to *monitoring*, otherwise it exits. If WCO receives a failed event from a TM, it proceeds to *failure processing*, and then ends.

The state transition of TMs is illustrated in Figure 5.5. The TM registers events, such as output events and status events, generated by its parent

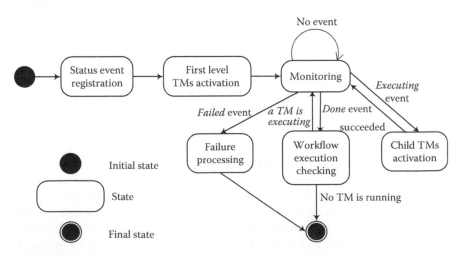

FIGURE 5.4 State transition of WCO.

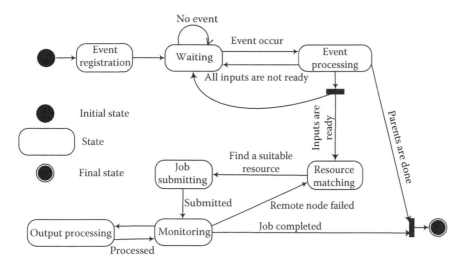

FIGURE 5.5   State transition of TM.

tasks and waits for the events to occur; when an event occurs, the TM goes to the *event processing* state. If all input data are available, it starts a new thread to process execution for a job; otherwise, the TM goes back to the *wait* state. A job is a unit of work that a TM sends to a grid node and one task may create more than one job. The job execution is started from *resource matching*, in which a suitable resource is selected from the resource group created by querying a directory service. If a suitable resource is available, the TM submits a job to the resource and then monitors the status of job execution on the remote resource. If the execution has failed, the TM goes back to *resource matching* and selects an alternative resource and then submits the job to it. If all parent tasks and execution jobs are completed, the TM ends.

5.3.3 Interaction

The interactions between the WCO, TMs, ESS, and remote resources are illustrated in Figure 5.6. First, the WCO needs to register to the ESS and to task status events. Then, the WCO activates TMS of first-level tasks of which, in this example, there is only one TM1. After TM1 finishes the pre-processing for the task execution, it sends a message to ESS saying "*I am executing the task.*" ESS informs the WCO and WCO activates TMs of the child tasks of TM1, namely TM2 and TM3, in this example.

The inputs of the task managed by TM2 and TM3 rely on the output of the task of TM1, so TM2 and TM3 register to ESS and listen to its output events. Once TM1 identifies a suitable resource, it submits a task to that resource. As soon as TM1 knows the output of the task, it informs TM2

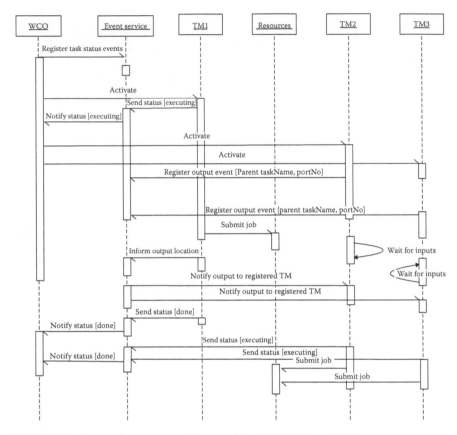

FIGURE 5.6   Interaction sequence diagram of the WCO, TMs, and ESS.

and TM3 through ESS, saying *"my output of port No. x is ready and its location is xxxx."* If all input data for TM2 and TM3 are ready, TM2 and TM3 report execution status to ESS, and then proceed to initialize the execution of their tasks. After WCO receives the notification of the execution of the tasks in TM2 and TM3, WCO will activate their child TMs, so that they can prepare for task execution. This process will be continued until the end of workflow execution.

5.4 Service Discovery

In a grid environment, many services with the same functionality and user interaction can be provided by different organizations. In addition, a service may be replicated and deployed in many locations. From the user's point of view, it is better to use a service that offers a higher performance at a lower price. Therefore, a method is required to allow users to find replicated services easily.

A directory service, called the grid market directory (GMD) [1], has been developed to support service publication and discovery in a market-oriented grid environment. GMD is an infrastructure that allows (1) the creation of one or more registries for service providers (2) the service providers to register the resources/application services that they wish to provide and (3) users such as workflow engines to discover resources/services and their attributes (e.g., access price, location, and usage constraints) that meet their QoS requirements.

Figure 5.7 illustrates service publishing and discovery in a grid environment through GMD. Service providers first register with the GMD and publish their static information such as location, service capability, and access methods. A grid user such as the workflow engine can query the GMD to find a suitable service. After that, the user can also query and subscribe to the service provider directly to obtain more dynamic information such as service execution status.

A provider can provide specialist applications for others to access remotely. Figure 5.8 shows an application service schema. An application service provider may also provide hosted machine information such as host name and host public key for remote secure access. Service providers also need to indicate middleware through which grid users can access the service.

The grid application model (GAM) is developed for application identification. GAM is a set of specifications and application programming interfaces (APIs) for a grid application. GAM can be published by service providers within the GMD, and the users can search GMD for services conforming to a particular GAM. Applications with the same GAM name provide the same function and API. In the case of the workflow system,

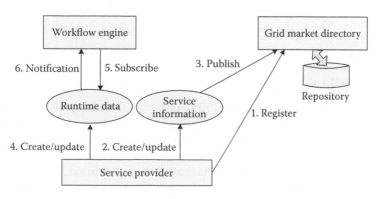

FIGURE 5.7 Service discovery using GMD.

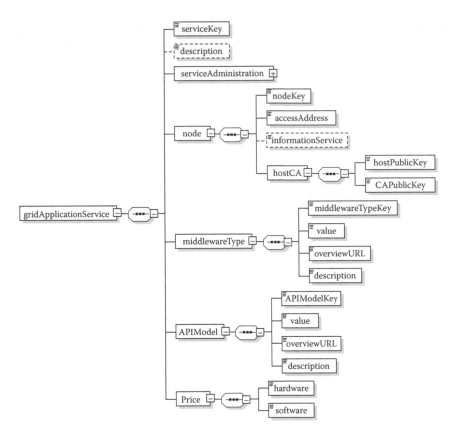

FIGURE 5.8   Grid application service schema.

if users do not specify a particular service for a task in the workflow description, the scheduler uses the GAM name associated with the execution of the task to query the GMD. The GMD will return a list of services. These services are all able to execute the task.

5.5 Workflow Language

In order to allow users to describe tasks and their dependencies, an XML-based workflow language (xWFL) has been defined. The workflow language provides the means to build new applications by linking stand-alone applications. Figure 5.9 shows the basic structure of the workflow language. It consists of two parts: task definitions defined in *<tasks>*, data dependencies defined in *<links>*, and QoS constraints defined in *<QoSconstraints>*.

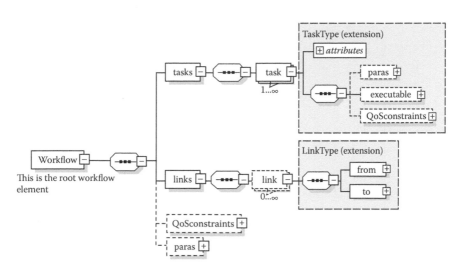

FIGURE 5.9   Structure of workflow language.

5.5.1 Tasks

The element *<tasks>* is a set of definitions of tasks that are to be executed. Figure 5.10 shows the schema of task definition. A task can be a single task or a parameter sweep task. A parameter sweep task is able to process a set of parameters. The parameters are defined in *<paras>*. The detailed design of parameter tasks is introduced later. The element *<executable>* is used to define the information about the application, input and output data of the task. The workflow language supports both abstract and concrete workflows. The user or higher workflow planner can either specify the location of a particular service providing a required application in *<service>* or leave it to the engine to identify providers dynamically at runtime. The middleware of the application is identified through a service information file by the GMD when dispatching tasks.

In the example that follows, task A executes the dock.exe program on the host *bellegrid.com* in the directory */services* and the executable *dock* has two input I/O ports: port 0 (a file) and port 1 (a parameter value). The example shows that task *A* has only one output.

```
<task name= "A">
  <executable>
  <name>dock</name>
  <service>
    <hostname="bellegrid.com" />
      <accesspoint value="/services/dock.exe" />
  </service>
  <input>
```

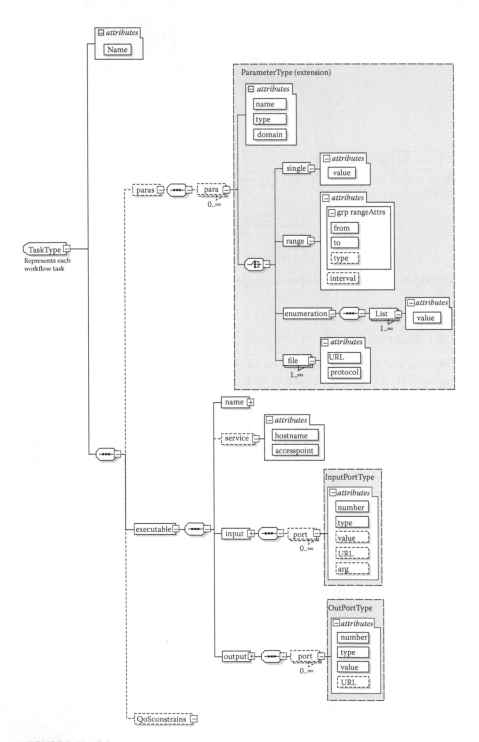

FIGURE 5.10   Schema of task definition.

```
<port num=0 type="file" url=http://www.gridbus.org/
dock.in value="dock.in"/>
   <port num=1 type="msg" value=1/>
</input>
<output>
   <port num=2 type="file" value="dock.out"/>
</output>
</executable>
</task>
```

5.5.2 Data Dependencies

A data link is used to specify the data flow between two tasks. The schema of the data link is defined in Figure 5.9. Figure 5.11 shows the example of a data flow description. The inputs of task *B* and task *C* rely on the output of *A*. The output of *A* needs to be transferred to the node on which tasks *B* and *C* are executed. Input could be a file, parameter value, or data stream.

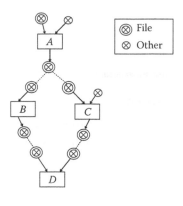

```
<workflow>
  <tasks>
    <task name= "A">
       ....
    </task>
    <task name= "B">
       .....
    </task>
  <task name= "C">
     ......
  </task>
  <task name= "D">
     ......
  </task>
  </tasks>
  <links>
  <link>
   <from task="A" port=2 />
   <to task="B" port=0 />
  </link>
  <link>
   <from task="A" port=2 />
   <to task="C" port=0 />
  </link>
  <link>
   <from task="B" port=1 />
   <to task="D" port=0 />
  </link>
  <link>
   <from task="C" port=2 />
   <to task="D" port=1 />
  </link>
  </links>
</workflow>
```

FIGURE 5.11 Flow diagram of tasks *A*, *B*, *C*, and *D*.

5.6 Parameterization

Supporting parameterization in the workflow language is very important for scientific applications. It enables scientists to perform experiments across a range of different parameters without being concerned about the detailed workflow description. A parameter defined in each task type is called a *local parameter*; when it is defined for the entire workflow, it is called a *global parameter*. As shown in Figure 5.9, multiple parameter types such as single, range, file, and enumeration are supported. An example for a single parameter type and a range parameter type is given in Figure 5.12.

Among these parameter types, range and enumeration types are used to define a range or a list of parameters valued that the task is required to be executed. This type of task is called a *parameter sweep* task [9] and is structured as a set of multiple execution jobs, each of which is executed with a distinct set of parameters. Figure 5.13 illustrates parameter sweep tasks. Two inputs are required by task *A* and task *B*; one is a local parameter and the other is a global parameter. The local parameter of *A* is a range type parameter whereas the local parameter of *B* is an enumeration type. At execution time, the global parameter value is combined with each local parameter value and generates ten subjobs for task *A* and three subjobs for task *B*.

5.7 I/O Models

As shown in Figure 5.11, a task receives output data from its parent as its input through a data link. However, for some tasks, more than one output could be generated by one output port. For example, such a task could be a data collecting task that continues to read information from a sensor

```
<paras>
      <para type= "single">
            <name>X</name>
            <value type=integer>10</value>
      </para>
      <para type= "range">
            <name>Y</name>
      <min>1</min>
            <max>20</max>
            <step>2</step>
      </para>
</paras>
```

FIGURE 5.12 Single parameter and range parameter.

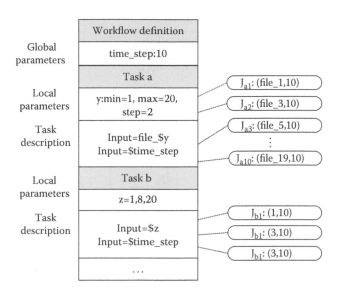

FIGURE 5.13   Illustration of workflow parameters.

device and generate corresponding output data or a parameter sweep task that generates multiple data based on various sets of parameters. These outputs can be produced at different times. Some successor tasks may not require to be processed until all these output data are generated. It can process the output once it is available. However, ancestor tasks can process the output data generated from a single parent differently, depending on their requirements.

Three I/O models have been developed in the workflow system to provide data-handling capabilities. These models are: *many-to-many, many-to-one*, and *synchronization*. In Figure 5.13, there are two tasks: task *A* and task *B*. They are connected by a data link. There are multiple subjobs in *A* and each job produces an output. For the many-to-many model, task *B* starts to process data and generates an output once there is an input available on the data link. As shown in Figure 5.14a, four outputs generated by four subjobs of *A* are processed individually by four subjobs of *B*. For the many-to-one model, task *B* starts to process data once there is an input available; however, the result is calculated based on the result generated by earlier subjobs. As shown in Figure 5.14b, subjob *B1* processes the output generated by sub-job *A1*. Once the output of *A2* is available, sub-job *B2* is created and processes the output of *A2* based on the output generated by *B1*. For the synchronization model, task *B* does not start processing until all the output is available on the data link. As shown in Figure 5.14c, there is only one subjob in task *B* and it processes all outputs generated by subjobs of *A* at one time.

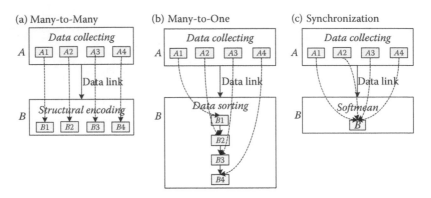

FIGURE 5.14   Input/output models.

5.8 Fault Handling

Two fault handling mechanisms are developed in the system, *retry submission* and *critical task replication*. The retry submission mechanism reschedules a failed job onto a current available resource, and also records the number of failed jobs for each resource. Once the failed job number exceeds a *warning threshold*, the scheduler decreases the number of jobs submitted to these resources. If the number of failed jobs exceeds a *critical threshold*, the scheduler terminates submission of jobs onto this resource. The critical task replication mechanism replicates a task execution on more than one resource. The result produced earliest is then used for the rest of the workflow. This mechanism is designed to execute a long running task when there are multiple spare resources.

5.9 Implementation

The WFEE has been implemented by leveraging the following key technologies: (1) IBM TSpaces [10] for supporting subscription-notification-based event exchange; (2) Gridbus broker [11] for deploying and managing job execution on various middleware; (3) XML parsing tools including JDOM [12]. The detailed class diagram and event server implementation are presented as follows:

5.9.1 Design Diagram

The class design diagram of the WFEE is shown in Figure 5.15. *XMLParsing ToModel* parses XML-formatted workflow description into Java objects

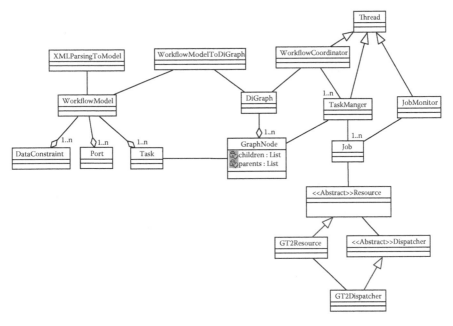

FIGURE 5.15   Class diagram of WFEE.

that are instances of class of *Task*, *Port*, and *DataConstraint*. These objects are passed on to *WorkflowModel*. *WorkflowModelToDiGraph* converts *Workflow Model* into a directed graph represented by class *DiGraph* which encompasses many *GraphNode* objects. An instance of *GraphNode* contains a workflow task and the references of *GraphNodes* of its parent and child tasks. *WorkflowCoordinator* creates and controls the instantiation of *TaskManager* according to the graph node dependencies. *Job* class represents a unit of work assigned to a grid resource. Every job has a monitor implemented by *JobMonitor* to monitor job execution status on the remote node. In order to extend WFEE to support multiple grid middleware, we abstract class *Resource* and *Dispatcher*, which provides interfaces that interact with grid resources.

5.9.2 Event Messages

We have utilized tuple spaces to exchange events. The tuple space model originated at Yale with the Linda system [13]. A tuple is simply a vector of typed values (fields). A tuple space is a collection of tuples that can be shared by multiple parties by using operations such as read, write, and delete. In our work, we have leveraged IBM's recent implementation of tuple spaces called TSpaces [10] to be the event server.

There are three types of event tuples whose format is shown in Table 5.1, *task status event*, *output event*, and *job status event*. The task status event is sent

TABLE 5.1

Format of Events

Event Name	Field 1	Field 2	Field 3	Tuple Template for Registration
Task status event	Task no.	"Status"	Value	New tuple (new Field(String), status, new Field(String))
Output event	Task no.	Port no.	Value	New tuple (taskNo, portNo, new Field(String))
Job status event	Job no.	Task no.	Value	New tuple (new Field(String), taskNo, new Field(String))

by TMs and WCO used to control TMs activation. The first field is the ID of the task, the second field is string "status" to indicate the type of tuple for the registration purposes, and the third field gives the value of status.

Child TMs need output events sent by the parent TMs to be informed if their inputs are available. The events have three fields: the task ID is given in the first field, the second field is port numbers, and the third field is the location of output. One task can have multiple jobs. Job status events are sent by the job monitor. Every job status event provides a job ID and its task ID with status value. TMs make decisions according to the job events. For example, when a job has failed, the TM can reschedule it on another resource in the resource group. The tuple templates are used for subscribing to the corresponding event. For example, task status events can be received by a tuple template with the second field as a String called "status."

5.10 A Case Study in fMRI Data Analysis

Magnetic resonance imaging (MRI) [14] uses radio waves and a strong magnetic field to provide detailed images of internal organs and tissues. Functional MRI (fMRI) [15] is a procedure that uses MRI to measure the signal changes in an active part of the brain. fMRI is becoming a major diagnostic method for learning how a normal or diseased brain work. fMRI images are obtained by scanning the brains of subjects as they perform cognitive tasks. A typical study of fMRI data consists of multiple-stage processes that begin with the preprocessing of raw data and conclude with a statistical analysis. Such analysis procedures often require hundreds to thousands of images [16].

5.10.1 Population-Based Atlas Workflow

Population-based atlas [17] creation is one of the major fMRI research activities. These atlases combine anatomy imaging data from healthy and diseased populations. These describe how the brain varies with age, gender, and demographics. They can be used for identifying systematic effects on brain structure. For instance, they provide a comprehensive approach for studying a particular subgroup, with a specific disease, receiving different medications, or neuropsychiatric disorder. Population-based atlases contain anatomical models from many subjects. They store population templates and statistical maps to summarize features of the population. They also average individual images together so that common features of the subgroup are reinforced.

Figure 5.16 shows a workflow that employs the automated image registration (AIR) [18] and FSL [19] suite for creating population-based brain atlases from high-resolution anatomical data. The stages of this workflow are as follows:

1. The inputs to the workflow are a set of brain images, which are 3D brain scans of a population with varying resolutions and a reference brain image. For each brain image, *align_warp* adjusts the position and shape of each image to match the reference brain. The output of each process is a *warp parameter set* defining the spatial transformation to be performed.

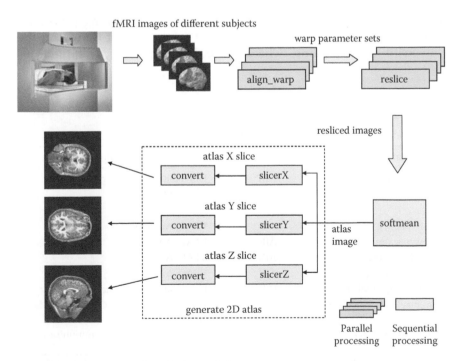

FIGURE 5.16 Population-based atlas workflow.

2. For each warp parameter set, *reslice* creates a new version of the original brain image according to the configuration parameters defined in the warp parameter set. The output of each reslice procedure is a resliced image.

3. *softmean* averages all the resliced images into one single atlas image.

4. The averaged image is sliced using *slicer* to give a 2D atlas along a plane in three dimensions (*x*, *y*, and *z*), taken through the center of the 3D image. The output is an atlas dataset.

Finally, each atlas dataset is converted into a graphical atlas image using *convert*.

5.10.2 Experiment

The experiment was conducted using the testbed provided by the University of Melbourne (Australia), Victorian Partnership for Advanced Computing (VPAC) (Australia), University of Electro-Communications (Japan), and University of Innsbruck (Austria). The configuration of all resources is listed in Tables 5.2 and 5.3. Table 5.2 shows the application software available on every resource. All application software cannot be installed on every resource, due to their varied capability and administration policy. The AIR application required for executing procedure *align_warp*, *reslice*, and *softmean* is installed on the sites of VPAC and the University of Melbourne, whereas the FSL application required for

TABLE 5.2

Applications Configuration of Grid Sites

Node/Details	Applications	Location
Manjra.cs.mu.oz.au	AIR* FSL Convert	University of Melbourne, Australia
vgtest.vpac.org	AIR	VPAC, Australia
Vgdev.vpac.org	AIR	VPAC, Australia
Brecca-1.vpac.org	AIR	VPAC, Australia
Brecca-2.vpac.org	AIR	VPAC, Australia
karwendel.dps.uibk.ac.at	FSL	University of Innsbruck, Austria
uuuu.maekawa.is.uec.ac.jp	FSL	University of Electro-Communications, Japan
walkure.maekawa.is.uec.ac.jp	FSL	University of Electro-Communications, Japan

*AIR package includes software for executing the task.

TABLE 5.3

Resource Attributes

Node/Details	CPU (type/#/GHz)	Middleware
manjra.cs.mu.oz.au	4/Intel Xeon/2.00 GHz	SSH/GT2
vgtest.vpac.org	1/Intel Xeon/3.20 GHz	SSH/GT4
ngdev.vpac.org	1/Intel Xeon/3.20 GHz	SSH/GT4
brecca-1.vpac.org	Intel Xeon/4/2.80 GHz	SSH
brecca-2.vpac.org	4/Intel Xeon/2.80 GHz	SSH
karwendel.dps.uibk.ac.at	2/AMD Opteron 880/2.39 GHz	SSH/SGE
uuuu.maekawa.is.uec.ac.jp	1/Intel Xeon/2.80 GHz	SSH/GT4
Walkure.maekawa.is.uec.ac.jp	1/Intel Xeon/2.80 GHz	SSH/GT4

executing procedure *slicer* is installed on other sites. The Convert application required to execute procedure *convert* is available only on the site of the University of Melbourne. Table 5.3 shows the processor capability and supporting middleware of each resource.

In the first experiment, the impact of the number of grid sites is investigated for the various numbers of subjects. Figure 5.17 shows the total execution times using 1–5 grid sites for generating an atlas of 25, 50, and 100 subjects. The size of the image file associated with each subject is around 16–22 MB. We can see that the total execution time increases as the number of subjects increases. Additionally, the larger the number of grid sites, the faster the execution time achieved. For example, the total execution time of generating an atlas of 100 subjects using one grid node is 95 minutes;

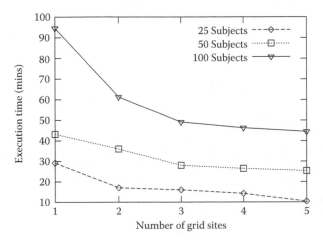

FIGURE 5.17 Total execution times of processing 25, 50, and 100 subjects over various Grid sites.

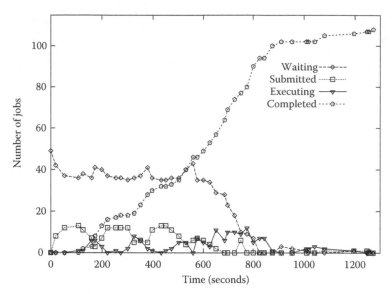

FIGURE 5.18   Execution progress for processing 50 subjects.

however, it takes only 45 minutes using five grid nodes. The speedup rate is over 50%. It shows that the performance of conducting fMRI data analysis can be significantly improved by using the grid.

Figure 5.18 shows the execution progress for processing 50 subjects. At the beginning of the workflow execution, 50 *align_warp* jobs are generated for the first step, and each job processes one subject image. Once a job in step 1 is completed, the task manager of step 2 is notified by the output event of this job. It then generates a new *reslice* job of step 2. Therefore, the number of waiting jobs does not continuously decrease when *align_warp* jobs are completed. In Figure 5.18, we can observe that the number of waiting jobs remains around 50 until 50 jobs of step 1 are completed. All the results of step 2 are processed once by the *softmean* task, and the completion of *softmean* generates three *slicer* jobs to produce 2D images along three dimensions. Therefore, there are only a small number of waiting jobs after 600 seconds.

Figure 5.19 shows the number of jobs of each task running over the execution. Available jobs of step 1 and step 2 can be executed in parallel. However, the jobs of step 1 have a higher priority than those of step 2, when they compete for resources. As we can observe, most of the execution jobs during 0–580 seconds are produced by *alignWarp* and only a few of the jobs of *reslice* were executed.

Table 5.4 shows the start and end time for each task in the workflow for 100 subjects. The start time we measured is the time when the stage-in of input data to the remote resource is started and the end time is the

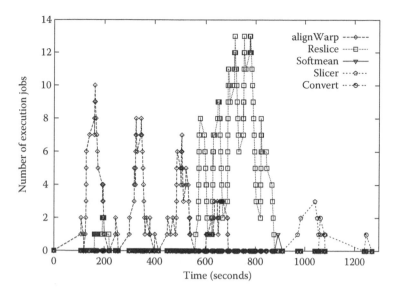

FIGURE 5.19 Execution tasks for processing 50 subjects.

completion time of a task. As we can see from Figure 5.16, there are multiple subtasks in task *align_wrap* and *reslice*. The task *reslice* process was started after the *align_wrap* process, once an output produced by a sub-task of *align_wrap* was available. However, the task *softmean* is started to process after all subtasks of *align_wrap* have been completed, because it requires all results produced by the subtask of *align_wrap* to generate a mean image. Theoretically, after the task *softmean* finishes, its child tasks should be submitted immediately; however, there are some time intervals

TABLE 5.4

Detailed Execution Times of the Tasks for Processing 100 Subjects

Task	Start Time (min)	End Time (min)	Duration (min)
align_warp	0	22.82	22.82
reslice	2.65	28.50	25.85
softmean	28.92	43.08	14.17
sliceX	43.1	44.1	1
sliceY	43.1	44.13	1.03
sliceZ	43.1	44.07	0.97
convertX	44.4	44.72	0.32
convertY	44.42	44.75	0.33
convertZ	44.07	44.43	0.37

between the parent tasks' end time and the child tasks' start time. This gap can be attributed to the overhead of running WFEE, including time involved in processing event notifications, resource discovery, and remote resource submission. However, compared with the running time of tasks, this gap is insignificant and less than 2%.

5.11 Related Work

The workflow engine presented in this chapter is an independent workflow execution system and takes advantage of various middleware services such as security, grid resource access, file movement, and replica management services provided by the Globus middleware [6,20], and multiple middleware dispatchers provided by the Gridbus broker.

Much effort toward grid workflow management has been made. DAGMan [21] was developed to schedule jobs to the Condor system in an order represented by a DAG and to process them. With the integration of Chimera [22], Pegasus [23] map, and execute complex workflows based on full-ahead planning. In Pegasus, a workflow can be generated from a metadata description of the desired data product using AI-based planning technologies. The Taverna project [24] has developed a tool for the composition and enactment of bioinformatics workflow for the life science community. The tool provides a graphical user interface for the composition of workflows. Other workflow projects in the grid context include UNICORE [25], ICENI [26], Karajan [27], Triana [28], and ASKLON [29].

Compared with the work listed above, the workflow engine provides a decentralized scheduling system by using the tuple spaces model, which facilitates deployment of different scheduling strategies to each task. It also enables resources to be discovered and negotiated at runtime.

A number of workflow languages [7,30,31] have been developed and most of them focus on the composition of Web services. However Web services are not the standard middleware used by the majority of today's scientific domains [32]. The workflow language proposed in this chapter is middleware independent and also supports parameterization [9], which is important to scientific applications.

5.12 Summary

In this chapter, a workflow enactment engine is introduced to facilitate composition and execution of workflows in a user-friendly manner. The engine supports different grid middleware as well as runtime service discovery. It is capable of linking geographically distributed standalone applications and takes advantage of distributed computational resources to achieve high throughput.

The event-driven and subscription-notification mechanisms developed using the tuple spaces model make the workflow execution loosely coupled and flexible. Supporting parameterization in the workflow language allows users to easily define a range and list of parameters for scientific experiments to generate a set of multiple parallel execution jobs. The engine has been successfully applied to an fMRI analysis application. The engine presented in this chapter facilitates users to build workflows to solve their domain problems and provides a basic infrastructure to schedule workflows in grid environments.

References

1. J. Yu and R. Buyya. A Taxonomy of Workflow Management Systems for Grid Computing. *Journal of Grid Computing*, 3 (3–4), 71–200, September 2005.
2. D. Hollinsworth. *The Workflow Reference Model*. Workflow Management Coalition, TC00-1003, 1994.
3. D.P. Spooner, J. Cao, S.A. Jarvis, L. He, and G.R. Nudd. Performance-Aware Workflow Management for Grid Computing. *The Computer Journal*, 48 (3), 347–357, 2005.
4. R. Buyya and S. Venugopal. The Gridbus Toolkit for Service Oriented Grid and Utility Computing: An Overview and Status Report. *Proceedings of the First IEEE International Workshop on Grid Economics and Business Models*, pp. 19–66, Seoul, April 23, 2004.
5. J. Yu and R. Buyya. A Novel Architecture for Realizing Grid Workflow using Tuple Spaces. *Proceedings of the 5th IEEE/ACM International Workshop on Grid Computing (Grid 2004)*, pp. 119–128, Pittsburgh, November 8, 2004.
6. I. Foster. Globus Toolkit Version 4: Software for Service-Oriented Systems. *IFIP International Conference on Network and Parallel Computing*, pp. 425–432, Beijing, China, December 2005.
7. T. Andrews, F. Curbera, H. Dholakia, Y. Goland, J. Klein, J. Klein, F. Leymann, et al. Business Process Execution Language for Web Services Version 1.1, 05 May 2003. Available at: http://www-128.ibm.com/developerworks/library/ws-bpel/, accessed February 2005.
8. B. Allcock, J. Bester, J. Bresnahan, A.L. Chervenak, I. Foster, C. Kesselman, S. Meder, V. Nefedova, D. Quesnal, and S. Tuecke. Data Management and Transfer in High Performance Computational Grid Environments. *Parallel Computing Journal*, 28 (5), 749–771, May 2002.
9. D. Abramson, J. Giddy, and L. Kotler. High Performance Parametric Modeling with Nimrod/G: Killer Application for the Global Grid? *Proceedings of the 14th International Parallel and Distributed Processing Symposium (IPDPS 2000)*, pp. 520–528, Cancun, Mexico, May 1–5, 2000.
10. P. Wyckoff. TSpaces, *IBM Systems Journal*, 37, 454–474, 1998.

11. S. Venugopal, R. Buyya, and L. Winton. A Grid Service Broker for Scheduling e-Science Applications on Global Data Grids. *Concurrency and Computation: Practice and Experience*, 18 (6), 685–699, May 2006.
12. JDOM. Available at: http://www.jdom.org, accessed December 2004.
13. N. Carriero and D. Gelernter. Linda in Context. *Communications of the ACM*, 32, 444–458, April 1989.
14. J. Mattson and M. Simon. *The Pioneers of NMR and Magnetic Resonance in Medicine: The Story of MRI*. Jericho & New York: Bar-Ilan University Press, 1996.
15. J. Van Horn. Online Availability of fMRI Results Images. *Journal of Cognitive Neuroscience*, 15 (6), 769–770, 2003.
16. Y. Zhao, J. Dobson, I. Foster, L. Moreau, and M. Wilde. A Notation and System for Expressing and Executing Cleanly Typed Workflows on Messy Scientific Data. *ACM SIGMOD Record*, 34, pp. 37–43, September 2005.
17. P. Thompson, M.S. Mega, and A.W. Toga. "Sub-Population Brain Atlases," in A.W. Toga and J.C. Mazziotta (Eds.) *Brain Mapping: The Methods* (2nd edn), Academic Press, USA, 2002.
18. R.P. Woods, S.R. Cherry, and J.C. Mazziotta. Rapid Automated Algorithm for Aligning and Reslicing PET images. *Journal of Computer Assisted Tomography*, 16, 620–633, 1992.
19. S. Smith, P. Bannister, C. Beckmann, M. Brady, S. Clare, D. Flitney, P. Hansen, et al. FSL: New Tools for Functional and Structural Brain Image Analysis. *Proceedings of the 7th International Conference on Functional Mapping of the Human Brain*, June 10–14, Brighton, 2001.
20. I. Foster and C. Kesselman. Globus: A Metacomputing Infrastructure Toolkit. *International Journal of Supercomputer Applications*, 11, 115–128, 1997.
21. D. Thain, T. Tannenbaum, and M. Livny. "Condor and the Grid," in *Grid Computing: Making the Global Infrastructure a Reality*. John Wiley, New York, 2003.
22. I. Foster, J. Vöckler, M. Wilde, and Y. Zhao. Chimera: A Virtual Data System for Representing, Querying, and Automating Data Derivation. *Proceedings of the 14th International Conference on Scientific and Statistical Database Management (SSDBM)*, pp. 37–46, Edinburgh, July 24–26, 2002.
23. E. Deelman, J. Blythe, Y. Gil, C. Kesselman, G. Mehta, S. Patil, M.H. Su, et al. Pegasus: Mapping Scientific Workflow onto the Grid. *Proceedings of the Across Grids Conference 2004*, pp. 11–20, Nicosia, Cyprus, 2004.
24. T. Oinn, M. Addis, J. Ferris, D. Marvin, M. Senger, M. Greenwood, T. Carver, et al. Taverna: A Tool for the Composition and Enactment of Bioinformatics Workflows. *Bioinformatics*, 20 (17), 3045–3054, 2004.
25. M. Romberg. The UNICORE Architecture Seamless Access to Distributed Resources. *Proceedings of the 8th IEEE International Symposium on High Performance Computing*, pp. 287–293, Redondo Beach, 1999.
26. S. McGough, L. Young, A. Afzal, S. Newhouse, and J. Darlington. Workflow Enactment in ICENI. *Proceedings of the UK e-Science All Hands Meeting*, pp. 894–900, Nottingham, September 2004.
27. G. von Laszewski and M. Hategan. *Java CoG Kit Karajan/GridAnt Workflow Guide*. Technical Report, Argonne National Laboratory, Argonne, 2005.
28. I. Taylor, M. Shields, and I. Wang. Resource Management of Triana P2P Services. *Grid Resource Management*. Kluwer, Amsterdam, 2003.

29. T. Fahringer, A. Jugravu, S. Pllana, R. Prodan, C.S. Jr, and H.L. Truong. ASKALON: A Tool Set for Cluster and Grid Computing. *Concurrency and Computation: Practice and Experience*, 17, 143–169, 2005.
30. T. Fahringer, S. Pllana, and A. Villazon. AGWL: Abstract Grid Workflow Language. *Proceedings of the International Conference on Computational Science, Programming Paradigms for Grids and Meta-Computing Systems*, pp. 42–49, Krakow, June 2004.
31. S. Krishnan, P. Wagstrom, and G.V. Laszewski. *GSFL: A Workflow Framework for Grid Services*. Technical Report Preprint ANL/MCS-P980-0802, pp. 1–12, Argonne National Laboratory, Argonne, 2002.
32. Y. Zhao, M. Wilde, I. Foster, J. Voeckler, T. Jordan, E. Quigg, and J. Dobson. Grid Middleware Services for Virtual Data Discovery, Composition, and Integration, *Proceedings of the 2nd Workshop on Middleware for Grid Computing*, pp. 57–62, Toronto, 2004.
33. J. Cardoso. *Stochastic Workflow Reduction Algorithm*. Technical Report, LSDIS Lab, Department of Computer Science University of Georgia, 2002.
34. I. Tayler, E. Deelman, D. Gannon and M. Shields (Eds.). *Workflows for e-Science: Scientific Workflows for Grids*. Springer-Verlag, London, December 2006.
35. J. Yu, S. Venugopal, and R. Buyya. A Market-Oriented Grid Directory Service for Publication and Discovery of Grid Service Providers and their Services. *Journal of Supercomputing*, 36 (1), 17–31, April 2006.

Part II

Grid Services

6

UK National Grid Service

Andrew Richards and Gillian M. Sinclair

CONTENTS

6.1 Introduction

The UK National Grid Service (NGS) [1] currently provides researchers within the UK and their collaborators with access to a wide range of computing and infrastructure resources using grid technologies that have been developing in the UK since 2001. The NGS is not the only grid activity within the UK and has developed alongside other activities such as GridPP, EGEE (Enabling Grids for E-SciencE), and other institutional-based activities such as campus grids. Working toward the future, the NGS is aiming to increase the collaboration with such existing activities and also look toward a sustainable future, in line with other activities such as the planned European Grid Initiative (EGI).

6.2 Background to the Creation of the NGS

The UK National Grid Service developed out of the much larger e-Science program funded within the UK from 2001. The e-Science program comprised a core program to develop and broker generic technology solutions and application-led Research Council programs to build communities and develop exemplar services running on top of the underlying grids. For the grid infrastructure development two main aspects were the establishment of regional e-Science centers [2] and the creation of the UK Engineering Task Force (ETF) [3], which oversaw, using expertise from each of the regional e-Science centers, the development of prototype national grids within the UK. In parallel, other activities such as the GridPP project were funded to specifically focus on developing e-infrastructure for the particle physics community and to be a part of the much larger LHC experiment at CERN and the Worldwide LHC Computing Grid Project (W-LCG) and, in time, the EGEE project.

The ETF oversaw the collaborative efforts of the regional e-Science centers in order to provide a grid test-bed for the deployment and evaluation of grid middleware and applications. These efforts culminated in the creation of the "Level 2" [4] grid in 2003, which paved the way for the creation of a production grid service in the UK that would underpin e-Research activities. In 2003 the ETF Production Grid (ETFp) was created, which was later renamed the National Grid Service. This first phase of the project was to establish, at four sites in the UK, resources to underpin the grid infrastructure. In the first phase the constituent sites chosen to be part of the NGS were the Science and Technology Funding Council (STFC; then called CCLRC), the University of Oxford, the University of Manchester, and the White Rose Grid site at the University of Leeds. Hardware was procured through a tender exercise from Clustervision and was deployed into service starting in April 2004.

6.3 Supporting the National Grid

Funded as part of the original e-Science core program by the Department of Trade and Industry (DTI), the Grid Support Center (GSC) was funded for two and a half years by the DTI Core e-Science Program. It was managed by the then Central Laboratory of the Research Councils (CLRC; now STFC) e-Science Center and run by four STFC staff plus one FTE (full time equivalent) from each University of Manchester Research Computing facility and Edinburgh (Edinburgh Parallel Computing Center) respectively. GSC also received direct funding from the Particle Physics Research Council (PPARC) Biotechnology and Biological Sciences Research Council (BBSRC), and Natural and Environmental Research Council (NERC), and collaborated with the University of Manchester Physics Department to provide support for the HEP Data Grid via the GridPP project.

The GSC initially helped sites to deploy Globus, Condor, and SRB and to learn how to manage and use the middleware. It produced the Grid Starter

Kit and associated Web sites, and its staff took leading roles in tutorials, courses, and workshops.

From October 2004 to September 2006 the Grid Operations Support Center, funded by the Engineering and Physical Sciences Research Council (EPSRC), continued the work of the Grid Support Center in providing end user support of grid activities within the UK. It was also part of the Grid Operations Support Center remit to directly support the activities of the then separately funded National Grid Service. The Grid Operations Support Center and the National Grid Service were combined under the one name of the National Grid Service for phase 2, which started in October 2006.

6.4 National Grid Service Today

Today the NGS manages and coordinates the generic grid infrastructure for e-research within the UK. Combining the expertise of a core set of sites providing access to computational and data resources, a support center providing end-user support, documentation, and training, materials the NGS currently supports over 700 individual users and a number of projects that directly access the NGS and partner resources.

The mission of the NGS is to provide coherent electronic access for UK researchers to all computational and data-based resources and facilities required to carry out their research, independent of resource or researcher location. The vision of the NGS is a robust, reliable, and trusted service providing electronic access to the full range of computation and data-based research facilities required by UK researchers. Access to resources is location independent and based upon integrated open standards. A range of sophisticated services to support coordinated collaborative and cross-resource activities is provided.

To achieve this, the NGS leads the deployment of a common grid infra-structure for combining services and information from multiple sources, reducing the need for specific arrangements for access, bespoke software, or intensive researcher effort. Through the NGS Partnership program, the NGS integrates services to access a growing number, scale, and variety of resources spanning the complete space from advanced real-time facilities such as synchrotrons and telescopes through to complex queries of historical data stored in national or institutional data centers.

The NGS operates the core services required to exploit local, national, and international partner facilities, providing support to the users of the service, and monitors and troubleshoots the services provided by all part-ners and the underlying infrastructure. This integration supports straight-forward migration of work from local to national or international resources as required and provides a standard set of interfaces and services whereby local systems can be shared across institutions to optimize use and provide a return on investment. The NGS aims to pioneer systematic arrangements that relieve the development load on both resource providers and resource users. The NGS also supports the policy, operational, and monitoring framework required to support the UK's grid infrastructure.

To oversee and manage the project the NGS appointed an executive director, who has overall day-to-day responsibility for the project and works in conjunction with the NGS Director (PI) and the NGS Technical Director. Separate boards were created to oversee the project (NGS Board), the project management (NGS Development Board), day-to-day management of member sites and operational policies (NGS Technical Board), and day-to-day technical implementation and system management (NGS Operations Board). The work of the various boards is encapsulated in the NGS project plan, which comprises nine work packages, plus a special work package to interact with and include input from the Engineering Task Force (ETF).

- WP1 project management: To oversee the overall planning and implementation of the NGS project. This includes financial management of the project and appropriate staffing.
- WP2 support center: To oversee the core NGS Support Center activities providing end-user support and operating auxiliary services as defined in the support center service-level definition.
- WP3 core node deployment and operations: To procure, deploy, and operate the hardware that comprises the core site resources at the four sites of University of Oxford, University of Manchester, White Rose Grid University of Leeds, and STFC Rutherford Appleton Laboratory.
- WP4 resource and partner integration: To coordinate and engage with resource providers and ensure support is provided so that partner and affiliate institutes are successfully joined into the NGS.
- WP5 sustainability and business plan: To plan for the future of the NGS and ensure a sustainable e-infrastructure that researchers can depend upon.
- WP6 application support: To deploy applications requested by users and develop as appropriate tools to support the NGS user community.
- WP7 outreach, training, and user documentation: To provide end-user training and to develop and coordinate outreach activities to engage with current and prospective users.
- WP8 core service development: To oversee the development of additional services, as required, for the successful operation of the NGS. This has included activities such as the provision of an accounting and user management system.

- WP9 road mapping: To ensure the NGS future development and deployment strategy is planned and accounts for changing technologies and user-requested services.
- ETF: To provide an interface between the NGS activities and those of the ETF. This ensures that the ETF undertakes work appropriate to that of the NGS and ensures the NGS is aware of technological changes.

6.5 Core Founding Member Sites

The current NGS, based around four cluster-based computer systems and the UK's High Performance Computing services, entered full production in September 2004, following a preproduction test phase that started in March 2004. During the period from March 2004 to June 2006, the NGS continued to grow in both numbers of users and partner institutions.

The NGS is made up of different physical types of nodes but it can be useful to classify them as one of two types, compute nodes or data nodes. Compute nodes are powerful in terms of computation whereas data nodes provide more storage. The additional storage supports database activities, through the provision of high-performance Oracle systems, and the ability to store other large datasets as requested by end users. Each node also has a useful software environment that enables high-performance parallel programs to be developed and executed. The base level of software required to be compatible with the NGS is defined in the NGS Minimum Software Stack (MSS). This document defines for prospective resource providers what middleware software they need to deploy, in order to provide interfaces that are compatible with the NGS. To ensure compatibility the NGS also monitors all certified sites on an ongoing basis, and regularly reviews partner sites to ensure they continue to meet their defined service-level definition.

6.6 Core Site Resources

In phase 1 of the NGS, the four core sites were split into two types of resource. The provisioned hardware was provided via a tender exercise by ClusterVision.

At the University of Oxford and the White Rose Grid, University of Leeds site, two "compute clusters" were installed. These were designated as compute clusters as they were provisioned to primarily focus on users who required large n-way compute jobs. In addition to the two compute clusters, at the University of Manchester and the STFC Rutherford Appleton Laboratory, two "data clusters" were installed. These were different from the pure compute clusters in that they had a smaller number of compute nodes (of the same specification as those at the compute cluster sites), but were complemented by a larger amount of physical storage and the provision of a high-performance Oracle-based database. These nodes were intended for users who had larger dataset requirements than compute power requirements, and for users looking to closely couple large datasets with high-performance compute power, without the issues of network performance affecting their ability to transfer large amounts of data at sufficient rates to meet their compute job requirements.

6.6.1 Phase 1 Configuration

Each compute facility comprised 64 dual CPU Intel 3.06 GHz nodes and 4 TB storage, while each data facility was configured with 20 dual CPU Intel 3.06 GHz nodes and 18 TB fiber SAN storage, giving a total of 44 TB of storage across all nodes available to users of the NGS.

6.6.2 Phase 2 Configuration

In phase 2 of the NGS, which started in October 2006, the four core sites were upgraded so that each core site had an identical compute cluster, while the two data sites at University of Manchester and the STFC Rutherford Appleton Laboratory were additionally upgraded with storage to provide a total of approximately 200 TB between the four core sites. Phase 2 deployed a mixture of dual and quad core technology using Opteron 64-bit processors. Forty-eight nodes with two dual-core Opteron 64-bit processors, 8 GB RAM and eight nodes with four dual-core Opteron 64-bit processors, 32 GB RAM and eight file server nodes were procured and deployed at each of the four core sites, giving a total of 1024 cores. ClearSpeed technology was also deployed as a test to enable users to experiment using such accelerator cards with their research applications.

6.7 National Grid Service Support Center

At the core of the NGS, the Support Center provides not only direct support to end users, but operates and maintains a set of core "auxiliary" services required to run the grid and to support user communities. The services delivered by the NGS Support Center are to provide a framework of support to UK Academia to assist and enable them in using the computation and data storage facilities delivered by the UK National Grid Service, its partners, and affiliates.

To clarify the set of services offered by the NGS Support Center, a service-level definition (SLD) has been created, similar to those written for each of the NGS partner sites, that details the types and level of services available. The SLD currently defines two sets of services, the Core Services and the Additional Services. The currently defined list of core services provided is as follows.

6.7.1 Core Services

6.7.1.1 Helpdesk

The NGS provides a helpdesk facility for users to interact with the NGS Support Center. Users may log tickets with the helpdesk via phone email or Web-form. All three methods are targeted to be available 24/7 with no unscheduled downtime. The telephone is targeted to be manned 10:00 am–4:00 pm Monday to Friday excluding public holidays and shutdowns as publicized on the NGS Web site. When not manned, there is a voicemail system.

The helpdesk is administered by a dedicated resource but tickets may be allocated to core NGS-funded staff or farmed out to other partner or affiliate sites. The support center will target the response to a ticket and/ or allocate a ticket to an agent within one working day and will target to resolve a ticket within five working days. However, the NGS helpdesk has no direct control over tickets that are passed to third-party sites, thus preventing an effective SLA being presented to an end user to whom the NGS resource appears as a single entity.

The helpdesk provides executive-level reports to the NGS Technical Board (such as number of tickets logged and average time to close tickets) and operational/tactical level reports (such as unresolved tickets per core/ partner site) to the NGS Operations Board when requested.

6.7.1.2 Certification Authority

The UK e-Science Certification Authority (CA) provides X.509 digital certificates for the UK e-Science community. The certificates are a method of ensuring secure access between authorized users and trusted computers on the grid. The UK e-Science CA is the CA for Grid and e-Science activities in the UK and it is accredited by the International Grid Trust Federation (IGTF), meaning it is internationally recognized to issue certificates to people and computers in the UK.

The CA authenticates users through a wide network of registration authorities (RAs). Users are initially authenticated via means of photo ID (such as a passport photo or university card). Requests for certificates are accepted through the NGS Web site with certificate signing (securely generating requested certificates) performed every working day by a member of the core NGS support staff. There are nominally five people at RAL authorized to/capable of signing certificates. Note, however, that the elapsed time between a request for an initial certificate and the certificate being generated and dispatched is usually greater than one day, due to the requirement of a face to face meeting between the applicant and their local RA.

6.7.1.3 Registration Authority

An RA is a body that approves a user's request for a digital certificate from the UK Certification Authority. An RA operator performs the role of validating a user's identity and passing that information to the certification authority. In summary, a user must present photographic ID to their local RA operator to verify their identity before their request for a digital certificate is approved. A process is followed by which the RA operator interacts with the certificate authority in order for the user to obtain a certificate.

The support center is responsible for the creation of new RAs, and ensuring the RAs understand their roles and responsibilities as well as providing training for new RA operators (see Section 6.7.1.7), and manages RAs and RA operators. A review of existing RAs is also performed by the support center.

6.7.1.4 NGS Resources

The NGS Support Center oversees the provision of core resources at the four founding sites and also has a major role in managing the information relating to each site, overseeing the site's service-level description review process, and collating and publishing site-specific information.

Information on each site's resources is available on the NGS Web site. Published information contains detail on the resources available at each site, hostnames, available applications, and details relevant for each site that they choose to publish. Each core and partner site must also publish a service-level description that is agreed by the technical board; these documents are available on the NGS Web site.

6.7.1.5 User Support

The support center provides the following services:

- Guidance and assistance for users applying for access and using the NGS resources from initial application for a certificate through to using an application/accessing data on the NGS. Users are subscribed to two mailing lists—a fortnightly news bulletin to inform them of software updates, training, events, and so on and also a "status" mailing list that informs users of service status, downtime, and so on. Both email lists are publicly available through the JISCmail service. Users also have a storage resource broker (SRB) account created for them. Note that users apply for use of NGS resources via an application through the NGS Support Center, which is fed through a peer review process to validate the research the user wishes to conduct on the NGS resources.

- Guidance and advice for new sites wishing to join the NGS through documentation (Joining Guide), verbal communication, and a "buddy" system.
- Details of applications and software installed at NGS sites.

6.7.1.6 Web Site

The support center provides content management for the NGS Web site, which holds a vast range of information on general information on the NGS, certification, site and user joining information, site resource details, tools for using and accessing the grid, and much more. The site is also linked to the helpdesk knowledge base repository.

The Web site is regularly updated to include latest information; for example, news articles, current research using NGS resources, events, vacancies, and reports on numbers of certificates generated. The target audiences/users of the Web site are new NGS users, experienced NGS users, resource providers, NGS partner and affiliate staff, and anyone interested in UK grid computing. In addition to "regular" Web site content management, the site also contains areas that are secured through the use of certificates and are used to host secure content such as minutes of board meetings.

6.7.1.7 Training

In addition to end-user training provided via the Training, Outreach, and Education (TOE) team based at the National e-Science Center (NeSC), the NGS runs regular courses designed for new RA managers and operators that offer comprehensive training on the procedures and tasks that RA operators will encounter.

6.7.1.8 NGS Portal

The support center hosts and develops, in conjunction with OMII-UK, the NGS portal (JSDL Application Repository). The NGS portal provides a range of standard (JSR168) compliant components that support user credential management, network resource information management, job execution, file transfer between distributed servers, SRB file management, and a new JSDL job repository and submission portlet.

The application provides three core functionalities:

- A browsable repository of personal and shared/public JSDL job profile documents (job submission description language documents). JSDL can be searched for by category of interest and published to be made available to other users by the portlet administrator(s).
- A grid job submission and monitoring interface (currently, only Globus but more grid middleware providers are being added, e.g., GridSam/WSRF).
- A JSDL GUI editor for constructing, validating, sharing, and uploading jobs described in JSDL.

Additional functionality is currently under development for the second portlet version, including support for additional middleware providers (GridSam, WSRF) and staging protocols (SRB, Webdav). In addition, the portlet is extensible in that application-specific interfaces can be added to the portlets client interface page-registry to provide more "application-friendly" interfaces for building JSDL for specific applications.

6.7.1.9 Documentation

The support center publishes and manages an extensive collection of documentation relating to the NGS, certificates, and grid computing. This documentation is available across the NGS Web site, and also in a dedicated "Documentation" section within the Web site. Documentation is provided to assist end users who want to use NGS resources and also NGS resource providers looking to implement NGS compatibility at their site.

6.7.1.10 User Account Management

The NGS Support Center develops and maintains the User Accounting Lifecycle System. This is used to manage all NGS user accounts and applications. It also automates the process of locking out users who exceed their current CPU quota and renewing accounts of those who reapply.

The support center is responsible for the LDAP VO (virtual organization) server used with the UAL system. The VO server provides user information which sites use to grant users access to resources (through the mechanism of "gridmap" files). The VO server will be replaced by a virtual organization management system (VOMS) server during NGS phase 2.

6.7.1.11 Promotion, Outreach, and Education

The NGS Support Center aims to promote the NGS and increase the number of users by increasing awareness of the services provided. This involves attendance at scientific and grid conferences as well as peer-reviewed conference submissions, a regular newsletter, press releases of cutting-edge research performed on the NGS, and articles in relevant literature from other bodies such as Joint Information Systems Committee (JISC) and other grid organizations. Direct liaison with interested research groups such as computational biologists is also undertaken and a great deal of interaction takes place to deliver a service of high relevance to research communities. The support center also works with the National e-Science Center (NeSC) to promote the NGS service and grid computing to UK academia.

6.7.1.12 Global Activities and Collaboration

The support center collaborates with other grid initiatives across the world. The purpose of such collaboration is to provide leadership in areas that we are advanced in, to leverage others' experience in areas that we are developing in, and to collaborate in areas of mutual benefit. This collaboration takes many forms, such as co-development of software/grid middleware, membership, and involvement in bodies such as Open Grid Forum (OGF) and EGEE in order to influence, and adhere to, emerging standards.

6.7.2 Additional Services Provided by the NGS Support Center

In addition to the core service activities of the support center, there are a number of additional or auxiliary services that support and facilitate the operation of the NGS. Some of these services, for example, the MyProxy service, VOMS, and GridSam, can be used independently by projects that may already have access to their own grid infrastructure and simply require such additional tools without the need for installing them at their own sites.

6.7.2.1 Monitoring Services

Monitoring services are provided to enable the overall health of the NGS to be tracked on a day-to-day operational basis and to enable certification of sites wishing to join the NGS.

6.7.2.1.1 *Ganglia monitoring service*

Ganglia is a scalable distributed monitoring system for high-performance computing systems such as clusters and grids. The support center hosts a Ganglia monitoring service that reports on the core and partner sites in real time via the NGS Web site.

6.7.2.1.2 *Inca monitoring service*

Inca is a flexible framework for the automated testing, benchmarking, and monitoring of grid systems. It includes mechanisms to schedule the execution of information gathering scripts, and to collect, archive, publish, and display data. The support center manages Inca software running on the NGS, and reports from the system are presented to the technical board and are also available on the NGS Web site. It is planned to include the monitoring service uptime within the Inca reports themselves.

6.7.2.1.3 *NGS Web site monitoring tools*

Reports on Web site usage, for example, hits per pages and the length of time visitors spend on the site, are recorded and used in management reporting.

6.7.2.2 GSI-SSH Terminal

The grid Security Infrastructure Secure Shell (GSI-SSH) is an application written in Java that enables a user to connect to a server running GSI-SSH using a UK e-Science grid certificate. It may be run as a signed applet from within a browser or as a standalone application. By default it should work with any UK e-Science grid where a user has an account (e.g., the NGS) using a UK e-Science certificate or a proxy certificate stored in the MyProxy server. The applet is provided through the NGS Web site.

6.7.2.3 MyProxy Upload Tool and MyProxy Server

The "MyProxy Upload Tool" is software that is used to upload a user's certificate into a MyProxy server. The MyProxy server acts on the user's behalf to provide authentication details to third parties, rather than the user having to use his certificate directly. NGS provides the "MyProxy Upload Tool" and also hosts a MyProxy server for use by the NGS user base. The MyProxy server reports its availability every 15 minutes. If a failure is detected during normal working hours, then a system administrator is contacted in order to investigate and resolve the situation.

6.7.2.4 NGS CPU Checking Utility

To keep users up to date with their account, the NGS provides a facility where users can check their allocated CPU time against what they have actually used. The purpose of this service is to enable users to monitor if they are approaching their allocated limit of CPU time and to allow them time to reapply for more hours if required.

6.7.2.5 BDII MDS Service

The support center runs a Monitoring and Discovery Service (MDS) for the National Grid Service using Berkley Database Information Index (BDII) from the EGEE project. It retrieves data from NGS sites by querying the Grid Resource Information Service (GRIS) at each site and makes it available via an LDAP server. The information is primarily for the automatic discovery of resources by a resource broker, but can be viewed on the NGS Web site it may be queried on the command line if an OpenLDAP client is installed. The BDII server reports its availability every 15 minutes. If a failure is detected during normal working hours, then a system administrator is contacted in order to investigate and resolve the situation.

6.7.2.6 Advanced Reservation

The support center hosts an advanced reservation form on the NGS Web site, for users to request resource reservation in the case of, for example, training and demonstrations at conferences. The support center manages the reservation process, coordinating the reservation with the sites as appropriate.

6.7.2.7 Wiki

The support center provides a Wiki for NGS stakeholders to share information. In order to add information to the Wiki, a user needs a recognized certificate (such as one provided by the NGS Certification Authority). The purpose of the Wiki is for the NGS community to share information that may not be contained on the NGS Web site, for example, users sharing their experiences of how to use certain applications, or explain the purpose of any virtual organizations that they may belong to.

6.7.2.8 Grid Operations Center DB

The Grid Operations Center DB (GOCDB) contains general information about participating Large Hadron Collider Computing Grid (LCG) sites, their various computing resources (nodes), any periods of scheduled maintenance, and their site administrators and other contacts. The database can be accessed by users holding X.509 digital certificates. Any user certificate generated by an LCG-recognized CA will work. The support center is responsible for ensuring that relevant NGS data are published to the GOCDB.

6.8 Partner- and Affiliate-Certified Sites

In addition to the core NGS sites at STFC and the Universities of Leeds, Oxford, and Manchester, the UK's high-performance computing (HPC) facilities HPC-x and HeCTOR are also partners in the NGS. In addition, five new partner sites have joined the NGS since October 2004 (Queen's University Belfast, the University of Glasgow, the University of Cardiff, Lancaster University, and the University of Westminster) and a number of affiliate sites (the University of Keele, the University of Southampton, the University of Bristol, and the University of Reading). Through the partner and affiliate program the NGS has grown to include a wide variety of homogeneous resources, united by the common grid interfaces. At present the NGS facilitates access to a range of different system resources from those similar to the core services running Linux or other flavors of UNIX to Windows-based Condor pools and Windows-based compute cluster services.

Following the specification of the NGS minimal software stack, the resource sites are integrated into the NGS following a period of certification to ensure that the required interfaces that should be compatible with the NGS are available. Sites that do not directly contribute resources for the use of general NGS users are termed "affiliates," while sites that provide a contribution of resources for use by all NGS users are granted partner status. As a partner they are expected to write a service-level definition that details the type and level of service offered to NGS users. As a partner site they are also entitled to direct representation, via a nominated person, on the NGS technical board. This enables partner sites to be directly involved with decisions relating to the operation of the NGS and for future policy decisions.

6.9 Documentation and Training

The National e-Science Center in Edinburgh was originally established to not only act as a regional center of e-Science specialism but also provide dedicated training activities for users across the spectrum of e-Science activities. Today the e-Science center hosts training and educational events and supports and leads training activities across the EGEE grid project. As a center specializing in training, NeSC was invited to participate within the NGS project. Through the provision of funding from the central NGS grant, NeSC has provided a number of training courses from general NGS induction courses, which are designed to enable a user new to the grid to be able to start to use NGS resources and understand what is available, through to more specific training courses focused on particular subject areas, such as using the OGSA-DAI and Oracle services for users who have specific data requirements.

Through NeSC's links with the EGEE project, the NGS has been able to leverage training and related materials developed for the EGEE project. This has proved more difficult than hoped due to a range of incompatibilities between EGEE and NGS. The EGEE infrastructure has initially focused on a rather uniform software and hardware architecture. This, combined with the tight EGEE timescales that are driven in part by the real needs of the particle physics community, has led to technical choices and software products that are difficult to deploy and operate in a heterogeneous production service. EGEE is working to overcome these deficiencies as its user base expands. Equally, EGEE and NGS experience is teaching us which of the EGEE decisions are absolutely vital for large-scale production grids. From the user perspective differences between NGS and EGEE are gradually reducing. Despite the detailed differences, a successful series of training events has been run at a number of locations in the UK and an increasing amount of NGS-specific training material is available. Beyond induction training, user documentation is also available on the NGS Web site. Adequate and appropriate level user documentation remains a challenge, however, in the rapidly evolving grid world. This is an issue that confronts all grid infrastructure projects and will continue to be a problem until the technological and complexity gaps between the underlying infrastructure and the end users can be bridged. Grid infrastructures are still complex to use and increased basic training materials will be required as the user base expands.

A dedicated outreach strategy has been put in place to encourage more users from all research areas to take advantage of NGS resources. This involved reaching out directly to the end users at subject-specific communities and integrating the NGS into the research community. To facilitate this the NGS appointed a dedicated outreach officer to oversee and coordinate these activities and to ensure a professional representation of the NGS at workshops, conferences, and other appropriate events.

An important factor in encouraging more users of the NGS is the presence of user case studies—real examples of research that has been carried out in their field using NGS resources. The NGS collects examples of research performed using the NGS and uses them to produce outreach and promotional material to reach out to new communities. Real examples of how the NGS can be used are essential in attracting new users, as even if the case studies are not directly in the same research area, the practices illustrated can often be used across a broad range of research areas and topics.

6.10 Future of the National Grid Service

The NGS currently represents development efforts within the UK to create a standardized infrastructure based on the requirements of today's users. For future development the NGS has to look beyond the classic e-Science communities and instead toward providing e-infrastructure that will facilitate researchers across all disciplines and within all institutes. To further this aim the NGS constantly works with funding bodies to understand the requirement of their user communities and how best to facilitate not only deployment of infrastructure but also embedding e-Science techniques into researchers' day-to-day activities. Beyond the UK the NGS also works to further its collaborative links with international grid efforts such as the TeraGrid and the EGEE and participates in future discussions over the role and requirements for initiatives such as the current European Grid Initiative (EGI) which aims to facilitate sustainable grid infrastructures throughout Europe.

In order to engage further communities, the NGS must also tackle challenges such as securing user data and green computing. For disciplines that use personal data such as medical data, being able to guarantee that work undertaken using national resources is secure and cannot compromise that data is becoming increasingly essential. Similarly, with ever-increasing costs associated with ownership and running of large-scale infrastructure, the NGS must look to pioneering green computing, in terms of how existing resources can be more fully utilized and how new resources can be implemented in such ways as to avoid duplication of effort and thus duplication of running costs, and how they can be deployed with the lowest carbon footprint achievable using today's available mechanisms. The main premise for the NGS will be to continue to drive standards and to deploy and adopt standard interfaces for grid services across institutes, thus making it easier for end users to make use of the grid.

6.11 Case Studies

The strength of the NGS is in the wide range of communities that are currently using NGS resources to undertake their research. The current end users of the NGS range cover the entire research spectrum from physical sciences to the arts and humanities and include individual researchers working alone, to multidiscipline, cross-institutional research groups.

The following case studies are more detailed examples of the current types of research being done using the NGS. These case studies are produced by the NGS to promote and illustrate to communities the types of research and the benefits of using grid resources such as the NGS. These and other case studies are regularly updated and made available on the NGS Web site.

6.11.1 Grid-Enabled Neurosurgical Imaging Using Simulation (GENIUS)

One of the largest community user groups of the NGS is the biosciences. For example, the NGS has played a major role in the GENIUS project (grid-enabled neurosurgical imaging using simulation), which is led by Professor Peter Coveney at UCL (Figure 6.1).

Many neurological conditions can be better understood and treated if the blood flow of the patient is studied, and studying the patient's blood flow around the brain with a system like GENIUS before surgery can greatly increase the chances of success. MRI scans are used to provide the data to reconstruct the neurovascular structure of real patients and a lattice-Boltzmann fluid flow solver called HemeLB is used to simulate the flow of blood around the reconstructed vessels. The GENIUS project makes use of distributed grid computing by running geographically distributed parallelization of simulations. This allows the simulations to run on a number of processors on several machines, decreases the turnaround time for simulations, and returns the results to the clinicians as fast as possible. This project demonstrated the collaboration that can

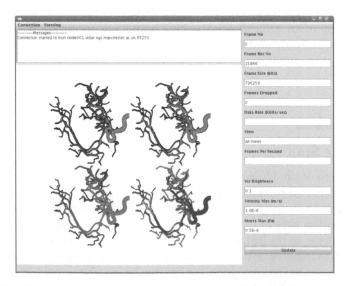

FIGURE 6.1   Visualizing blood flow around a patient's brain in real time.

take place between different national grids as this simulation was run on the NGS, the TeraGrid (the US national grid), and the Louisiana Optical Network Initiative (LONI). Real-time visualization is achieved by incorporating a ray tracer into the HemeLB algorithm. Dedicated optical fiber links connect the supercomputers to the clinicians' desktop so clinicians can then easily view the live visualization on their desktops. They can also steer the simulation by moving the model around to get a better look at areas of interest and altering input data.

6.11.2 Modeling and Simulation for e-Social Science (MoSeS)

Planning for services such as transport, health, schools, and housing requires information about the human population, their residences, and activity. The UK census, last carried out in 2001, has become more detailed and more used in activities such as public service planning. It can provide important statistics, including estimates of how many people live in an area, how old they are, their current occupations, and the health and provision of care to others.

Dr. Andy Turner and colleagues at the University of Leeds are aiming to take the use of the census and other population data in public service planning a step further by developing national demographic simulations and to enable their use by social scientists and policy planners to help answer specific questions and analyze scenarios; for example, estimating demand for services such as transport schemes or for access to health and social care services for specific areas.

The basis of the demographic modeling are data from the 2001 census and various census datasets are used to generate individual and household-level datasets for 2001. These data for 2001 are then projected at the individual and household level for small areas on a yearly basis up to 2031. It is hoped that a scenario-based forecasting approach will allow social scientists to examine possible effects of environmental change and better plan for the likely effects of an aging population. All of the modeling code has been written by the project members using Java and parts of it have been parallelized to take advantage of high-end computers like those available on the NGS (Figure 6.2).

6.11.3 RTGrid

Radiotherapy, an extremely effective form of treatment for many cancers, requires a great deal of careful planning for each individual patient as radiotherapists need to define the target area and amount of radiation each patient will receive. The shape and the direction of the radiation beam are also important considerations as it is essential to ensure the least amount of damage to the healthy tissue surrounding the treated area.

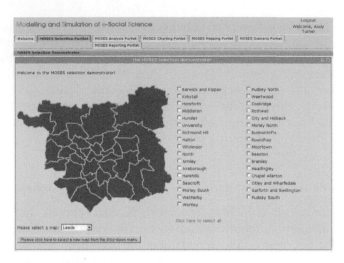

FIGURE 6.2   Demographic modeling on the NGS.

In order for radiation-dose simulations to be clinically relevant, they need to be completed in a matter of hours. Unfortunately this is often not the case, with many simulations taking days, which is where the NGS has been brought in. Dr. Mary Chin from the University of Surrey and the Welsh e-Science Center have worked together to tackle this problem.

Radiation doses in patients are calculated using the Monte Carlo method to simulate the creation, path, and destruction of each particle. The statistical sampling of the Monte Carlo method means that the greater the number of simulated events, the better the confidence level of the result as the entry of trillions of particles into the body can be simulated. Using the NGS has enabled the simulation time to be reduced from almost a week to 3 hours. This obviously has implications for the clinical usefulness of such simulations. Medical physicists can verify treatment programs before and during the treatment. As complex radiotherapy modalities become increasingly common, verification of the treatment is vital.

Acknowledgments

We would like to acknowledge all those who have contributed to and made possible the various grid activities within the UK, especially those who have worked for the Grid Support Center, the Grid Operations Support Center, the National Grid Service, and the Engineering Task Force.

References

1. UK National Grid Service. Available at: http://www.ngs.ac.uk.
2. UK e-Science Core Program. A National e-Science Center linked to a network of Regional Grid Centers. Available at: http://www.rcuk.ac.uk/cmsweb/downloads/rcuk/research/esci/centres.pdf.
3. UK e-Science Engineering Task Force Reports. Available at: http://www.ngs.ac.uk/ETF/.
4. Allan, Rob; Baker; David; Boyd, David; Chohan, Dharmesh (2003) "Building the e-Science grid in the UK: Middleware, applications and tools deployed at level 2," presented at the UK e-Science All Hands Meeting 2003, Nottingham, September 2–4, 2003.

7

Grid Resource Allocation

Zhou Lei, Zhifeng Yun, and Gabrielle Allen

CONTENTS

7.1 Introduction

Grid computing concepts and technologies were developed for coordinated resource sharing globally. The shared resources include accessible computers, software, data, services, a QoS network, and other resources, as required by applications in industry, science, and engineering. Grid applications address complex problems, ranging from jet engine design and reservoir simulation to bioinformatics, biomedical imaging, and astrophysics. They are usually compute-intensive, data-intensive, knowledge-intensive, and/or collaboration-intensive. A grid application consists of a number of jobs associated with workflow. Each job presents different requirements on how to utilize and coordinate shared resources for problem solving. In general, single sites are simply no longer efficient for meeting the resource needs of a grid application.

Grid scheduling is critical to effectively manage grid application execution across a grid. It is defined as the process of making scheduling decisions involving resources to meet various requirements of applications. It selects proper jobs of an application, and searches multiple domains to use a single machine or scheduling a single job to use multiple resources at a single site or multiple sites. Basically, there are two kinds of activities involved in grid scheduling: job selection and resource allocation. Workflow management puts a lot of effort into job stream arrangement, while resource allocation focuses on efficient resource utilization. One of the primary differences between grid resource allocation and local resource scheduling is that grid resource allocation does not "own" the resources at a site (unlike local resource scheduling) and therefore does not have control over them. This lack of ownership and control is the source of many problems to be solved in this area. In this chapter, we do not address the situation of workflow management. We instead discuss grid resource allocation in terms of application needs, resource allocation stages, and major approaches to efficient resource allocation.

In the next section, we categorize the needs of resource allocation from an application perspective. We then describe the general process of grid resource allocation. Major approaches to improve resource allocation are illustrated in the final section.

7.2 Resource Allocation Needs

From an applications perspective, resource allocation needs are classified into four categories: high-throughput computing, resource co-allocation, time-critical needs, and computing synchronization.

7.2.1 High-Throughput Computing

For many experimental scientists, scientific progress and quality of research are strongly linked to computing throughput. In other words, most scientists are concerned with how many floating-point operations per month or per year they can extract from their computing environment rather than the number of such operations the environment can provide them with per second or minute. High-throughput computing (HTC) [1] environments focus on large amounts of processing capacity over long periods of time.

The key to HTC is effective management and exploitation of all available computing resources. The main challenge in a grid environment is how to handle resource characteristics of autonomy, heterogeneity, and geographic distribution and maximize the amount of resources accessible to an application.

Large-scale parameter sweep [2] is a classic scenario requiring HTC. A parameter-sweep application model is a combination of task and data parallel models, and applications formulated to use this model contain a large number of independent jobs operating on different datasets. A range of scenarios and parameters to be explored are applied to program input values to generate different datasets. The programming and execution model of such applications resembles the Single Program, Multiple Data (SPMD) technique. The execution model essentially involves processing N independent jobs (each with the same task specification, but a different dataset) on M distributed computers where N is, typically, much larger than M. Fortunately, this high-throughput parametric computing model is simple, yet powerful enough to formulate distributed execution of many application areas such as radiation equipment calibration analysis, searching for extra-terrestrial intelligence, protein folding, molecular modeling for drug design, human-genome sequence analysis, brain activity analysis, high-energy physics events analysis, *ad hoc* network simulation, crash simulation, tomography, financial modeling, and Mcell simulations.

7.2.2 Resource Co-Allocation

Resource co-allocation in a grid refers to the simultaneous assignment to application multiple resources belonging to possibly different administrative domains [3]. Basically, two types of applications need resource co-allocation: (1) computationally intensive applications that may require more resources than a single computing machine can provide, and (2) applications requiring several types of computing capabilities from resource owners in other administrative domains. For example, a scientific instrument, five computers, and multiple display devices were used for collaborative real-time reconstruction of X-ray source data [4]. While such simultaneous allocation can in principle be achieved manually, resource co-allocation strategies are required to support not just the management of individual resources but also the coordinated management of multiple resources.

Grid applications typically operate on ensembles of resources that span administrative domains of control, with resources in the ensemble being independently operated. Furthermore, access to resources is in general unreliable, due to either competing demands for resources or outright failures. In a grid environment, it is not reasonable to assume a centralized global scheduler that resolves resource co-allocation. Two features of grids complicate the co-allocation process, rendering ineffective existing approaches based on centralized control and a strategy of aborting on the failure of any resource request.

Some techniques were used to improve co-allocation effectiveness. One approach is to enhance the local resource management system. For example, by incorporating advance reservation capabilities into a local resource manager, a co-allocator can obtain guarantees that a resource will deliver a required level of service when required. Alternatively, the resource management system can publish information about current queue contents and scheduling policy, or publish forecasts of expected future resource availability. This information can be used to improve the success of co-allocation by constructing co-allocation requests that are likely to succeed. The co-allocator may use information published by local managers to select from among alternative candidate resources, or it may attempt to allocate more resources than it really needs.

7.2.3 Time-Critical Needs

Time-critical applications, such as medical simulation and weather forecasting, have special requirements with respect to quality of service (QoS) [5]. For such applications it is crucial to know at which point in time the results of remote simulation tasks executed on some grid resources will be available.

Let us take coastal and environmental modeling [6] as an example to explain time-critical needs for resource allocation in a grid environment. The economically important Louisiana Coastal Area (LCA) is one of the world's most environmentally damaged ecosystems. Beyond the economic loss, LCA erosion has devastating effects on its inhabitants, especially in New Orleans whose location makes it extremely vulnerable to hurricanes and tropical storms. To effectively model the LCA region, an integrated framework has been developed for coastal and environmental modeling capable of simulating all relevant interacting processes from erosion to storm surge to ecosystem biodiversity, handling multiple time (hours to years) and length (meters to kilometers) scales. This requires dynamically coupled models and the invocation of algorithms based on streamed sensor or satellite data, location of appropriate data and resources, and creation of necessary workflows on demand, all in real time. Such a system needs to collaborate various relevant resources, such as sensors, data storage, computing capacity, with real-time responses to improve ecological forecasting, sensor placement, control of water diversion for salinity, or predict/control harmful algal blooms, and support sea rescues and oil spill responses. In extreme situations, such as approaching hurricanes, results from multiple coupled ensemble coastal models, dynamically compared with observations, could greatly improve emergency warnings.

The described scenarios were included in the emerging field of dynamic data-driven application systems [7], which identifies new, complex, and inherently multidisciplinary application scenarios where simulations can dynamically ingest and respond to real-time data from measuring devices, experimental equipment, or other simulations.

7.2.4 Computing Synchronization

Grid scheduling usually needs the dispatch of components of a grid application to different resources. As described above, resource co-allocation strategies are developed to deal with the interaction among application components in order to improve application execution performance. Furthermore, there is a kind of multistep application, such as grid-enabled inverse modeling, in which resource co-allocation is needed for each step (or iteration) and computing synchronization is necessary among steps since each step depends on the results of the previous steps [8].

Ensemble Kalman filter (EnKF) subsurface modeling, extensively adopted by science and engineering, is a good example to explain the requirements of computing synchronization [9]. In petroleum engineering, model inversion is important for the value determination of model parameters and for making accurate predictions. It is used to calibrate subsurface properties (e.g., porosity, permeability, and hydraulic conductivity) in a subsurface simulation model. In this way, the computed values of observables, such as rates, pressures (or head), and saturations, at different observation locations are in reasonable agreement with the actual measurements of these quantities. Nowadays, the increase in sensor deployment in oil and gas wells for monitoring pressure, temperature, resistivity, and/or flow rate (i.e., smart/intelligent wells) has added impetus to continuous model updating. Instead of simultaneously using all recorded data to generate an appropriate reservoir flow model, it has become important to capture reservoir flow information by incorporating real-time data. Automatic and real-time adjustment procedures are needed for efficient model inversion.

The EnKF method reduces a nonlinear minimization problem in a huge parameter space by changing objective function minimization with multiple local minima to a statistical minimization problem in the ensemble space. This method is extensively adopted not only for model inversion but also for uncertainty assessment, optimization, control, and so on. However, the EnKF method is processing intensive due to iterative simulations of a large number of subsurface models. Large datasets (comprising model and state vectors) must be transferred from ensemble-specific simulation processes to Kalman gain or uncertainty assessment processes, and back again to the ensemble process at each assimilation iteration.

Computing synchronization in a grid is used to synchronize all the components of each iteration of inverse modeling and then initiate the next iteration. Without effective computing synchronization, application execution reliability would be seriously impacted by the slowest computing resource(s) due to the nature of grid resources; that is, geographical distribution, heterogeneity, and self-administration.

7.3 Resource Allocation Stages

Generally speaking, grid resource allocation can be broken into three stages: resource discovery, resource selection, and job submission [10]. We will now discuss these separately, but they are inherently intertwined, as the decision process depends on available information.

7.3.1 Phase 1: Resource Discovery

Resource discovery involves determining which resources are available to a given user. At the beginning of this phase, the potential set of resources is the empty set; at the end of this phase, the potential set of resources is some set that has passed a minimal feasibility requirement. The resource discovery phase is done in three steps: authorization filtering, job requirement definition, and filtering to meet the minimal job requirements.

The first step of resource discovery for grid scheduling is to determine the set of resources that the user submitting the job has access to. Computing over the grid is no different from remotely submitting a job to a single site: without authorization to run on a resource, the job will not run. At the end of this step the user will have a list of machines or resources to which he or she has access. In the second step, to proceed with resource discovery, the user must be able to specify some minimal set of job requirements in order to further filter the set of feasible resources. Given a set of resources to which a user has access and at least a small set of job requirements, the third step in the resource discovery phase is to filter out the resources that do not meet the minimal job requirements. At the end of this step, the user acting as a grid scheduler will have a reduced set of resources to investigate in more detail.

7.3.2 Phase 2: System Selection

Given a group of possible resources (or a group of possible resource sets), all of which meet the minimum requirements for the job, a single resource (or single resource set) must be selected on which to schedule the job. This selection is generally done in two steps: gathering detailed information and making a decision. In order to make the best possible job/resource match, detailed dynamic information about the resources is needed. Since this information may vary with respect to the application being scheduled and the resources being examined, no single solution will work in all, or even most, settings. The dynamic information gathering step has two components: what information is available and how the user can access it. The available information will vary from site to site, and users currently have two main sources: the grid information service (GIS) (e.g., MonaLisa [11]) and the local resource scheduler. Details on the kind of information a local resource scheduler can supply are given in [1]. With the detailed information gathered, the next step is to decide which resource (or set of resources) will be used.

7.3.3 Phase 3: Job Submission

Strictly speaking, job submission does not belong to resource allocation; we briefly mention it just for completeness. Once resources are chosen, the application can be submitted to the resources. Job submission may be as easy as running a single command or as complicated as running a series of scripts and may or may not include setup or staging.

7.4 Major Approaches

In general, there are five main approaches to achieve efficient grid resource allocation: performance prediction, resource matching, economic scheduling, peer-to-peer method, and advanced reservation.

7.4.1 Performance Prediction

Grid resource performance delivering to any given application fluctuates dynamically in any time duration. A grid scheduler needs to predict what the deliverable resource performance will be for the time period in which a particular application component will eventually use the resource [12].

In order for a scheduler to determine which resource is most valuable in terms of performance, there must be some kind of resource evaluation or characterization methods for the scheduler to predict the performance of each computation node. Moreover, some particular formations of performance information, such as quantifiable metrics, are needed to fit the requirement of grid scheduling.

There is static and dynamic information on a resource. Static information changes slowly with respect to program execution lifetimes. Due to quick changes, application execution time on the resource and job waiting time in the queue are dynamic characteristics. Since grid schedulers do not have central control over resources, these dynamic fluctuations make scheduling complex and difficult. In order to accurately predict the performance, people developed several ways to model this information.

7.4.1.1 Runtime Predictions

Many techniques have been proposed to predict application execution time. The type of prediction techniques can generally be separated into statistical and analytical approaches. Statistical approaches use statistical analysis of previous applications that have completed to conduct the prediction, whereas analytical approaches construct equations to describe application execution time. Statistical approaches can be further classified into three classes: time series analysis [13], categorization [14], and instance-based learning [15]. Analytical approaches develop models by hand [16] or use automatic code analysis or instrumentation [17]. Figure 7.1 shows that the runtime prediction approaches taxonomy.

However, studies showed that the statistical properties of these characteristics are difficult to model since time series of measurements for a given characteristic have a slowly decaying auto-correlation structure. Despite these statistical properties, predictions based on historical data are more general since no direct knowledge about applications is required.

The categorization approach derives execution time prediction from historical information of previous similar runs. This approach is based on the observation that similar applications are more likely to have similar runtimes than applications that have nothing in common [18]. In this

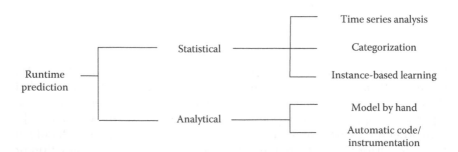

FIGURE 7.1   Runtime prediction approaches taxonomy.

approach, a template of attributes needs to be defined to identify the category where the application will be assigned. Once the set of templates has been defined, new applications can be added to the categories containing similar applications, and the execution time and a confidence interval can be calculated. Different statistical techniques, such as mean or linear regression, are applied to generate prediction. A prediction is formed from each similar category of applications, and the prediction with the smallest confidence interval is selected to be the prediction for the application. Instance-based learning (IBL), which is also called the locally weighted learning technique, is another approach for runtime prediction. In this approach, a database of experiences is used to make a prediction. For each prediction, a query point is inputted into the database, and the data point in the experience base that is most relevant to the query will be selected. A proper distance metric is defined to measure the distances between data instances. In fact, the IBL algorithm is a generalization of the template approach, in which distances are simplified to binary values.

7.4.1.2 Queue Wait-Time Predictions

Jobs submitted to a resource normally cannot be executed immediately. They are placed in the queue waiting for resource allocation. Predictions of queue wait times can guide a user in selecting an appropriate resource. Wait-time predictions are also useful in grid environments when trying to submit multiple requests so that the requests all receive resources simultaneously. The third use of wait-time predictions is to plan other activities in conventional supercomputing environments.

One approach to predicting wait time is using the wait times of applications in the past that were in a similar scheduler situation. This approach uses the same mechanism as the approach to predict application execution time. Another approach is to simulate scheduling algorithms such as first-come first-served, least-work-first, and backfilling to obtain predictions for queue wait times, where application runtimes used in simulations are estimated using the template approach. Although this approach can provide accurate wait-time predictions in some cases, it needs a detailed knowledge of the scheduling algorithm to make the prediction. By performance prediction, a scheduling system can capture future expected resource behavior, guide the scheduler in selecting appropriate resources, and therefore make the execution more efficient.

7.4.2 Resource Matching

In conventional resource management systems, a system model is established, which contains information about the underlying resources. A centralized allocator schedules the resources and maintains the allocation information. Although such a resource management strategy works well in high-performance scheduling regimes, it cannot be directly applied to distributed environments because of their dynamics, heterogeneity, and lack of central control. More sophisticated resource management strategies need to be proposed to address this issue.

A matchmaking paradigm can be used to address this issue in a flexible and robust manner. The idea of matchmaking is simple. In the matchmaking scheme, providers advertise resource descriptions to the matchmaker, and customers send the request to the matchmaker. These advertisements contain the constraints and preferences on the entities they would like to be matched with. The matchmaker finds compatible providers and customers and notifies the matched entities. A claiming protocol is activated to mutually confirm the matched entities. Figure 7.2 illustrates the matchmaking process.

Notice that the advertisements are periodically refreshed to keep the latest information about participating entities. All policy considerations are contained within the advertisements themselves: the matchmaker merely provides the mechanism to interpret the specified policies in the context.

Four steps are involved in the matchmaking process in the Condor system [1]: (1) advertising, (2) matching, (3) notification, and (4) claiming.

1. A semi-structured data model called classified advertisements (classad) specifies the characteristics, constraints, and preferences of the resources and requests. A classad is a mapping from attribute names to expressions. Each expression may consist of simple constant or arithmetic and logic operators. It can also contain combinations of variables and built-in functions. Providers and

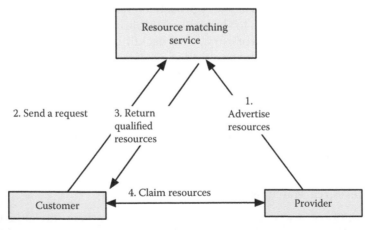

FIGURE 7.2 Grid matchmaking process: (1) providers advertise the resource; (2) customers send request to matchmaker; (3) matching services find compatible pairs and notify the matched entities; and (4) matched entities confirm and complete the allocation.

customers of services use classad to advertise their capabilities in the resource matching systems.

2. A matchmaking algorithm is used by the matchmaker to create matches. Matchmaking takes effect in two classads, evaluating one with the other. Each classad must define *Requirements* and *Rank* to express the constraints and preferences of the entities. Only when the *Requirements* expressions of both classads are evaluated to be trues can these two classads be defined as matched. The *Rank* expression measures the desirability of a match, with larger value denoting better match.

3. After the matchmaker finds compatible advertisements, it sends copies of the matched entities to every agent whose advertisement is involved. The match registrar and collectors also receive notification information so that they can update the status of the pool and discard the matched advertisements, respectively.

4. When the matchmaker notifies the match, a separate claiming protocol is activated, and notified agents need to mutually confirm the match. However, this claiming procedure allows entities to decline the match if they do not want to proceed with the claim. Claiming protocols avoid duplication problems. If the claiming protocol is not completed, the entities can abort the match process and readvertise their availability.

In the matchmaking paradigm, resources are used as soon as they become available, and applications are migrated from resources that become unavailable. By using this opportunistic scheduling paradigm, problems associated with defining system models can be thoroughly avoided.

7.4.3 Economic Scheduling

Unlike matchmaking, where matchmakers make the selection of the compatible entities based on their own advertisements, market economy approaches provide a different solution for grid scheduling, where each participant can make decisions on its own behalf. Resource scheduling is totally individual economic behavior of all the participants. Some problems that occur in grid computing, such as site autonomy, decentralization, and heterogeneous substrate, can be solved by these economic methods.

The Enterprise [19] is a market-like scheduler for distributed computing. Each idle computer sends estimated task completion time as a bid to the client, whereas the client chooses the one that best fits the requirements. However, this model is limited to a single user preference and is not practical for distributed autonomous resources. The Spawn [20] system utilizes idle computational resources in a distributed network of heterogeneous workstations. Although this method works well for heavily loaded systems, it cannot obtain good performance in lightly loaded system. Ernemann and Yahyapour [21] proposed a supply-and-demand mechanism that enables the system to optimize different objectives for different participants. It provides a robust and flexible approach for handling failures as well as allowing adaptivity during changes.

Figure 7.3 shows the general flow of economic scheduling. According to the site-autonomy policy, the total area is separated into a local domain and a remote domain. In each domain, there is a MetaManager controlling the resources. It can act as a broker with the MetaManagers in the remote domain, and is able to explore them by directory services or peer-to-peer strategies. All users or clients send requests to a local MetaManager. Each request can specify the requirements, job attributes, and objective functions. The request is described by request-and-offer description language, which is similar to classad, but has its own parameters. After the local MetaManager received a request, it first analyzes the requirements of a job and finds the matched local resources. Local offers are generated, whereas the best one is kept for further processing. The local scheduler then forwards the request to other MetaManagers. The remote schedulers search their domains and send back their best offers to the original domain. All the offers are compared and the final offer is selected. Finally, the client receives the feedback on the resource allocation and execution of jobs will be initiated. The MetaManager can reallocate the schedule to optimize the objective or to recover from system changes. In order to prevent permanent forwarding of the request, the maximum number of hops are restricted. The original request is eliminated if no offer is received in some particular deadlines.

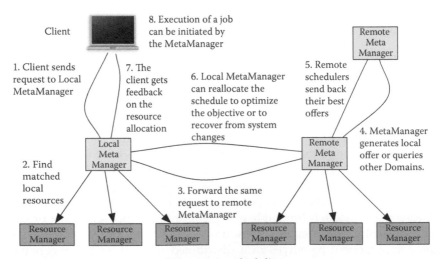

FIGURE 7.3 Architecture of market economy scheduling.

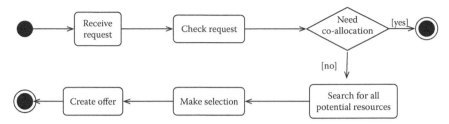

FIGURE 7.4 Creation of local offer.

Figure 7.4 describes how to generate local offers. Users or clients send out a request to the MetaManager. The necessary scheduling requirements and job attributes, such as the run time, the required number of resources, the earliest start time, and the latest end time, are included in each request. When the local MetaManager receives this request, it extracts this information and searches the local domain to find whether the local resources can meet the job's requirements and whether the user's budget is enough to process the job. If there are sufficient resources in the local domain to support the job, the scheduler will find the best-fitted resource to execute the job; otherwise, multisite scheduling will be initiated.

The scheduler first collects all the free-time intervals of each resource. Since one free-time interval normally cannot satisfy the job's requirements, combinations of intervals need to be considered. The deviation of bucket sort is used to combine the intervals. For example, all the intervals with the same start time can be collected into one bucket. Then the selection is made again to collect the intervals with the same end time from that bucket. As a result, a combination of intervals with the same start time and end time is generated. If there are several combinations fulfilling the requirements, the scheduler can select the one with the highest utility value based on some heuristics.

When there is no single resource able to execute the application by itself, the local scheduler sends the request to the remote MetaManager to generate multisite co-allocation. In this case, the job will be divided into several small parts as specified by some parameters. However, several aspects of the grid, such as the site autonomy and cost management, must be taken into account. This economic scheduling approach provides support for individual access and services. It combines the independent domains and limits the failure impact in the local area. With this system, co-allocation of resources from different sites with different scheduling is possible.

7.4.4 Peer-to-Peer Method

Peer-to-peer (P2P) systems and grids are two resource-sharing environments. Although they differ in their usages, they both link resources and people with physically distributed locations within virtual organizations (VOs).

Moreover, they seem likely to converge. Some of the characteristics of P2P systems, such as unreliable resources and intermittent participation, provide important experiences for grid computing and will constitute an even more significant share of resources. Consequently, services, particular in resource discovery, will have to tolerate failures and adapt to dynamic resource participation. At the same time, grid scalability is increased and more communities are allowed to participate.

Four components of P2P resource discovery solution in grids are defined in [22], membership protocol, overlay construction, preprocessing, and request processing, in order to allow the joining and leaving of nodes at any time and to recognize the unexplored regions. A commonly used mechanism in P2P systems is adopted for the membership protocol: a node contacting its member node to allow joining. The joining member learns about the neighbors from the membership information sent back by the contacted nodes.

Overlay construction is built by selecting the collaborating nodes from the membership list based on desirable characteristics. Much research has been done [23–25] to construct overlay topology to make efficient broadcasting with good performance.

Preprocessing is a technique that enhances search performance before actually executing jobs. Rewiring overlay construction to update topology is one kind of preprocessing. Dissemination of resource descriptions to out of the local area is also a preprocessing technique. However, the preprocessing strategies may not work well in grids since resources and users change frequently, making the results of preprocessing inaccurate.

Requests are first processed by local components. Unless a perfect match is found, the request will be forwarded to remote components. A request is dropped where either there is no further forwarding to be pursued or time to live (TTL) is equal to zero.

Since there are no central controllers, how to efficiently propagate requests becomes a bottleneck. Figure 7.5 shows the network organization taxonomy of P2P systems. Generally P2P systems can be categorized as structured or unstructured networks. For unstructured networks, each request is flooded to the whole network. However, since the large amount of query messages generated by flooding, this kind of approach does not scale well. Structured systems use distributed hash tables (DHTs) to maintain a structured overlay network among peers and use message routing to forward the query messages. In a system with N peers, each peer in DHTs only needs to maintain an index for $O(\log N)$ peers and can guarantee to resolve a look-up request in $O(\log N)$ time.

The P2P scheduling infrastructure [26] is based on economic scheduling; however, there is no MetaManager to help forward the request:

1. A consumer submits the job request to a local node.
2. Once the request is received, the node estimates whether it can finish the job before the deadline. If it meets the requirement, this

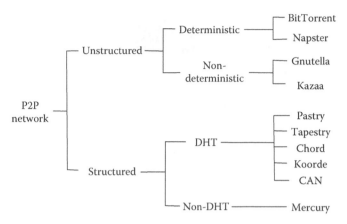

FIGURE 7.5   Peer-to-peer network taxonomy.

node sends back a bid that contains the charge for the job to the original node; otherwise, it forwards the request to other nodes based on the above propagation approaches.

3. The consumer collects all the bids and chooses the one charging the least, and then sends the job offer as well as job parameters to the selected node.

4. The selected node inserts the job into its job queue once it receives the offer and sends back the results.

A grid emulator was proposed to evaluate the performance of a combined environment: a large-scale grid with P2P systems' dynamism and intermittent participation properties. Results give a quantitative measure of the sharing environment and can be found in [22].

How to discover and allocate the resources in this combined resource-sharing environment is not trivial. Some issues, such as integrating task scheduling into grids and efficient request propagation approaches, are currently active research topics.

7.4.5 Advanced Reservation

There are two scenarios for resource allocation. One is for metacomputing applications such as TeraGyroid [27] and SPICE [28]. This kind of application requires all the resources to be available at the same time. Another scenario is for workflow applications such as Pegasus [29]. In this scenario, resources need to be scheduled at different times to guarantee task turnaround. Each task cannot be executed unless all its dependent tasks are completed. Resource reservation is one solution to these allocation concerns. It allows users to request resources in a proper order and guarantees access to sufficient resources for their applications.

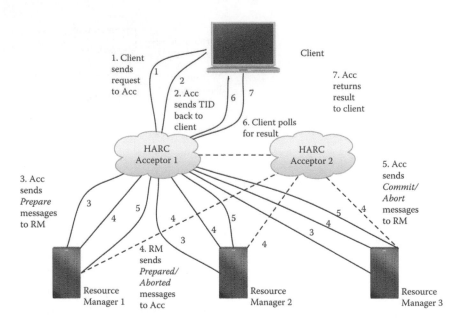

FIGURE 7.6   HARC architecture and resource reservation procedure.

Although a local resource manager can be used for local reservation, it is inconvenient to manually reserve geographically distributed resources for applications that might simultaneously have access to those distributed resources. HARC (the highly available resource co-allocator [30]) was developed to reserve multiple types of resources in a single step. It reserves multiple resources in an all-or-nothing fashion (i.e., atomically) and provides a co-allocation service supporting both metacomputing and workflow applications.

The HARC architecture, shown in Figure 7.6, consists of clients (requesting resource co-allocation), acceptors (making reservations), and resource managers (talking with local schedulers). The clients first contact the resource manager (RM) for timetable information. The timetable gives information such as what resource is free and when it is free, and so on. The client analyzes the information, selects a set of resources, and sends the requests to acceptors (Acc) asking for co-scheduling. The Acc sends back the transaction identifier (TID) to clients to differentiate the co-allocation.

Then Acc sends *Prepare* messages to all the RMs asking for scheduling of the resources based on a client's request. Each RM evaluates the request and responds with a *Prepared* message if it can meet the requirement.

However, since the timetable will be slightly out of date, there is a possibility that the request will be *Aborted*. Acc collects all the responses from RMs, returns a *Commit* message if all RMs can meet the requirements, or sends an *Abort* message if one of the RMs cannot make the schedule. The next time, the client uses the TID to poll the results from Acc. Acc will return a reservation ID if the co-allocation succeeds, or an error message if it fails. Note that several acceptors may exist simultaneously to provide redundancy. The acceptor that the client first contacts is the leader. When the leader fails, one of the other acceptors will automatically take over as leader for that transaction.

The client can send request messages to acceptors to make, modify, or cancel the reservations. Each *Make* action request message needs to define three elements:

1. "Resource" specifies where to make the reservation.
2. "Schedule" specifies when to make the reservation.
3. "Work" specifies what is to be made.

Modify and *Cancel* actions share the similar structure of request message. The *Modify* action contains the Resource and Ident element, as well as new Schedule and/or Work elements, depending on the part to be changed. The *Cancel* action contains the Resource and Ident element, which specifies the reservation to be canceled.

In HARC, all messages are written in XML and sent directly over HTTPS, which is called transport-level security. All HTTPS connections are initiated by X.509 certificates, which means that each Acceptor and each HARC RM must have its own X.509 credential. HARC can work with local batch schedulers, such as PBSPro, TorqueMaui, TorqueMoab, LSF, and so on.

There are some other reservation tools developed in industry and academia, such as the Generic (Grid) Universal Remote (GUR) [31]. However, GUR does not support workflow reservations and cannot be used to reserve multiple types of resources. HARC provides a co-scheduling framework suitable for any resource and is shown to be well suited to the scheduling of large scientific workflows.

7.5 Summary

This chapter has discussed grid resource allocation in terms of application needs, allocation processes, and major approaches to efficient allocation. Resource allocation needs include high-throughput computing, resource co-allocation, time-critical needs, and computing synchronization.

The resource allocation process has three stages: resource discovery, system selection, and job submission. In order to achieve effective resource allocation, there are five major approaches presented: performance infor-mation, resource matching, economic scheduling, peer-to-peer method, and advanced reservation.

Acknowledgments

This work was sponsored by the U.S. Department of Energy (DOE) under Award Number DE-FG02-04ER46136 and the Board of Regents, State of Louisiana, under Contract No. DOE/LEQSF(2004-07). We also thank the support from the Louisiana Optical Network Initiative (LONI).

References

1. Condor High Throughput Computing. Available at: http://www.cs.wisc.edu/condor/htc.html.
2. R. Buyya, M. Murshed, D. Abramson, and S. Venugopal. "Scheduling parameter sweep applications on global grids: A deadline and budget con-strained cost-time optimization algorithm." *Software: Practice and Experience*, 35, 491–512, 2005.
3. K. Czajkowski, I. Foster, and C. Kesselman. "Resource co-allocation in com-putational grids." In: *Proceedings of the Eighth IEEE International Symposium on High Performance Distributed Computing (HPDC-8)*, Redondo Beach, California, August 3–6, pp. 219–228, 1999.
4. K. Sim. "Relaxed-criteria G-negotiation for grid resource coallocation." *ACM SIGecom Exchanges*, 6 (2), 37–46, 2006.
5. I. Brandic, S. Benkner, G. Engelbrecht, and R. Schmidt. "QoS support for time-critical grid-workflow applications." In: *Proceedings of the First International Conference on e-Science and Grid Computing (e-Science 05)*, Melbourne, Australia, December 5–8, pp. 108–115, 2005.
6. P.S. Bogden, T. Gale, et al. "Architecture of a community infrastructure for predicting and analyzing coastal inundation." *Marine Technology Society Journal*, 41 (1), 53–71, 2007.
7. Dynamic Data Driven Application Systems. Available at: http://www.dddas.org/.
8. X. Li, Z. Lei, C. White, G. Allen, G. Qin, and F. Tsai. "Grid-enabled ensemble subsurface modeling." *Proceedings of the 19th IASTED International Conference on Parallel and Distributed Computing and Systems (PDCS 2007)*, Cambridge, MA, November 19–21, pp. 57–62, 2007.
9. G. Evensen. "The ensemble Kalman filter: Theoretical formulation and prac-tical implementation." *Ocean Dynamics*, 53 (4), 334–367, 2003.
10. J.M. Schopf. "Ten actions when grid scheduling: the user as a grid sched-uler," In: *Grid Resource Management: State of the Art and Future Trends*,

J. Nabrzyski, J.M. Schopf, and J. Weglarz (eds), Kluwer Publishing, Massachusetts, pp. 15–23, 2004.

11. I. Legrand, H. Newman, R. Voicu, C. Cirstoiu, C. Grigoras, M. Toarta, and C. Dobre. "MonALISA: An agent based, dynamic service system to monitor, control and optimize grid based applications." *Proceedings of International Conference on Computing in High Energy and Nuclear Physics 2004*, Interlaken, September, pp. 33–36, 2004.

12. R. Wolski, L.J. Miller, G. Obertelli, and M. Swany. "Performance information services for Computational Grids." In: *Grid Resource Management: State of the Art and Future Trends*, J. Nabrzyski, J.M. Schopf, and J. Weglarz (eds), Kluwer Publishing, Massachusetts, pp. 193–213, 2004.

13. R. Wolski, N. Spring, and J. Hayes. "Predicting the CPU availability of time-shared Unix systems." In: *Proceedings of the Eighth IEEE International Symposium on High-Performance Distributed Computing (HPDC-8)*, Redondo Beach, California, August 3–6, pp. 105–112, 1999.

14. A. Downey. "Predicting queue times on space-sharing parallel computers." In: *Proceedings of the International Parallel Processing Symposium (IPPS)*, Geneva, Switzerland, April 1–5, pp. 209–218, 1997.

15. N. Kapadia, J. Fortes, and C. Brodley. "Predictive application performance modeling in a computational grid environment." In: *Proceedings of the Eighth IEEE International Symposium on High-Performance Distributed Computing (HPDC-8)*, Redondo Beach, California, August 3–6, pp. 234–239, 1999.

16. J. Schopf and F. Berman. "Performance prediction in production environments." In: *Proceedings of Fourteenth International Parallel Processing Symposium and the Ninth Symposium on Parallel and Distributed Processing*, Orlando, FL, 30 March–3 April, pp. 647–653, 1998.

17. V. Taylor, X. Wu, J. Geisler, X. Li, Z. Lan, M. Hereld, I. Judson, and R. Stevens. "Prophesy: Automating the modeling process." In: *Proceedings of the Third International Workshop on Active Middleware Services*, San Francisco, August 6, pp. 3–11, 2001.

18. W. Smith. "Improving resource selection and scheduling using predictions." In: *Grid Resource Management: State of the Art and Future Trends*, J. Nabrzyski, J.M. Schopf, and J. Weglarz (eds), Kluwer Publishing, Boston, Massachusetts, pp. 237–253, 2004.

19. T.W. Malone, R.E. Fikes, K.R. Grant, and M.T. Howard. "Enterprise: A market-like task scheduler for distributed computing environments." In: *The Ecology of Computation: Volume 2 of Studies in Computer Science and Artificial Intelligence*, Elsevier Science, Cambridge, MA, 1988, pp. 177–255.

20. C.A. Waldspurger, T. Hogg, B.A. Huberman, J.O. Kephart, and W.S. Stornetta. "Spawn: A distributed computational economy." *IEEE Transactions on Software Engineering*, 18, 103–177, 1992.

21. C. Ernemann and R. Yahyapour. "Applying economic scheduling methods to Grid environments." In: *Grid Resource Management: State of the Art and Future Trends*, J. Nabrzyski, J.M. Schopf, and J. Weglarz (eds), Kluwer Publishing, Boston, Massachusetts, pp. 491–506, 2004.

22. A. Iamnitchi and I. Foster. "A peer-to-peer approach to resource location in Grid environments." In: *Grid Resource Management: State of the Art and Future Trends*, J. Nabrzyski, J.M. Schopf, and J. Weglarz (eds), Kluwer Publishing, Boston, Massachusetts, pp. 413–429, 2004.

23. Y. Chawathe, S. Ratnasamy, L. Breslau, N. Lanham, and S. Shenker. "Making Gnutella-like P2P Systems Scalable." In: *Proceedings of ACM SIGCOMM 2003: Conference on Computer Communications*, Karlsruhe, Germany, August 25–29, 2003.

24. B. Krishnamurthy and J. Wang. "Topology modeling via cluster graphs." In: *Proceedings of the First ACM SIGCOMM Internet Measurement Workshop: IMW*, San Francisco, California, November 01–02, pp. 19–23, 2001.

25. V.N. Padmanabhan and L. Subramanian. "An investigation of geographic mapping techniques for internet hosts." In: *Proceedings of ACM SIGCOMM 2001—Applications, Technologies, Architectures, and Protocols for Computers Communications*, vol. 31, San Francisco, California, November 01–02, pp. 173–185, 2001.

26. L. Xiao, Y. Zhu, L.M. Ni, and Z. Xu. "GridIS: An incentive-based grid scheduling." In: *Proceeding of the Nineteenth IEEE International Parallel and Distributed Processing Sysposium (IPDPS'05)*, April 3–8, Denver, Colorado, pp. 40–47, 2005.

27. R.J. Blake, P.V. Coveney, P. Clarke, and S.M. Pickles. "The teragyroid experiment supercomputing 2003." *Scientific Computing*, 13 (1), 1–17, 2005.

28. S. Jha, M. Harvey, and P. Coveney, "Spice: Simulated pore interactive computing environment—using grid computing to understand DNA translocation across protein nanopores embedded in lipid membranes." In: *UK e-Science All Hands Meeting*, Nottingham, UK, September 19–22, pp. 56–62. Available at: http://www.allhands.org.uk/2005/proceedings, 2005.

29. E. Deelman et al., "Pegasus: A framework for mapping complex scientific workflows onto distributed systems." *Scientific Programming*, 13, 219–237, 2005.

30. J. MacLaren. "Introduction to HARC document: version 1.9". En-LIGHTened Project Report, Center for Computation and Technology, Louisiana State University Technique report. Available at http://www.cct.lsu.edu/site54.php, 2007.

31. K. Yoshimoto, P. Kovatch, and P. Andrews. "Co-scheduling with user- settable reservations." In: *Proceedings of the 11th Workshop on Job Scheduling Strategies for Parallel Processing*, Kendall Square, Cambridge, MA, June 19, pp. 146–156, 2005.

8

Grid Services Orchestration with OMII-BPEL

Liang Chen, Wolfgang Emmerich, and Bruno Wassermann

CONTENTS

8.1 Introduction

The service-oriented paradigm continues to gain traction in grid computing. Open-grid services architecture (OGSA) has defined a service-oriented grid-computing environment based on existing Web service standards and technologies. As more and more grid resources are made available as grid services in practice, it has become commonly understood that a solution that can combine and coordinate existing services into useful application logic is needed. Considered to be a key element toward a fully integrated service-oriented architecture, business process execution language (BPEL) is the *de facto* standard for Web service orchestration in industry, providing the models and language specification for defining executable application workflows based on service invocations. It is in our interest to explore the applicability of using BPEL in grid applications.

Use of workflows allows the abstraction of process logic in a more human-friendly manner. To a certain extent, consideration of implementation details will no longer be required. Instead, people can focus on the development of a process model and the specific workflow interpretation. In grid computing, workflows are useful for the manner in which they facilitate the modeling and the automation of large-scale concurrent processes. A variety of workflow systems, or similar, exists, targeting different technologies, levels of abstraction, application domains, formalisms, and so on. A service-oriented paradigm is deemed to change the mindset of distributed computing in general and will provide us with new possibilities for problem solving with architectural approaches.

OMII-BPEL contributes to the initial attempts of using BPEL for grid workflows modeling and execution. The project studied the language expressiveness for grid-specific requirements, looked at existing BPEL resources and how they may be leveraged in a grid environment. An integrated BPEL development toolkit has been developed under the OMII (Open Middleware Infrastructure Institute) stack for the UK e-Science community. Through user feedback and conducted research, the project is devoted to improving user experiences on working with BPEL in scientific grid computing.

In the rest of this chapter, we will first get to learn some important features of BPEL and focus on how they are related to grid workflow modeling. This is then followed by an introduction to the OMII-BPEL software bundle, through which we highlight the specific requirements and characteristics of grid service orchestration and how users are supported by the tools and environments made available. Finally, a case study that will demonstrate how everything has come to work together in a real-life project will be given before we conclude.

8.2 BPEL and Grid Service Orchestration

In a service-oriented architecture, computing logic and resource are demarcated in a unit of *service*. Services as autonomous self-administrative modules are loosely coupled in highly distributed computing environments where service interactions wholly depend on the exchange of self-containing messages across networks, while a service itself exposes nothing but functional interface to the outside world. It is an intrinsic requirement for a service-oriented architecture (SOA) that services as composable and reusable building blocks should be coordinated in high-value logic to achieve advanced application objectives, while a service on its own is simply less likely to be capable of meeting such requirements. Enforcement of open standards of SOA helps eliminate implementation diversities, like differences between platforms, implementation languages, network protocols, security mechanisms, open or proprietary technologies, and so on, in the heterogeneous computing environment. It is now possible for services to be treated equally in a standardized and consistent manner to serve application requests despite underlying differences. *Service orchestration* simply provides the means of connecting the services through a central point of control and communication. An orchestration engine drives the execution of application workflow and service invocations according to the orchestration logic that has been defined. BPEL is a standard way of documenting such orchestration logic.

By focusing on service-level activities, flexibility, and simplicity, BPEL accommodates the creation, modification, merging, and division of large, complex application logic through the control of single or multiple process definition documents, which is much preferred by developers, architects, or domain experts whoever hold overviews of applications but rather not to drill down to implicit programmatic levels. The fact is, BPEL, as a purpose standard, reflects the intrinsic nature of interoperability, composability, reusability, and extensibility of servic orientation. It is an inseparable facet of designing and building a service-oriented architecture. To understand the context in which BPEL exists, the problem scope it targets, the essential models of service orchestration it promotes, as well as those technical backgrounds it relies on, helps us appreciate the essence of service-orientated architecture, around which grid services are eventually built. We will develop our discussions with these essential principles in mind throughout the chapter.

Working right on top of Web services, BPEL relies on the primitive Web service standards, the Web service description language (WSDL), which unifies service interface description, and the simple object access protocol (SOAP), which regulates service messaging, to engineer service orchestration. Grid services, for example, those developed with Globus toolkits or available with OMII distribution, can be orchestrated with BPEL in the exact same way as standard Web services given conformance to the same standards.

The proposal of BPEL was initially submitted to the OASIS (Organization for the Advancement of Structured Information Standard) in April 2003 by IBM, Microsoft, and other industrial partners, while the origin of the language can actually be traced back to IBM's Web service flow language (WSFL) and Microsoft's XLANG, both of which were part of the process management solutions built into the companies' commercial servers. At present, BPEL is mainly seen in two versions: the early BPEL4WS 1.1 and the latest WS-BPEL 2.0 published on 11 April 2007. The name change voted in the autumn of 2004 aligned with the Web service standard naming conventions. From then on, significant enhancements have been able to merge into the language with requirements and features raised by the BPEL community over the years since its first release. Companies like ActiveEndpoints, Adobe, Deloitte, JBoss, TIBCO, WebMethods, Oracle, and others, which have contributed to the standardization of the new specification, have further geared up the technology's moving into the mainstream industry. Products, solutions, services, and applications are continuously seen to spring out and evolve.

At the core of BPEL is a set of language constructs. For an XML-based language, the specification gives syntactic definitions to the XML elements, and explains how they should be used and structured to document valid BPEL processes according to the semantics and the models established behind. The key set of constructs is known as the *"activities."* They abstract the behavioral and structural capabilities of BPEL, dealing with service interactions (e.g., *invoke, receive, reply*), flow control (e.g., *if, while, repeatUntil, pick, sequence, flow, foreach*), data manipulation (e.g., *assign*), and fault handling (e.g., *throw, rethrow*). Others have to do with advanced scope management (e.g., *faultHandler, compensationHandler, eventHandler*), declarations (e.g., *variable, partnerLink, correlation*), and so on.

8.3 Interaction Modeling

The primary use of BPEL is to model service interactions that comprise distinct application behaviors. The hub-and-spokes orchestration model has decided that the elementary interaction activity BPEL can describe is based on a peer-to-peer model that involves exactly the process itself and one of its partner services. Complex interactions are built with more of these basic activities. Given the important fact that a BPEL process becomes a Web service as soon as it has been deployed, these interactions will appear with no difference from those between standard Web services. Thus we infer that the primitive interaction patterns that BPEL supports are consistent with the interfacing models implied by WSDL, since a WSDL description is the only contract of communication of a Web service.

Depending on the availability of an *output* message of a service *operation* definition in a WSDL description, two invocation patterns can be derived accordingly: the *request–response* pattern, in which a reply is always explicitly sent by a service in response to a request; and the *one-way* pattern, where a reply is not expected. By further applying these patterns to all two peers in an interaction, we have four combinations that, from a process prospective and considering the role the process plays as either a service provider or a service consumer, we can quickly summarize as the following: (1) process receives a request, and replies with the processed result (requestor blocks until the reply is received); (2) process receives a request but does not reply (requestor does not expect one either); (3) process initiates an invocation to a partner service and waits for a reply; and (4) process initiates an (e.g., a notification) invocation without waiting for any reply.

The four are considered the minimal conversation patterns a BPEL process needs to implement, and upon which complex patterns can be built. To enable the modeling, BPEL introduces the corresponding language notations that directly map to these interaction activities, namely *invoke*, *receive*, and *reply*. *Invoke* denotes an initiative invocation to a partner service. *Receive* denotes a receipt of service request. *Reply* denotes an explicit reply to a matching request. Then, for example, to model a request–response style of interaction, a *receive* and a *reply* can be placed at both the beginning and the end of a process respectively; an *invoke* without an output variable can be used to simulate a one-way invocation:

```
<bpws:invoke    inputVariable="SubmitJobRequest"
    name="jobmanager"    operation="submitjob"
    outputVariable="GetJobStatusResponse"    partnerLink=
    "jobmanagerPartner"    portType="jman:jobmanager"/>
<bpws:receive    createInstance="yes"    operation="submitjob"
    partnerLink="jobmanagerPartner"    portType=
    "tns:jobmanager"    variable="SubmitJobRequest"/>
<bpws:reply    operation="submitjob"    partnerLink=
    "jobmanagerPartner"    portType="tns:jobmanager"
    variable="GetJobStatusResponse"/>
```

In the snippets above, a *partner link* is an abstraction of a communication link between a process and its partner, whose roles in the target interaction are defined. A partner service must be imported through a partner link declaration that extends the original WSDL description before it can be referenced in a BPEL process. Accordingly, *PortType* and *operation* correspond to those in the WSDL interface exposed by the partner service. A *variable* is an instance of a WSDL message used in an interaction.

In grid computing, asynchronous interaction is often found to be advantageous for system optimization that better meets scalability requirements that push the system to its limits. Negligible concerns in system design and configuration that are based on synchronous communication could be well magnified or even turned into bottlenecks in systems when massive processes are spawned onto the grid. For instance, keep-alive signaling for long-running invocations generates unnecessary network traffic; blocked process accumulating memory footprints and acquiring extra CPU cycles complicate the process and system maintenance; services end up building long-lived dependencies that tend to be less tolerant to failures with resulting chained effects. Asynchronous interaction addresses partly the problem by further decoupling the services in interaction; that is, the service requestor will no longer stay in the communication with the service provider as soon as the request has been successfully submitted; the service provider calls back at a later time only when it is ready with the response data.

To illustrate in BPEL in detail, we will look at an example of job submission. At some point in a process, a job needs to be submitted to the grid via a dedicated job submission service. The process then queries a monitor service for the final job status and decides whether a resubmission is necessary or not. Suppose the submission service always replies with a unique job ID as soon as a job is received, synchronous invocation is simply adequate to describe such behavior. However, as it is unknown in advance how long it takes for a job to complete, are thus how long a job status report will take, it is more appropriate for the query with the monitor service in this case to be asynchronous, especially if the job is meant to be long-running. In order to model this in BPEL, we can first use an *invoke* activity to send the job description to the submission service and in return to obtain the job ID. Then we use another *invoke* activity but this time without specifying the *output* variable to perform the one-way invocation with the job ID to the monitor service, which will be calling back with the final job status in, preferably, another one-way invocation. The callback message will then be captured by an event listener that must be defined subsequently in the process. There are a couple of ways to set this up in BPEL. For example, *receive* is good enough for a simple case like this; *pick* is a selective event listener, which selects exactly one event out of all those having been defined; for more complex situations, *event handlers* can be used as a container of event-triggered routines alongside the normal process logic; as long as they are in an active state, events will be continuously picked up by the corresponding handlers. *Events* supported in BPEL can be inbound messages; for example, our job status notification, from the process service port, or customer timers set off after specific durations or at deadlines. Finally, in the process WSDL description, the use of a listener must be exposed as an available *operation* of the process, to which the asynchronous service provider can refer in the callback. In this way, there will be no blocking of the process whatsoever.

Apparently asynchronous interaction tackles certain issues that may be raised by synchronous designs, nevertheless at the expense of complication in process modeling. Suppose the BPEL process in our example is called many times, if we need to submit more jobs. Each resulting callback message must then be directed to the correct process instance that has initiated the asynchronous service requests in the first place, so that the matching notification listener can close the session of listening when the message arrives. However, this is not as straightforward as it appears. From the point of view of the monitor service in our example, the only information regarding callback invocations is the service endpoint reference bound in runtime to the callback port that we have exposed. It is shared by all process instances in the same way as they were created on service requests made at the same service port at the beginning. WSDL, of course, does not provide any information about the process instance as it is not designed to be aware of any extension standards like BPEL at all. On the other hand, BPEL services are not meant to be considered any more special than standard Web services in service interactions, so neither should their WSDL descriptions. The asynchronous service provider, just like any external service partner and client, is not concerned about whether the service it is communicating with is a BPEL process or not. In a word, the service endpoint reference available to the monitor service cannot be used to distinguish process instances, and all callback messages will be unfortunately sent to the same destination. There must be a way to reassociate these messages with the initial process instances they are intended for. The notion of *correlation* in BPEL allows the use of application data delivered with SOAP messages to construct process identification information dynamically in runtime and, moreover, independent of any implementation-specific mechanisms. From a service-oriented perspective, application data embedded in the messages exchanged with service partners constitute the context of application significance that pure implementation-level mechanisms cannot convey. The application-level protocol is required to address such a requirement, while at the same time avoiding competition with any possible low-level implementations. Starting with the extension to the initial WSDL description, it is possible to declare rules for how messages should be parsed against the creation of identity information in XML, before labeling the process document when and how the declared correlation should be performed; for example, when a callback message is received by the listener. It is then up to the BPEL engine to implement the correlations in practice: messages will be intercepted and checked at arrival before they are reassociated and passed to the matching instances. Partner services and original messages defini-tions, however, are not required to be altered in any way when correla-tions are performed transparently on the other side of a conversation. In our example, a job ID qualifies as a simple process identifier. Together with the job status, it should be included in the callback message to distin-guish a process instance. Given that it will be the same instance of the monitor service that allocates the job ID, this will not be difficult. As we can see, message correlation is essential for maintaining application con-sistency in extensive service interactions.

Lastly, we have not yet pointed this out explicitly, but any services participating in a BPEL process, like the submission service or the monitor service in the example, can themselves be BPEL processes as opposed to implementations in Java, .net, or by other technologies. Thanks to the intrinsic composability, BPEL processes deployed as Web services can be seamlessly integrated into the Web-service architecture pretty much effort-lessly. As a matter of fact, when a BPEL process becomes reasonably com-plex in practice as one continues streamlining the initial design for performance, scalability, deployment, and other reasons, the top-down modeling approach usually ends up breaking down the process into sub-process modules so that they themselves in turn can be further optimized and mixed up with BPEL processes and normal services in coordination. An in-production BPEL application is hence usually made up of a number of workflows that establish the hierarchical dependencies with each other to enable sophisticated interaction modeling.

8.4 Flow Control

Process modeling is about structuring activities into proper control flows that map to an application process. BPEL supports common structure con-trols similar to those found in modern programming languages, like *while*, *repeat until*, *if*, and so on. *While* is a notation of controlled looping, whose enclosing process loops through all iterations one after another until the condition is no longer held true. *Repeat until* is similar to *while*, only the first iteration is always executed regardless of condition. *If* specifies how to branch a workflow on conditions, including an optional default branch. Other BPEL controls like *sequence* and *flow* prescribe the order in which their containing activities should be executed.

Proper support for parallelization is fundamental in grid computing. *Flow* allows the embedding of concurrent procedures in processes. Typical use cases are found in business process modeling. Small business logic can be rearranged in parallel to maximize the execution efficiency of the overall business process. Dependencies between sublogics can be synchronized using *links* that connect them with optional *transition condition*s and/or *join condition*s at the start or the end of a link, respectively, to evaluate and control flow transitions. The limitation of *flow*, though, is the lack of expressiveness to define repeated parallel processes. Typically in grid computing, application processes or computation models are repeatedly executed in parallel. Differences between individual executions are merely those parameters or data being used. Apparently, *flow* will be too tedious or even impossible* in this case to duplicate, say, thousands of subprocesses just to implement concurrency. As difficult as it is to construct a workflow like this, changes to parallel branches that must be applied frequently in development cycle or due to experimental requirements are neither manageable. In WS-BPEL 2.0, a new construct has been added to cater to the exact parallel looping like this. In the default mode of sequential looping, *for each* is a functional equivalent to *while*, whereas in the parallel mode it allows the spawn of iterations all at once. A user can define how each iteration should differ from the other with reference to built-in counter or custom expressions or variables, based on a same procedure template. In the example below, all ten *wait*s will start timing for a period of *Counter* seconds:

```
<bpws:forEach counterName="Counter" name="ForEach"
    parallel="yes">
        <bpws:startCounterValue>1</bpws:startCounterValue>
        <bpws:finalCounterValue>10</bpws:finalCounterValue>
        ...
        <bpws:wait>
                <bpws:for>concat('PT', $Counter, 'S')</bpws:for>
        </bpws:wait>
        ...
</bpws:forEach/>
```

A common concern in the design of concurrent systems is that concurrent access to shared resources may cause race conditions leaving resource status undeterministic due to corrupted read/write operations. Problems like this can happen for BPEL processes too. Take variable access for example: the state information of a BPEL process is stored in *variables*; for example, invocation messages, counters, constants, user-defined data models, and so on. Attempts of updating variables from within concurrent procedures are almost certain to lead to unexpected results unless proper synchronization has been explicitly implemented as part of complete process logic. BPEL addresses the problem by forcing serialization of these concurrent accesses using synchronized scopes. Being similar to the concept in procedural programming languages, *scope* is such a context in which local activities and declarations are defined. *Isolated scope* is one that has implemented concurrency control in particular. For example,

---

* Related work on parallel looping using *flow* is discussed in a separate chapter.

a procedure **P** defined in a parallel *for each* statement modifies a variable **V** in the outer scope **S**. When **P** is iterated in parallel, there are chances that **V** will be modified at the same time. To avoid the situation, an *isolated scope* **IS** can be inserted to *for each* and wraps up exactly where in **P** variable **V** is modified, indicating a forced serialization of that access. The BPEL engine is responsible for the actual implementation of serialization. An *isolated scope* can be activated by setting the corresponding attribute in a normal scope.

```
<bpws:scope isolated="yes" name="aScope">
    ...
</bpws:scope>
```

8.5 Data Handling

The Web service framework is built on the foundation of XML supports. XML schemas and documents ensure the consistency, interoperability, integrity, and validity, as well as the flexibility of data sharing across information systems. Inherently, XML suggests a strong typing system for defining structure application data, which we find to be rather important for grid computing. Typical grid applications produce a massive data volume that is critical either as computational results on their own or as a shared database across organizations. A weakly typed system does not give enough confidence on the validity of data, especially if the computation is experimental or by any means capable of producing exceptional results in unpredictable manners. Syntactic validation gives early opportunity to expose apparent mistakes in data, and possibly to define error-recovering strategies in advance. Faulty data can be kept from further processing, wasting more time and resources, or even putting systems in an unstable state. For a complex system, especially one operating in a service-oriented architecture in which services as distributed modules utterly reply on the accuracy of application data received to participate in application logic, the ambiguity of information must be avoided. The bottom line is, some forms of validation will be required sooner or later to build a reliable application system, if not systematically, then probably programmatically.

BPEL inevitably needs to deal with XML data. Service interaction is about delivering service data in XML, either sending or receiving, so they must be modeled in the BPEL process. BPEL needs to access these data as defined in their WSDLs and schemas to extract, modify, compose, and validate application information with fine-grained details. *Assign* is a powerful activity of BPEL designed for such purposes. *Assign* allows the encapsulation of single or multiple *copy*s to specify how XML data can be assigned *from* a source *to* a destination. For example, an *assign* can indicate an XML data of the reply message of an invocation to be copied to the input message of another invocation. Effectively this will be implementing the interaction between the two services through the controlling BPEL process. *Assign* is often used to update variables and other process properties.

XML data models of real applications, of course, tend to be more complex. Support for XPath in BPEL will literally allow any node of a given XML structure to be precisely located using proper queries or expressions; for example, to look for the element that has a particular attribute, or an element in an array at a particular position, or an element that has a particular child node, or a set of nodes that hold particular text values, and so on. XPath functions, as a subset of the XPath specification, provide further operability for XML data manipulation, such as strings concatenation, rounding of numeric values, returning position information, arithmetic operations, boolean evaluation, and so on. The example below shows how the value of an input variable is copied to the output using a simple XPath query. The full XPath support should be sufficient in most cases for most users. In more complex situations, BPEL also supports XSLT transformation that basically allows complete transformation of the XML document according to provided external XSL style-sheets, in which transformation templates have been defined by a user in advance.

```
<bpws:assign name="initialisation" validate="no">
    ...
    <bpws:copy>
            <bpws:from>$input.payload/tns:input</bpws:from>
            <bpws:to part="payload" variable="output">
                    <bpws:query>tns:result</bpws:query>
            </bpws:to>
    </bpws:copy>
    ...
</bpws:assign>
```

Assign is one of the most frequently used activities in BPEL. The decent support of XML processing has made it a crucial element for defining data flows in process modeling with full control of XML data and definitions. Given the flexibility of using it in BPEL, people are likely to find themselves more than a single solution for data handling in practice.

8.6 Error Handling

Any system can unexceptionally encounter errors. Like any distributed system, BPEL processes, which involve service interactions beyond administration boundaries, are typically error-prone. Error handling in BPEL, in general, is to allow the definition of error-handling procedures along with the normal process definition to be activated in case of process failures. Error-handling logic, as critical as the main process logic in a fault-tolerant workflow design, should be explicitly defined by a process modeler. BPEL pursues ACID transactions at the application level to ensure integrity of process executions. This means, in case of errors, that any committed operations that have caused changes of system status, including both logic and data, must be undone to reverse the effects of having been passed on to not only the process itself but also the partner services that have been interacted within the same transaction that has failed and must be abandoned. In an example of database service, committed submission of persistent data can be deliberately deleted from the database through a normal service invocation but at a different, dedicated reversing operation port; a snapshot of the process state at the point of failure in the transaction will be captured and used for assembling reversing invocation messages. In BPEL, such a reverse of committed enactment is called *compensation*. The container of compensation logic is called a *compensation handler*. A compensation handler can be attached to the same *scope* in which compensatable forward-working logic resides. Since *scope* is usually used to demarcate a sublogic unit in BPEL, it is sensible to allow various extension handlers of complementary logic that have only constrained influences to the companion main process in scope to be associated with it. Including *compensation handler*, there is *message exchange handler*, *event handler*, *termination handler*, and *fault handler*. *Fault handler* can capture explicit faults, for example during a service invocation, and then activate corresponding fault-handling logic in reaction. Therefore, an explicit call of compensation is usually incorporated as part of the fault-handling routine to enable compensation in time. BPEL supports both WSDL faults associated with service operations, and customer-defined faults. A class of implicit faults has also been defined in the specification.

8.7 Extensibility

A service-oriented architecture is an open and extensible platform ready to accommodate dynamic and agile developments of modern information systems. All Web service standards, including BPEL, support extensibility to some extent. BPEL allows user-defined *activities* to be implemented and works side by side with standard constructs. In a way, user extensions like the macro process, embedding of programming codes or scripts, and so on can be integrated to a BPEL implementation seamlessly. More likely, vendor extensions implemented on customer requests will appear commonly in commercial BPEL products.

8.8 Web Service Extensions

Surrounding the primitive Web service standards like WSDL and SOAP is an ever-evolving family of Web service standard extensions (WS-*), which represent the collective thoughts of defining and improving the Web service framework. For example, standards like WS-Addressing, WS-Notification, WS-Policy, WS-Security, and others are considered as beneficial complements to the primitive Web services framework in their own target areas of concerns. In a layered environment that will continue to expand and evolve like this, BPEL will consistently work with Web services through its direct dependencies on WSDL and SOAP standards orthogonally to the WS-* extensions. Eventually, it will still be the application data peeled off the SOAP messages that BPEL is really concerned about.

8.9 OMII-BPEL and BPEL Environment

BPEL in itself is only a specification that does not come with any concrete implementations. Being an open standard, it instead encourages vendor-diverse solutions. OMII-BPEL has bundled up some best-of-breed open-source software to deliver a fully integrated BPEL environment that includes a client-side process editor, which eases the effort of BPEL generation, the BPEL Designer, and a server-side BPEL engine, which ensures reliable process hosting and enactment, the ActiveBPEL. The tailor-made environment developed under the OMII middleware infrastructure is released with other OMII-managed services and tools on a regular basis.

8.10 Process Modeling and BPEL Designer

The BPEL specification is a nearly 300-page-long document, including some 40 pages of XSD schema and another 12 pages of static analysis rules. All must be applied to validate process documentation. For proper modeling, the level of complexity is well beyond any existing text editors or XML editors. It is tedious and error-prone even for experienced users if no further assistance is available. The provision of GUI editors that support visual representation of the process model is surely more than simply having an interactive user interface, nice self-explaining icons, or an easy drag-and-drop style of composition. Most importantly, through the minimal one-to-one mappings between BPEL activities and their visual equivalences, the transition of abstract logic to concrete process model can be facilitated without having to unveil the actual translation taking place under the cover. Furthermore, in response to frequent changes of process definitions, for example, in scientific experiments, visual editing is much more efficient. It must be noted that the target users of BPEL are not only computer professionals who have the expertise of developing BPEL processes in every detail, but also those domain experts who have the real cases of workflow logic but rather limited technical background by way of contrast. This is particularly true for *Abstract Processes* introduced in WS-BPEL 2.0 as nonexecutable processes that only focus on logics in large instead of implementation details. All in all, the design of primary importance of a productive modeling tool is to ensure the modeling itself is well supported at all times. We will see what useful functionalities BPEL Designer has provided us in this section.

With the release of the home-grown Sedna editor, OMII-BPEL has made available one of the first visual BPEL editors for the community. It was a BPEL4WS1.1-compatible editing tool based on the Eclipse platform. In addition to standard BPEL supports, Sedna has integrated advanced features like parallel looping with *flow*, macro process extensions, deployment with ActiveBPEL, and so on. Being a fully functional BPEL editor, Sedna was available to the public with early OMII distribution. Some first-hand user feedback and requirements were therefore able to be gathered and studied for successive developments.

Work on bringing BPEL to scientific grid computing materialized with the cooperation of the Eclipse-BPEL project in 2006. Largely led by IBM and Oracle with volunteer contributions, the project was aiming to add comprehensive support of BPEL to the Eclipse IDE, around which a set of key functionalities can be offered to support the open-source development of the advanced BPEL modeling environment. This includes a visual editor, a BPEL data model, a validation framework, a runtime framework, and a debug framework.

Being very similar to Sedna, which has since been replaced, BPEL Designer was developed as an Eclipse plugin, and builds on well-known Eclipse plugin platforms like GEF, EMF, WTP, and so on. The consistent look and feel of BPEL Designer should appear familiar to existing Eclipse users. A screen dump is shown in Figure 8.1.

The main editing area in the middle is where a process is composed using constructs, mainly standard BPEL activities, which can be picked up from the palette alongside. Constructs are connected and aligned automatically. Container-constructs like *scope*, *if*, and so on can be collapsed or expanded for a better spacing and viewing experience. Summary tips are available when the mouse pointer hovers above. Declarations of *variables*, *partner links*, *correlation sets*, and other process properties can be managed

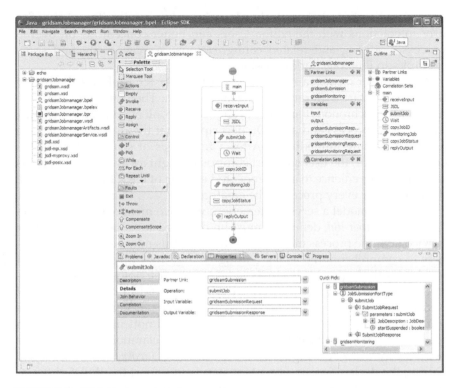

FIGURE 8.1 BPEL Designer interface.

in the panel to the right. The main editor interface is also divided with a number of windows (or *view*s in the Eclipse terminology). Apart from the common ones like *project explorer* and *outline* views, the *properties* view at the bottom is extensively used by BPEL Designer to provide further information and interactive configurability. For instance, a click on the blank canvas will have the *properties* view refreshed with the basic process information, like name, associated namespace, documentation, and so on. We will see more examples in the coming paragraphs. Other views like *problem* view, *servers* view, and so on are usually docked alongside *properties* view, which can be activated explicitly from the menus.

8.11 Usability

By reduction of user involvements in producing process documents, BPEL Designer has managed to integrate a number of usability features like automations, wizards, auto-completion aids, templates, forms, quick access menus, and so on to reduce the burden of operational details and simplify user interactions with the editor. Users can better concentrate on overall process modeling and related configurations. We can easily find some examples to illustrate our discussions.

When a user creates a new process, BPEL Designer allows the use of templates to start with rather than from scratch. There are both synchronous and asynchronous templates to choose from, all of which will generate the skeleton process and a WSDL descriptor that the user does not have to bother editing immediately. After a few steps of setup following the creation wizard, the user can start composing right away. Apparently, this is of great convenience to avoid repeated editing of basic BPEL and WSDL documents for every process to be created.

In order to model a service interaction, we have mentioned that a corresponding *partner link* definition must be declared to associate the partner service with the process model. As difficult as this may be for BPEL beginners to comprehend, the actual editing of a partner link declaration that involves both the process document and WSDL descriptor(s) could be frustrating. BPEL Designer has been able to simplify this by bringing a series of wizard pages that walk through all required steps. A partner link will be unmistakably declared in the end without the user noticing any changes in files at all. The wizard can be started from the declaration panel explicitly or on the fly when defining relevant activities (e.g., *invoke*, *receive*, and *reply*). During the declaration, an important step is to import the WSDL description of the partner service so that the information derived can be used to define service interactions in the process. BPEL Designer supports the import of WSDL files from a file system, Eclipse project space, or remote locations with given URLs. Once a WSDL description is imported, the service interface as well as any associated schema definitions will be stored and made available throughout the process project. BPEL Designer will render them into an expandable tree controller. Users can simply pick up the corresponding nodes on the tree to associate with the current partner link declaration without having to go through the actual WSDL description. A similar use of tree controllers can be found in a number of occasions in BPEL Designer.

For example, in order to define a service invocation, one must provide the information of partner link, port type, operation, and input/output variables altogether for an *invoke* activity. In the *properties* view (Figure 8.2), candidate values can be selected from the drop-down lists respectively. Or preferably, simply picking the intended *operation* node on the tree will have all the fields filled up automatically without further operation from the user. Input/output variables will be created on the user's behalf if they do not exist already by silently resolving the *operation* and the associated *message* definitions of the imported WSDL. The *variable* panel will be updated with the new declarations accordingly. This is more than a useful feature as users tend to put off variable declarations until they are actually needed.

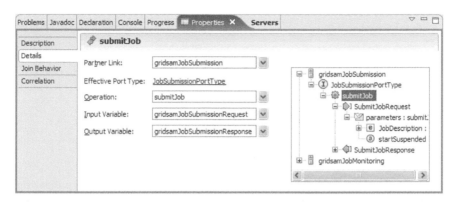

FIGURE 8.2 The use of a tree in the *properties* view of an *invoke* activity.

Users will no longer need to figure out WSDL and XSD definitions themselves to make matching declarations for new variables. The burden of XML operation is greatly removed.

Similarly in the *properties* view of an *assign* activity, available variables and their XML structures will be listed for both source and target specifications. The user can select the XML element of a variable from the tree and immediately obtain the corresponding XPath query that will locate the select XML element in runtime (as shown in Figure 8.3). This is extremely helpful for non-XML experts to get a grip on the XPath query quickly. Advanced editing may be applied in addition whenever appropriate.

To deal with expressions in BPEL Designer, for example, to extract XML data in an *assign*, or compose iteration conditions of a *while*, and so on, autocompletion aid is available to suggest candidate elements on the go as the user keys into the editor. Results are filtered out instantly until the right one is selected. XPath functions and XML elements are differentiated with different colors. Figure 8.4 is a dummy example for using input variable to determine the final value of a *for each* counter.

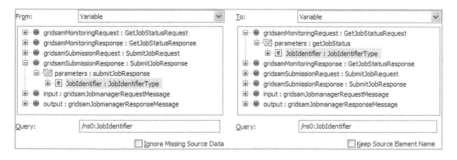

FIGURE 8.3 XPath query editor of an *assign* activity.

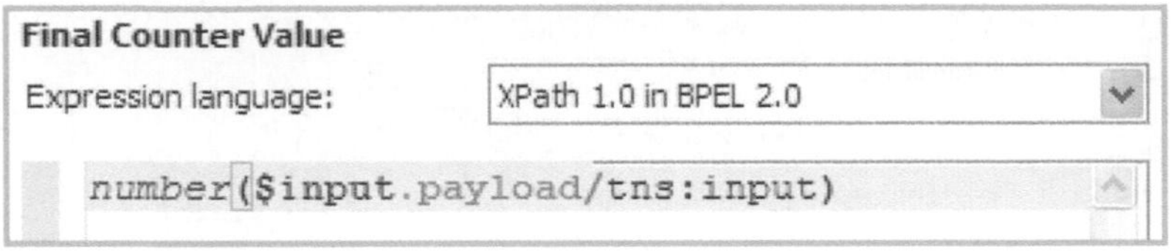

```
Final Counter Value
Expression language:        XPath 1.0 in BPEL 2.0                    ▼

    number($input.payload/tns:input)                                ▲
```

FIGURE 8.4   An example of XPath expression.

Last but not least, users of BPEL Designer always benefit from existing features of Eclipse IDE out of the box. Eclipse is one of the most influential development platforms in recent years. The open architecture fosters fruitful plugin developments to integrate extension functionalities to the classic platform. It is a handy toolbox where software components can be easily assembled and customized in the way that best facilitates user developments. Putting any third-party software aside, there are already a good set of features worth mentioning: error reports, CVS control, Ant scripting, server integration, help system, online update, and many more.

8.12 Validation

A trustworthy modeling environment should guarantee the correctness of BPEL documents being generated as a result of user interactions. While usability features may vary from tool to tool, validation in BPEL is strictly required by the specification. Besides the XSD schema that can be used for syntactic validation just like any XML language, static analysis must be performed additionally to detect semantic violations in the process definition. Since users are mediated by graphical interface and kept from direct access to text documents when using a visual editor, syntactic violations are physically impossible under normal situations. Nevertheless, this does not make a syntactically valid document being very semantically mistaken any less possible. For example, a static analysis rule says that the *from* specification and the *to* specification of an *assign* activity must be of compatible types. Syntactically, there will be no way to prevent an implementation of the very opposite. Therefore, BPEL Designer enforces validations whenever a process is updated and saved with confirmation. Immediate feedbacks will be made available in the form of error, warning, or info messages populated in the *problem* view window if any violation has been caught. Line numbers of faulty definitions in the process document will be given to locate the sources of violation. Users should then consider the fixes before continuing.

Conformant BPEL engines perform static analysis in deployment time too. This is to ensure that vanilla BPEL documents developed by any client editors can be validated independently. Note that BPEL also allows additional analysis beyond the standard set in extension to the specification. Even though this *should* be configurable to be disabled, potential issues, under the circumstances, are still up to the specific implementation and users to settle, while BPEL Designer only sticks to the basic standard.

8.13 Server Runtime Integration

The server runtime framework of BPEL Designer is an advanced facility that allows diverse BPEL implementations to be plugged into the editor. Vendors are free to extend the open interface exposed by the framework to hook up an installation of their engines so that the processes developed within BPEL Designer can be attached to a running server instance with which they can be tested against the engine before production deployments. A similar feature is widely used for J2EE developments in Eclipse: mainstream application servers have integrated (e.g., IBM WebSphere, JBoss, Apache Tomcat, and so on) to support professional application implementation, deployment, and debugging. Servers can be controlled and configured in the editor. A BPEL server runtime can be declared easily in the *server* view window of BPEL Designer by indicating the server type, the installation location, and some other information using a wizard. Available operations will then be available from the associated context menu within mouse clicks. For example, the ActiveBPEL runtime developed by OMII-BPEL is able to transparently generate a deployment archive by parsing existing process documents and resource files without user interference. Users assign process modules to be published to the server and start the server subsequently. Once published, the process will be synchronized with the running server. Changes will be updated immediately in the deployment archive, keeping the publishing up to date.

8.14 Simulation and Debug

Although this is an ongoing development, with the existing BPEL model and server runtime framework it will be possible to simulate process execution from within the editor. The debug framework will allow stepping through a process and toggling inspection breakpoints during simulation. Errors can be reported in real time and advice for corrections can be given.

As the development goes on, we are interested in learning about the following: import of simulation data, preselection of branching, timer control, and moreover how typical grid requirements can be supported in cases where scaled-down versions are not sufficient to evaluate a workflow design.

8.15 Web Service Client

BPEL Designer comes with a general Web service client. This is of immediate demand after process deployment. The Web Tools Platform (WTP) is an extension platform of Eclipse for developing Web and J2EE applications. It provides essential data models, APIs, and tools that are relied on by many derivative Eclipse projects and products, including BPEL Designer. The Web Standard Tools (WST) subproject supports the processing of a wide range of popular Web standards, like HTML, XML, XSD, XSLT, SOAP, WSDL, UDDI, and so on. The Web Service Explorer (WSE) is a tool with which users can interact with Web services and UDDI registries through Web interfaces.

In order to invoke a Web service, WSE uses the full WSDL description of a service to dynamically generate the Web form that needs to be filled out by a user to initialize the service request body. WSDL files may be located either locally or remotely. Associated XML schemas will be parsed to support form generation. The user confirms and submits a form by clicking a button. Request data will be validated before sending out. Errors detected during the invocation will be reported in the status window in which formatted response data of successful invocation will be displayed. Both request and response messages in raw SOAP can be viewed in a separate page, in which SOAP segments can be edited directly, or even saved or loaded if necessary. OMII-BPEL has integrated OMII security support into WSE. OMII users can invoke services hosted by OMII servers, whether BPEL processes or not, seamlessly and safely.

In order to work with UDDI registries, WSE allows users to specify the registry they want to create, update, delete, and query for services. Without going into too much detail, a typical scenario of using the UDDI registry is that a user can log in to the registry to search for services; for example, using keywords of service descriptions. The WSDL file of the returned service can then be saved directly into the current project space, and used in the partner link declaration as we have mentioned before. OMII distribution ships its own UDDI registry, Grimories, which implements the standard UDDI specification. It can be simply deployed to register and manage service information, including BPEL processes. The registry itself is a Web service, and the access from within WSE is secured in a similar way. Finally, WSE may be loaded externally to Eclipse in popular Web browsers such as Internet Explorer or Firefox.

8.16 Process Enactment and ActiveBPEL

A BPEL enactment environment is where application logic in the form of standard BPEL descriptions is finally executed, which takes account of process deployment, monitoring, and other aspects related to process execution. We focus on the discussions of these aspects in this section.

ActiveBPEL was one of the first BPEL implementations found to be appropriate for proper academic research. The software is freely distributed by Active Endpoints under a Lesser GPL license. It deploys Apache Tomcat as a Web application, while commercial products that run on mainstream application servers with enhanced performance and integration of enterprise solutions are developed around the very same codebase. Comprehensive documentation and community supports are available for open-source users, while ActiveEndpoint also provides training, consultancy, certification, and other professional services. Because of the early availability as a full-function BPEL enactment engine, ActiveBPEL had been widely adopted and evaluated, whose quality-assured software development and extensive user supports have been found to be difficult for general academic bodies to catch up with.

8.17 Deployment

A BPEL process is given a Web service front end at deployment time with extra binding and service information appended to the initial process WSDL description created in design time, which are important to regulate service messaging and identify service location. No differences are exhibited by any process after successful deployment from the standpoint of standard Web services to serve process requests. The BPEL processor at the core of an enactment engine understands and performs the mapping between the process description and the equivalent programming statements and procedure calls that can be executed in the server container. WSDL and XSD definitions need to be parsed beforehand to construct data models required by both predeployment validation and process execution. Additionally, process deployment is likely to require more than a single process document besides WSDL and XSD descriptors. Extra engine configurations and deployment-time information, like the locations of partner services, may be needed for an engine to customize individual deployment. Depending on proprietary implementations, this may be achieved declaratively with a deployment descriptor. BPEL Designer handles the generation of the deployment descriptor (as well as other file formats) transparently with the ActiveBPEL server runtime that we have mentioned.

Once a process archive is uploaded to ActiveBEPL, deployment files are unpacked to the working directory that the engine monitors periodically to pick up new deployments and make changes to existing ones. ActiveBPEL supports *hot deployment*, which avoids interruption of ongoing processes or other hosted applications on the server when a process needs to be modified and redeployed. In particular, this is of benefit to long-running processes that cannot afford unmanaged terminations.

Nevertheless, some practical issues emerge with process deployment. A BPEL engine could spend considerable time and computer resources, particularly memory resources, on deploying large process descriptions depending on the complexity that is roughly in proportion to the size of a plain BPEL document. It is possible that system reliability may be threatened so that stable process execution will not be guaranteed as a consequence, or the process deployment could not even complete in the first place. In less extreme scenarios, there are indeed great chances (when processes are becoming too complex and believed to be larger than necessary), in which cases decomposition that can distribute process complexity to coordinated subprocess modules is always suggested and likely to be applicable normally. In fact, the divide-and-conquer approach permits better control of process logic through the separate optimizations of individual subprocesses. It is good practice to perform service-oriented analysis and design in general anyway, even if memory constraint is not a concern of a particular system setup. Consequently, subprocess units may be further distributed onto multiple server instances whenever necessary to decentralize associated computing stress, without affecting the initial process. Because of the deliberate separation of service interface and binding (which only takes place at deployment time), service relocations, which are completely transparent and irrelevant at design time, can be simply managed as altering subprocess service endpoint references in the deployment descriptor, leaving the process document unmodified completely. Conventional strategies such as server clustering, which is supported by Tomcat, are still feasible to enable load balancing.

8.18 Execution

Process execution is clearly far more complex. Except for the static contents of a process like the process map deployed on an engine that must be consumed, a large amount of dynamic information, which is only determinable in runtime but crucially in charge of the process state and enactment, such as messages, queues, alarms, counters, faults, correlations, and so on, must all be coordinated. Not only that, executable processes as normal computer programs are influenced by a wide spread of system variables, from hardware configuration to operating system support, from hosting server to external services, which ought to be looked after. Likewise in grid computing, traditional concerns of those causing abnormal system behaviors when both hardware and software are under enormous stress remain themselves in a BPEL-enabled environment. We will try to discuss these in common with regard to large-scale and concurrent executions.

All computer programs consume memory. As memory usage begins to grow with processes in a system, memory-management mechanisms start to kick in to release memory slots that are no longer in use. Taking Java for instance, garbage collection is built into the Java Virtual Machine (JVM) and managed automatically. Optimization policies and related configurations can be passed directly at server (e.g., Tomcat) startup, and all hosted applications will be affected accordingly. With long-running executions, allocated memory is persistently held by processes that are waiting for conditions of continuing, for example, replies from partner services, to be satisfied. Deallocation noticeably lags behind, and available memory has to keep decreasing until the next major collection. Then, as the execution proceeds, a process itself can get more complex such that dependencies may establish and complicate themselves both internally and with other processes and services. It is almost certain, under such circumstances, that the general processing speed of the system will suffer a steep cut-down, not to mention the growing need of system resources of both existing and new processes that keep feeding in. The longer the processes stay in the system, the more the memory is needed by maintenance and the more the deallocation will be delayed, and in the end, less memory will be available overall for further executions. Having said that, an exponential climb-up of memory use may eventually occur and start to take down the system until a complete exhaustion under extreme conditions. Or, more likely, the virtual memory that has been swapped into the system may seriously slow down the machine even more due to demanding disk access. System stability will be seriously compromised such that a functional state cannot be maintained.

Another concern falls within threading. In a burst of spawn of massive concurrent processes, a large amount of threads are acquired to hold subprocess execution states. Typically, application servers implement their own native thread pools and threading managements to support high-density serving requests. With no exceptions to any large multithreaded systems, fine-grained tuning is required to meet concurrency requirements for particular application deployment. ActiveBPEL replies on Axis/Tomcat to handle low-level communications with external services, while its own scheduling of concurrent processes (e.g., via *flow* and *for each*) is implemented independently, to which extent process enactment can be internally optimized. Users can decide the range of number of threads to be managed by the built-in work manager of the engine, the maximum threads to be allowed for each process, and so on in order to control the workload to be placed on the server in runtime. As for Tomcat, dedicated tools and methods exist for professional diagnosis and rectifications. Nevertheless, it is still the fundamental system resources that will be in intensive demand when more threads are allocated in process executions: the operating system will not have an unbounded number of threads; memory management will be challenged again as we have just discussed; disk space may be subsequently exhausted; delays around the networks may result from high volume data transfer, and so on. It is important to point out that it is not only the critical components like the operating system, the hosting server, and the BPEL engine that must be sufficiently scalable, but also any orchestrated services as self-administrative components themselves must be aware of possible grid requirements in dedicated application scenarios. For example, a service invocation may be failed simply because of nothing but the service running out of file descriptors for opening files in the operating system. Extra synchronization must be implemented to avoid deadlocks or arbitrary data access; a service design should take potential performance requirements into consideration, so it does not refuse service requests made at a reasonable rate, for example. In other situations, extensive service interactions may result in communication deadlocks when both memory and threads are under tight constraints relative to large-scale executions. A process will be blocked on the reply of the partner service that is stuck in the thread queue.

BPEL is a specification that sums up a high-level agreement on the normalization of a solution to a specific problem. Process executions are still largely dominated by low-level implementations. Existing solutions and methods that have been efficient in resolving typical grid computing problems will still be applicable and play important parts in designing and implementing BPEL-enabled grid applications, given that BPEL and BPEL implementations are merely a thin layer on top.* Note that it is always a systematic consideration to put together pieces of best-fitting technologies and practices to achieve both functional and nonfunctional requirements. As far as process modeling is concerned, the responsibility of a process modeler is to incorporate possible optimizations into the process design, for example, with appropriate algorithms, interaction models, parallelization, fault handling, service-oriented analysis, and so on. The modeler must express the process logic not only accurately, but also efficiently, for the benefit of both process development and the underlying implementations.

---

* Benchmark data are available in separate publication.

The asynchronous interaction modeling may ease the network and memory tension for long-running invocations.

A BPEL process can be invoked by any valid Web service client that creates and delivers SOAP messages. Support has been given from many modern technologies since Web services gradually merged into mainstream IT. Various tools have become widely available. OMII-BPEL users benefit from the simple yet powerful Web Service Explorer so that they can work seamlessly and efficiently within the OMII environment.

8.19 Monitoring

It is obviously important that BPEL users should be able to keep track of their processes. ActiveBPEL provides the monitoring facility through its one-stop administration console that is deployed side by side with the core engine. An administration console is shown in Figure 8.5. The Web portal allows access to various information and configurations of the engine, while process monitoring is only one of the functionalities it provides.

Starting from the navigation index on the main Web interface, entries can be found for engine administration, deployment status, as well as process status in groups, which are followed by a link to a separate thorough user guide to the console. A short list of configurable engine settings can be found, from elements as simple as logging style through to validation control and thread management. The engine can be stopped and started without killing the host server to enforce lightweight reset of engine configurations and states; for example, to reset database login information. Upon the deployment of a process, apart from the deployment log given as expected along with a list of deployment files, full WSDL desertion of the deployed process is made available with a unique URL and made ready for use. Once a service request is made at the given WSDL service endpoint, for example, using WSE, the resulting process instance will be assigned with an entry in the console and labeled with a sequential process ID, a process name, a start date (and an end date if competed), and one of the main statuses of *running, completed, faulted,* or *terminated.* Every entry is linked to a separate page of full process details, where users can find out about variable data, partner links, execution progress, faults, log, and other state information of the process, using the given graphical process tree structure that maps to the process definition. A running process may be suspended, resumed, or terminated on the same page. The optimized JSP implementation of the detail page is capable of rendering large process instances (e.g., with *foreach* in parallel) in reasonable time.

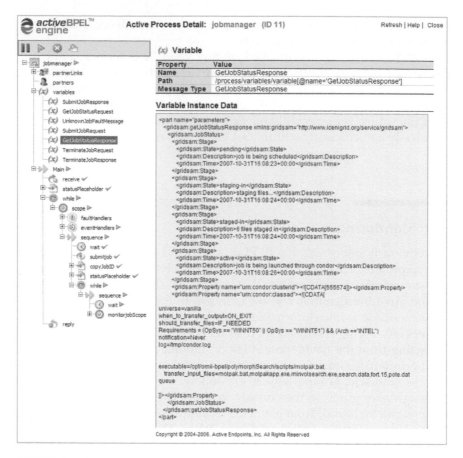

FIGURE 8.5   An active process in ActiveBPEL administration console.

8.20 Persistence

In-memory persistence is a major disadvantage as the process information lost in server shutdown is unrecoverable. A server may be shut down on purpose (e.g., for system maintenance) or, in most cases, due to unexpected failures. The persistence framework of ActiveBPEL can be configured to integrate with various RDBMS; for example, MySQL, Oracle, SQL Server, and DB2. PostgreSQL, as default choice of database of OMII infrastructure, is also supported. A large amount of information will persist in order to restore the engine state from corruptions such as process states, logs, variables, alarms, message queues, and so on. The engine will then be brought back to the state when the last transaction in database has completed in the event of failure. The admin console will also be restored and can be used as it was before the restart.

8.21 Security

As a center point of communication in service orchestration, a BPEL engine interacts with services scattered on networks, some of which may even be beyond organization boundaries. With exceptions, traditional security requirements of protecting private data and restricting access to logic still need to be handled by applications in a service-oriented environment. We will now explain the well-received security standard designed for the Web service framework, in particular the Web Service Security (WSS).

Conventionally, the protection enforced at the transport layer, like the secure socket layer (SSL), is no doubt a popular means of securing communication channels along which conversations are maintained. WSS, on the other hand, which protects the confidentiality and integrity of SOAP messages by digitally signing and encrypting XML elements in a message, is of importance for achieving end-to-end security in the application layer, which is vital to secure a message-centric interaction model that SOAP depends on.

Over HTTPS channels, OMII has implemented its WSS framework. A security header that contains the message signature and sender identity is inserted in every outbound message. Verification is performed at the intended recipient using the public key embedded in the X.509 certificate sent alongside. The certificates are issued for every instance of OMII server and client during the installation. WSS standardizes how this security information should be attached to the SOAP messages. All OMII-hosted services, which conform to the security policies, can be seamlessly orchestrated by ActiveBPEL, which in turn has been integrated into the framework. WSE in BPEL Designer, as the client-side program, can invoke any of these services safely as we have mentioned earlier.

8.22 A Polymorph-Search Case Study

The case study we have used is in the area of theoretical chemistry. Chemists look for different crystal structures, or polymorphs, of given organic molecules because the resulting characteristics that vary under different physical conditions are essential to determine solid states of, for example, pharmaceutical substances, industrial materials, and so on (shown in Figure 8.6). Among the available approaches, such as laboratory experiments, one has to computationally exhaust all theoretic possibilities using established scientific models, the results of which can then be cross-checked against those produced by companion approaches for final confirmation that scientists can reply on for subsequent study.

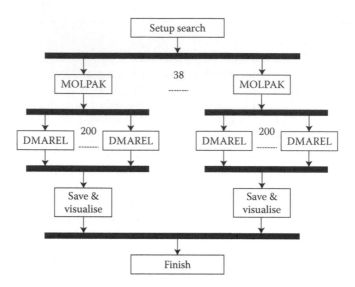

FIGURE 8.6   Overview of the polymorph-search process.

In a nutshell, the chemists rely on two Fortran application programs to perform this two-stage search. MOLPAK can generate up to 200 molecule packings in accordance with all 38 packing types known to scientists beforehand. Each of the packings will then be further optimized by DMAREL to determine whether a hypothetical structure is thermodynamically possible. During the execution of a full-scale search, a data volume in the region of 6 GB is usually produced. All MOLPAK and DMAREL calculations are independent of each other with separate input files, so full parallelization can be engineered respectively for all 38 and 7,600 executions (although no DMAREL jobs should go forth without the preceding MOLPAK calculation having been completed).

The initial system of the polymorph search was not built on the grids. Instead, it relied on a dedicated high-end server. The idea was straightforward. Around the two executables, a number of shell scripts could be coded to enable the automation of the search process, although the degree of manual interactions required was still so high that PhD students or chemical engineers were often assigned to maintain searches. Depending on the individual molecule, a MOLPAK calculation normally takes a few hours for completion, while in contrast a DMAREL optimization only lasts a few seconds or, occasionally, a couple of minutes. Because of the limited computing power and the lack of full automation and parallel processing of the system, it could easily take a few weeks to run a full-size search, which can be done now in hours with grid computing and BPEL enabled. Later on, it was possible for a user interface to be implemented for scientists to provide initial search parameters for the backend workflow rewritten in Java. Despite major improvements, the major limitation of such a system was how scientists lost the control and ownership of their workflow, so that whenever they wanted to make changes, they had to come to the software engineers to get the hard-coded workflow modified and recompiled.

In order to fully describe the polymorph search process in BPEL, five workflows have been developed so that a composite orchestration hierarchy is formed, as shown in Figure 8.7. Two of them correspond to the MOLPAK and DMAREL processing. Each includes the detailed description of how executable jobs should be prepared and submitted to the grid, and subsequently how returning data should be processed. The entry-level workflow coordinates the two with the visualizer and a utility service (e.g., to delete temp files or to package output data) provides the primary outline of the entire search process, which is the only workflow scientists must be concerned about to submit searches to the system. For example, the WSDL description of the process can be used within the Web Service Explorer of BPEL Designer to automatically generate the service request form that needs to be initialized with search parameters, and submitted by the user. In the production system, the standalone client is replaced with a more compact and lightweight Web client in the consideration of general usability, which also integrates the ActiveBPEL admin console and the access to the search repository altogether. Scientists can log in remotely with a single URL to submit, monitor, and manage searches. For each completed search, the system will create an output archive and a summary report accordingly in the search repository along with the initial input files uploaded to the server as well as other historical search data that have been submitted to the system. The archive will be downloaded later and the data inside will be retrieved by scientists for further optimization and study, while an overview of the search results can be found in the summary page, where a sorted table of key optimization data for each packing and the corresponding

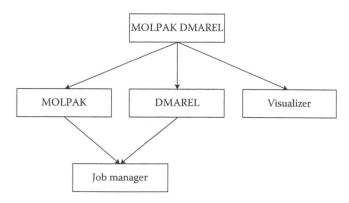

FIGURE 8.7 The hierarchy of polymorph-search workflows.

scatter plot showing the distribution of all polymorphs and pinpoints possible discovery of crystal structure are produced. The visualizer workflow coordinates everything from the creation of the basic summary page, to the dynamic update at the end of each batch of DMAREL executions. A Web service that assists the visualizer, like performing XSLT transformation of the data from chemical markup language (CML) to HTML, or storing and merging visualization data, and so on, has been developed and orchestrated. The last workflow is the job manager that is in charge of job submission and monitoring. The core interactions of the workflow with the underlying grid resources are supported by invocations to the middleware service called *GridSAM*. GridSAM is another OMII-managed program product, which mediates the interactions with the grid resources managed by Condor, Globus, or other infrastructures that may be bridged into the software architecture, by exposing a general interface of job submission, monitoring, and control. Job descriptions that are expressed in the GGF (Global Grid Forum) standard, job submission description language (JSDL), provided by the users, will be translated by GridSAM to the actual submission scripts used by the in-operation resource manager in runtime. A variety of data staging methods, like FTP, SFTP, Webdav, and so on, is supported to fetch files on grids. File paths to be specified in the job description can be dynamically determined and composed in a BPEL process using proper XPath queries and expressions with *assign* activities, while other information such as user account, file names, flags, operating system requirements, and so on can be statically assigned. In any case, the job manager takes any valid JSDL descriptor as its input and builds advanced process logic, such as fault handling, error detection, resubmission, timing control, and so on, around the core orchestration of GridSAM services to guarantee quality service provision that is largely relied on by both MOLPAK and DMAREL workflows. During the parallel job submissions, a vast number of job managers will be instantiated, running independently until the return of a completed execution. A detailed breakdown of the execution progress and status can be examined separately in the ActiveBPEL admin console. Considering the fact that the actual number of machines available in the Condor pool at any one time for the system is usually limited due to the concerns of fairness and administration, the top-level workflow has constrained the number of parallel DMAREL submissions to a number of three batches (i.e., 600 jobs by three DMREL workflow instances) at any one time based on a FIFO policy. Whichever MOLPAK has completed first,' the following DMAREL processing will be dispatched as soon as the token acquired by one of the previous batches is released. Synchronizations have also been carefully wired into the workflows to prevent corrupted updates to shared holder variables, like the one that holds all DMAREL optimization summaries that are used in the visualization. MOLPAK and DMAREL workflows also rely on another Web service to perform data processing in between executions; for example, to extract and transform intermediate

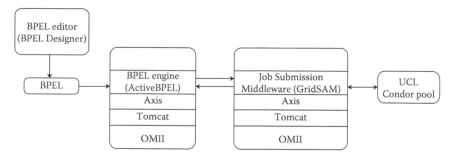

FIGURE 8.8   Architectural overview of grid job submission with BPEL.

data into usable format, to create and clean up temporary files, to set up dynamic execution parameters, and so on.

All BPEL workflows were developed with BPEL Designer installed on the client machine. ActiveBPEL and GridSAM were deployed within the OMII stack together with the other Web services used in the system. Multiple instances of servers may be installed to distribute potential workload. The gnuplot Web service implementation used by the visualizer for scatter plots is hosted by OMII at the University of Southampton. The supporting grid infrastructure of the system is the UCL Condor pool of approximately 1200 machines deployed in the student cluster rooms, almost all of which run Windows XP. Searches can be submitted and monitored from anywhere with an Internet connection using the Web interface. Figure 8.8 shows an architectural overview of grid job submission with BPEL.

The tailored ActiveBPEL engine and GridSAM can be selected and installed using the text installer of the OMII Server straight away. BPEL Designer can be installed similarly with OMII Client. The server deployment requires JDK (Java Development Kit) and has been tested on SuSE and Redhat series servers, although it should also be deployable on a wider selection of modern operating systems with minimal amendments. OMII Client has been tested on both Redhat Linux and Windows officially.

8.23 Conclusions and Future Work

The essential benefit of using BPEL is its seamless integration of Web service architecture that fundamentally promotes service orientation. In the research into service-oriented grid computing, BPEL introduces an agile solution to building a grid application framework, based on service composition and orchestration. With the rapidly developing technologies and long-existing industry supports, professional BPEL solutions will continue to be available and will benefit related grid developments.

OMII-BPEL has provided the first integrated BPEL environment suitable for scientific grid computing. Even though there is still room for improvements as user requirements grow constantly, it does not stop the adoption of BPEL as an executable workflow modeling language in grid computing in general. We are already seeing fruitful choices of BPEL implementations in the market: organizations and companies like Apache, IBM, JBoss, Microsoft, Oracle, and Sun have all built their own professional BPEL solutions. OMII-BPEL, when it continues to keep up with the quality software releases under OMII infrastructure, is looking forward to applying BPEL technology to more extensive scientific applications, through which we carry out an in-depth research and study of service-oriented grid computing with BPEL in a wider domain.

9

A Data Stream View of Scientific Workflow

Chi Yang and Jinjun Chen

CONTENTS

9.1 Introduction

e-Science is a buzz word when it comes to connecting different kinds of sciences and communities with each other to share scientific interests, data, and research results. This connection is the trend of scientific and technological development that augurs a rapid increase in the number of computations being employed by e-scientists. Consequently, scientific workflow, a new special type of workflow often underlying many large-scale complex e-science applications such as climate modeling, structural biology and chemistry, medical surgery, or disaster recovery simulation, deserves intensive investigation. Compared with business workflows, scientific workflow has special features such as computation, data or transaction intensity, less human interaction, and a large number of activities. Some emerging computing infrastructures such as grid computing, with powerful computing and resource sharing capabilities, present the potential for accommodating those special features. Some work both theoretically and empirically has been done toward this research frontier such as GredbusWorkflow, Kepler, Taverna, and SwinDeW-G series [1]. Each piece of work highlights different aspects of scientific workflow with different emphasis.

However, the research of scientific workflow management is still at its early stage in terms of issues to be solved and its development history. Furthermore, due to the wide application of the scientific workflow, the underlying environments vary from one to the other. All of the above problems with the scientific workflow give a new platform for people from the business process management community to investigate.

9.1.1 Research Scope of Scientific Workflow

Currently, many efforts are being focused on this new workflow area and the research work is fruitful. Below are some main topics related to scientific workflow; however, the scope of scientific workflow is not limited to these [2,3]:

- Special features of scientific workflows and their hints on new techniques
- Scientific workflow modeling, execution, and scheduling in support of computation, data, and transaction intensity
- Formal representation, scientific workflow patterns
- Control flows and data flows in scientific workflows
- Web/grid services-based scientific workflows
- P2P/decentralized architecture for scientific workflows
- Application programming interface and graphical user interface
- Scientific workflow verification and validation
- Exception handling, quality of service, performance and security issues in scientific workflows
- Underlying infrastructures targeting scientific workflow support
- Real-world scientific workflow applications

9.1.2 Scientific Workflow on a Grid

Scientific workflow is mainly concerned with the automation of scientific processes in which tasks are structured based on their control and data dependencies. Due to the high computation performance requirement, grid computing is one of the good infrastructures for its deployment [3–5]. Currently, some work toward this area can be seen. Specifically, the advantages of grid computing include the ability to build dynamic applications that orchestrate distributed resources, utilizing resources that are located in a particular domain to increase throughput or reduce execution costs, execution spanning multiple administrative domains to obtain specific processing capabilities, and integration of multiple teams involved in the management of different parts of the experiment workflow, thus promoting interorganizational collaborations.

9.1.3 Some Related Work on Scientific Workflow

A number of research projects have developed tools and runtime infrastructure to support composition and execution of scientific workflows [3,4,6–8]. Due to the limit of space, we briefly review some of the related work here to help understand the scientific workflow background. More projects can be referred to in [3].

- Swinburne Decentralized Workflow for Grid (SwinDeW-G) is a decentralized workflow system [1]. Since in e-Science, a workflow is naturally computation and/or data intensive, it is believed by SwinDeW-G that decentralized/P2P-based execution with modest centralized management is one of the good solutions. In this sense, SwinDeW-G was developed. In SwinDeW-G, workflows are executed by distributed peers. In grid computing, a peer can be just a grid service. The interaction between peers is performed in a P2P fashion [9] rather than via the centralized engine.

- The Chimera system implements support for establishing virtual catalogs that can be used to describe and manage information on how a data product in an application has been derived from other data [10]. This information can be queried, and data transformation operations can be executed to regenerate the data product.

- The Pegasus project develops systems to support mapping and execution of complex workflows in a grid environment. The Pegasus framework uses the Chimera system for abstract workflow description and Condor DAGMan and schedulers for workflow execution. It allows construction of abstract workflows and mappings from abstract workflows to concrete workflows that are executed in the grid.

- The Kepler project develops a scientific workflow management system based on the notion of actors. Application components can be expressed as actors that interact with each other through channels. The actor-oriented approach allows the application developer to reuse components and compose workflows in a hierarchical model.

- Adapting Computational Data Streams (ACDS) is a framework proposed by Isert and Schwan [11]. It addresses construction and adaptation of computational data streams in an application. A computational object performs filtering and data manipulation, and data streams characterize data flow from data servers or from running simulations to the clients of the application.

- Dv is a framework proposed by Aeschlimann et al. [12]. It is based on the notion of active frames. An active frame is an application-level mobile object that contains application data, called frame data, and a frame program that processes the data. Active frames are executed by active frame servers running on the machines at the client and at remote sites.

- JOpera for Eclipse has been developed, which supports grid work-flows fully integrated with the Eclipse user experience. It is extensible with users' own plugins to call any kind of service.

- Taverna, as a part of MyGrid project, uses the developing XML language Scufl to support the scientific workflow application.

- Teutaa is a system developed using JDK 1.4, in which the UML diagram is mapped to abstract grid workflow language (A-GWL) and then to concrete C-GWL.

9.1.4 Scientific Workflow Management by Database Management

Recently, in order to cope with the data-intensive scientific workflow, database-management-based techniques have been proposed. A certain Database Management System (DBMS)-centric architecture has been developed [6], in which conventional database technology can provide a quite satisfactory functional performance as a desired workflow management system. To deploy DSMS techniques for scientific workflow management, we can use different views to analyze workflows.

9.1.4.1 Transactional and Objective Views of Workflows

In conventional well-recognized conceptualization and modeling of work-flows, the research emphasis is based on activities. However, workflows can be treated as database transactions as shown in Figure 9.1 [6,13]. Except for the input and the output, all the processes belonging to workflows will be described as transactions in database systems.

If we separate the transactional part from the graph in Figure 9.1, we can obtain three sections: *output, workflow,* and *output.* Using the concept from the object database, we can abstract the transactional view into object view of the workflow by modeling input, *output* and *workflow,* into processing objects of a DBMS that is depicted in Figure 9.2.

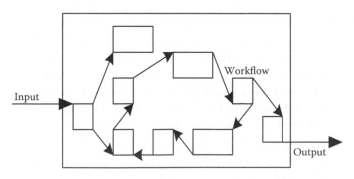

FIGURE 9.1   Transactional view of workflows.

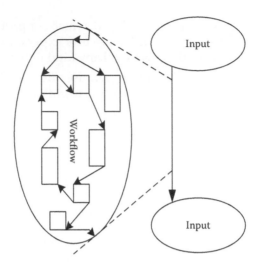

FIGURE 9.2   Objective view of workflows.

The processes that create the objects are expressed using the links between them. Both transactional view of workflows and objective view of workflows can be transformed to classical WFMSs that are deployed under most of current applications such as commerce and research types. Whatever view is adopted, a DBMS itself can be treated as a WFMS, offering much of the needed functionality with respect to process management without external software support [6]. Furthermore, almost all of the critical functionalities missing in the conventional DBMS can be easily provided with minimal, natural extensions that remain faithful to the philosophy of database technology.

The main benefits of this DBMS-based technique can be summarized as follows:

- Reduced implementation effort: In terms of many of the required functionalities, it does not require any external workflow tools.
- Real-time information availability: All the aspects of the work-flow are treated using database queries expression directly.
- Consistency in workflow management: In any DBMS, the access to the database systems depends on a unified access language that is used throughout the definition, execution, and monitoring. The same applies to the DBMS-based system.
- Increased optimization opportunity: The whole control and manipulation are managed by a DBMS like WFMS, the optimizer will have a global view of the entire system.

9.1.4.2 How to Build up a WFMS with DBMS Views

Based on the introduction of the transactional and objective views for workflows, below is a procedure for implementing a WFMS through the objective database view according to [6].

- A workflow is described as an object-oriented database schema
- An instance of the workflow schema is created during execution
- Invocation of workflow processes is captured and activated by rules from a constraints form
- External applications implementing workflow processes are scheduled through updates to system catalogs
- Status information and other kinds of information about running or finished workflow processes are obtained by database queries
- Information on the workflow data is obtained by database queries

9.1.5 Data Stream and Scientific Workflow

There exist overlapping research issues not only between the scientific workflow management systems and DBMS, but also between the scientific workflow management system and the data stream management system (DSMS) [14,15]. For example, the workflow execution and scheduling are somehow similar to the generation of the data stream query plan. Furthermore, the control flow and data flow [16] themselves can be treated as data streams, whose tasks focus on raw data prereduction, analysis, calculation, selection, and filtering. Even within the range of the pattern and representation, the processing techniques from DSMS can enrich the content in scientific workflow. Finally, the real-world impacts for scientific workflow applications breathtakingly resemble those for data streams such as meteorology, astronomy, and monitoring sensor network. Spontaneously, it comes into the picture that a DSMS view can be taken into the scientific workflow area.

9.1.5.1 Data Stream Brevity

For the strangers to DSMS with skeptical eyes, the first question is what is a data stream and what is a DSMS? Generally speaking, a data stream can be described as multiple transient relations or structures, each of which consists of infinite tuple sequence or defined data items. The data update method of data streams is append only. With respect to query processing, queries on data streams are usually continuous with approximate answers. In addition, query evaluation takes the form of one pass, and the query plan is adaptive. A DSMS is designed to manage or filter a large body and high speed of information. The management system also focuses on online processing mechanisms for manipulating information such as inserting and updating. In addition, it must provide features such as consistency, integrity, security, and so on [17] like in a DBMS.

A major purpose of a data stream management system is to provide users with an abstract view of the streaming data. That is, the system hides certain details of how the data are selected and maintained. Thereby, useful information data will not be lost, which offers an efficient retrieval, yet users see a simplified and easy-to-use view of the data. On the other hand, a scientific workflow management system provides users with an abstract view of the scientific activities, processes, and their external logic, hiding the details of how they are arranged and executed. Considering that under the theme of the scientific computation, most activities and processes are concerned with dealing with raw data, the processing of the scientific workflow is actually the processing of the data flow or data stream.

9.1.5.2 Data Stream View for Scientific Workflow

Inspired by the DBMS view for the scientific workflow in [6], the similarity between data stream management and scientific workflow management can also be found. In this subsection, the conceptual-level mapping relationship between data stream management systems and scientific workflow management systems will be presented.

For a given model of a certain scientific workflow, the execution of the workflow means the WFMS will manipulate and coordinate instances from its creation to its cancellation. In other words, a model determines a bunch of instances. This can be seen in Figure 9.3. As shown from the vertical axis, the upper flow is the workflow model or workflow schema. Under the same workflow model, during the execution time or runtime, there are a series of instances processing different activities defined by the model/schema simultaneously. But from the horizontal axis, it is obvious that the input data flows through the DSMS and the DSMS will match the ongoing data with reference to the query pattern given at the top of Figure 9.3. So it is clear that a scientific workflow schema can be treated as a data stream query expression during the processing, which indicates that the DSMS query processing technique can be applied to scientific workflows.

We previously mentioned that in any DBMS, the access of the database system depends on a unified access language that is used throughout the definition, execution, and monitoring. This is also the case in the DBMS-based system. In this sense, both DBMS and DSMS can improve scientific workflow management.

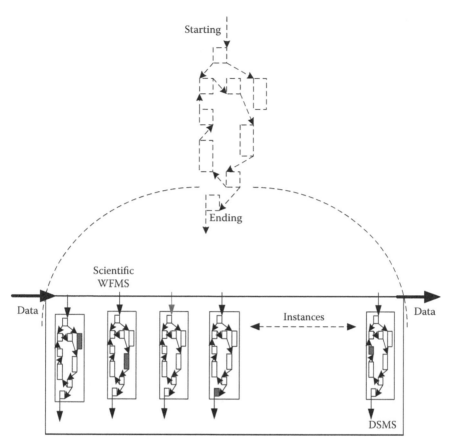

FIGURE 9.3 The data stream view for the scientific workflow.

9.1.6 Measures for Scientific Workflow Management Systems

To design a scientific workflow, the following problems need to be solved with the help of mature techniques from the database community:

1. Data-intensive and computation-intensive applications
2. Logical sequence of workflow requirement
3. Lower resource or inexpensive services consumption

In the rest of this chapter, we will explain the technique details according to the roadmap for deploying the techniques from DSMS to scientific workflow management. Briefly, the roadmap can be depicted as *schema → processing techniques → theoretic work → scheduling and optimization.*

9.2 Stream View of Scientific Workflow Schema

As discussed previously, the scientific workflow can be treated as data streams. Sometimes, the models of the scientific workflow can be described as query expressions of XML data streams. In fact, in many research fields, workflow models are also called as schemas [18,19]. The following section gives the definition of a workflow schema.

9.2.1 Workflow Schema

A workflow schema defines a state machine (deterministic finite automation—DFA) [18,19], consisting of

- States, including a marked initial state
- Transitions
- State variables

Under the general scheme, the workflow schema (*S*) is defined through a directed graph consisting of nodes (*N*) and flows (*F*) [14]. Flows show the control flow of the workflow. Thus $S = <N, F>$ is a directed graph where N is a finite set of nodes, F is a flow relation $F \subseteq N \times N$. Nodes are classified into tasks (*T*) and coordinators (*C*), where $C \cup T, C \cap T = \phi$.

Task nodes represent atomic manual/automated activities or subprocesses that must be performed to satisfy the underlying business process objectives. Coordinator nodes allow us to build control flow structures to manage the coordination requirements of business processes. Basic schema modeling patterns supported by these coordinators include sequence, exclusive or-split (choice), exclusive or-join (merge), and-split (fork), and and-join (synchronizer) [14]. An instance within the workflow graph represents a particular case of the process. Figure 9.4 gives an example of the graph-based workflow schema.

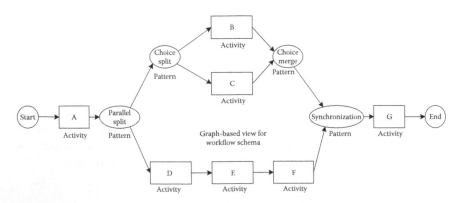

FIGURE 9.4   The workflow schema with basic patterns.

It should be mentioned here that the graphic patterns of workflows are not limited to the above structures. According to *Workflow Patterns* by van der Aalst [14], there are in total 20 patterns of workflow schema such as sequence, parallel split, and so on.

From the definitions of those patterns, we can deduce that a workflow schema can be simulated by the DFA [18,19]. It indicates implicitly that if there is any technique highly related to DFA and automata theory it may be used for processing the issues over the workflow. On the other hand, the technique should have a good capability for data processing because scientific workflow makes its reputation for the high data intensity. Consequently, combining the data processing ability and automata theory result from data streams can provide an alternative view for scientific workflow from a new angle.

9.2.2 Schema in DBMS and DSMS

Generally speaking, the schema describes the structure of a database system in a formal language supported by the DBMS. In a relational database, the schema defines tables, fields in each table, and the relationships between fields and tables. Schemas themselves are generally stored in a data dictionary by DBMS. Although a schema is defined in text database language, the term is often used to refer to a graphical depiction of the database structure.

However, whatever the system is, DSMS or DBMS, under the relational data structure, all the data are arranged using tables and fields, which constrains the patterns of the relational schema. It is obvious that the relational schema is totally different from the graphic/automata-based workflow schema. The data processing technique cannot be easily changed and deployed for the scientific workflow processing.

9.2.3 XML Schema and Semistructured Data

Nowadays, XML-format Web data are getting popular and XML data sources are of marvelous diversity [16,20,21]. Available XML data ranges from small Web pages to ever-growing applications, such as biological data, astronomical data, commercial data, and even to rapidly changing and possibly unbounded streams that are often encountered in Web data integration and online publish–subscribe systems. In May 2001, the W3C released the XML schema standard [10]. This standard had a very long gestation and this is not surprising, as the aim was to create a single modeling language that would please all interested parties.

In terms of features, XML schema models are backward compatible with DTDs. This is very important for the practical reason that it eases the transition from DTD modeling to XML schema modeling. It is always possible to convert a DTD into an XML schema model automatically. At the same time, the XML schema can offer the complement to the XML

DTD; the content includes functionality, namespace sensitivity support, and data exchange applications. Finally, the syntax of XML schema is self-description, which stands for using an XML document to describe the XML schema.

The XML schema is typically graph like, has a tree-like structure, and can be described by automata, actually by a nondeterministic finite automaton (NFA) [22].

```
<schema...>
  <!--ANY NUMBER OF FOLLOWING-->
    <include.../>
  <import>...</import>
  <redefine>...</redefine>
  <annotation>...</annotation>
<!--ANY NUMBER OF FOLLOWING DEFINITION-->
    <simpleType>...</simpleType>
  <complexType>...</complexType>
  <element>...</element>
  <attribute/>
  <attributeGroup>...</attributeGroup>
  <group>...</group>
  <annotation>...</annotation>
</schema>
```

Above is an example of an XML schema for an XML document structure and element content modeling. For element definitions, XML schema "Element" is used to define an element. The capability of "Element" can offer more constraints than the XML DTD can do. The specific power of element definitions include simple content, complex content, empty elements, child element content, namespace, occurrence options, choices, group occurrence options, embedded groups, mixed content, and so on. For a detailed description of XML schema application, see [23].

There are also patterns in an XML schema. Due to the resemblance of the workflow schema and XML schema, the overlapping patterns from both schemas can be found. This overlapping means that we can reuse the pattern techniques from XML data processing, especially XML data streams under the scientific workflow scheme, for data flow processing.

9.2.3.1 Pattern Analysis of the XML Schema

According to the work of van der Aalst, there are 20 workflow patterns [14]. In the XML schema, there are lots of pattern constraints and even quantity and data range constraints. The following examples demonstrate similarity, but are not exhaustive.

9.2.3.1.1 Sequence Control

```
<complexType>
   <sequence>
    <element ref= "DOC: a" />
    <element ref= "DOC: b" />
    <element ref= "DOC: c" />
   </sequence>
</complexType>
```

As shown in Figure 9.5, the sequence rule "(a, b, c)" within XML schema element declaration indicates that element "a" is followed by element "b," which in turn is followed by element "c," as shown above. All three elements will be indispensable in the document. This pattern in the XML schema is similar to the *Sequence* in the workflow schema patterns [14].

9.2.3.1.2 Choice Control

```
<complexType>
   <choice>
    <element ref= "DOC: a" />
    <element ref= "DOC: b" />
    <element ref= "DOC: c" />
   </choice>
</complexType>
```

The choice rule "(a|b|c)" indicates a choice between the elements "a," "b," and "c." Figure 9.6 describes this type of relationship. Because the choice control is exclusive, the selection of "a," "b," and "c" will cause the neglect of others. This pattern XML schema is similar to the *Exclusive Choice* workflow schema pattern in [14].

9.2.3.1.3 Occurrence Control

The XML schema author can also dictate how often an element can appear at each location. If the element is required and may not repeat, no further

FIGURE 9.5 Sequence control from XML schema.

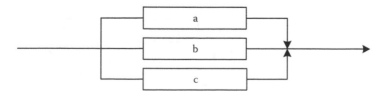

FIGURE 9.6 Choice control from XML schema.

information is required. All of the previous examples indicated a required presence. It is also a simple matter to specify a fixed number of reoccurrences of an element. For example, if every article in a magazine had three authors, the following model would ensure that three names are present:

```
<simpleType>
   <sequence>
      <element ref= "DOC: a" minOccurs= "0"
                                    maxOccurs= "3"/>
   </sequence>
</simpleType>
...author, author, author,...
```

But it is also possible to make an element optional, to allow it to repeat any number of times, and even to make it both optional and repeatable.

9.2.3.1.4 *Optional and Repeatable Element*

If an element is optional, and also repeatable, shown in Figure 9.7, the element name is followed by an asterisk, "*." The "*" may be seen as equivalent to the combination "?+." For example, "(a, b*)" indicates that element "b" may occur any number of times, the following XML schema demonstrates this type of constraint.

```
<complexType>
   <sequence>
     <element ref= "DOC: a" />
     <element ref= "DOC: b" minOccurs= "0"
                                    maxOccurs= "∝"/>
   </sequence>
</complexType>
```

An article element may contain any number of author elements, including none. To take another example, a chapter may have preliminary paragraphs, but may not always do so. The following XML fragment shows this XML schema phenomenon.

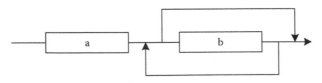

FIGURE 9.7 Optional and repeatable element.

```
<chapter>
    <para>...</para>
    <para>...</para>
    <para>...</para>
    <section>...</section>
</chapter>
<chapter>
    <section>...</section>
</chapter>
```

For detailed declarations of XML schema and all its functions and constraints, the W3C site and *The XML Companion* [16] can be used as references.

9.3 Scientific Workflow and XML Data Streams

We have previously discussed the possibility for the data stream view of the highly data-intensive scientific workflow. In Section 2.3.1, we have explored the similarity between XML schemas and workflow schemas. In this section, we propose the application of techniques from XML data stream processing to the scientific workflow domain [24–26].

9.3.1 XML Data Streams Brevity

XML data streams [19,22,24,25,27–30] combine data stream characters and XML data formats together to form a unique research frontier. The automata-based filtering technique is the current tool for processing XML data streams [17,20].

9.3.2 Event-Based Filtering Techniques

An event-based algorithm cannot select the data positively. Besides, it reports intercepted events (such as the start and end of elements in XML streams or critical data for ending or starting an activity in the scientific workflow) directly to the application through rollback, and does not usually build an internal data presentation. The application implements handlers to deal with the different events, which is much like handling events in a graphical user interface.

These event-based filtering algorithms sometimes are quite efficient. Consider the task: *Locate the record element containing the word "Ottawa."* If your XML document were 20 MB large (or even just 2 MB), it would be very inefficient to construct and traverse an in-memory parse tree just to locate this one piece of contextual information; an event-based filter would allow you to find it in a single pass using very little memory.

To understand how an event-based API can work, we consider the following sample document:

```
                    "1.0"?>
<doc>
<para>Hello, world!</para>
</doc>
```

An event-based interface will break the structure of this document down into a series of linear events, such as those listed below.

```
start document
start element: doc
start element: para
characters: Hello, world!
end element: para
end element: doc
end document
```

An application handles these events just as it would handle events from a graphical user interface; there is no need to cache the entire document in memory or secondary storage.

In scientific workflow management, data-intensive applications are typical. Should all data be calculated using some inner storage or filtered to reduce expense? We prefer the latter to build a filter-like processor for scientific workflow.

9.3.3 XML Data Stream View for Scientific Workflows

Specifically, we can take an XML data stream view for scientific workflows from three different points: query tree, theoretic bounds, and event-filtering techniques.

9.3.3.1 Similar Tree Pattern Schema

It is well known that under the XML data format, the query can be expressed as a tree structure [31]. Actually, a query tree is a part of an XML schema for the XML documents or the XML streams. It has (or partially has) functions like the schema for workflow.

Because a workflow schema is not a tree, to migrate the technique from XML data streams we should segment the workflow schema into different layers in order to obtain schema trees and to apply our tree-based algorithm to process it.

Figure 9.8 shows that there should be many strategies that can be adopted for segmentation. Which one is a more suitable candidature depends entirely on the applications and techniques to be applied. After segmentation, we can use tools for processing the tree-structured queries toward the scientific workflow such as automata (NFA/DFA).

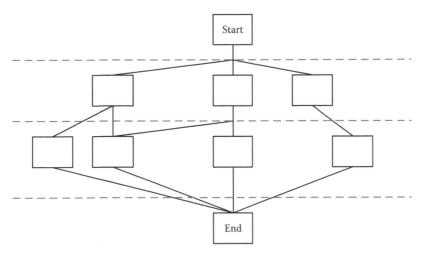

FIGURE 9.8 Sample segmentation of workflow schema.

9.3.3.2 Automata for XML Streams and Scientific Workflow

To process the tree structure or schema, under the XML schema, the most popular technique is automata theory. For query processing over XML data streams or a workflow schema after the tree-like separation as mentioned previously, lots of finite-state machines (FSMs) or automata-based filtering algorithms [19,28,32,33] can be introduced. FSMs are a natural and effective way of representing and processing path expressions. Elements of a path expression are mapped to states. A transition from an active state is fired when an element is found in the document that matches that transition. If an accepting state is reached, then the document is said to satisfy the query. For example, XFilter [19] has focused on the efficient evaluation of path expressions over streaming XML data, using indexed FSMs to allow many structure-based queries to be processed simultaneously. However, XFilter makes no attempt to eliminate redundant processing for similar queries. Figure 9.9 shows the DFA-based techniques for processing multiple queries. The query trees are finally transformed into some definite states for processing.

YFilter [22,34] combines multiple queries into a single NFA. The use of a combined NFA allows a dramatic reduction in the number of states needed to represent the set of user queries and greatly improves filtering performance by sharing execution work. YFilter also extends this NFA model to efficiently handle predicates within path expressions. Figure 9.10 shows that a set of queries (a) has been transformed into an NFA (b) for query evaluation. The NFA will then be indexed as shown in Figure 9.11, which is used to match the elements from the incoming stream [22,34].

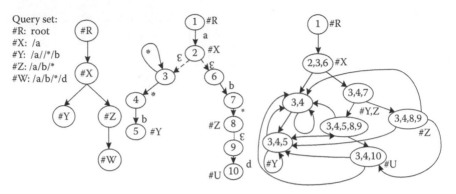

FIGURE 9.9   DFA for XPath query evaluation.

The SPEX system progressively matches all the elements with query one by one, to form the output query result [35]. The SPEX system will evaluate the predicates during the matching of the incoming elements. As shown in Figure 9.12, the query node (b) can be visited several times and a data node (a) can be visited once at most by a query node. In addition to the above techniques, there are several important theoretic bounds that should be introduced after some modification for the scientific workflow.

9.3.3.3 Theoretic Lower Bounds for Scientific Workflow Scheduling

As an important standard for the system consumption and expense, the lower-bounds problem is critical. It is also a measurement for us to conceive

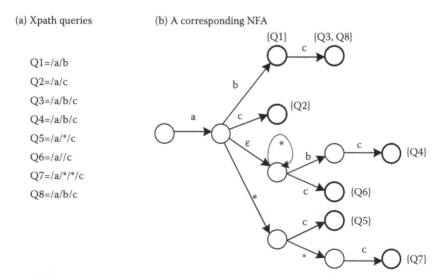

FIGURE 9.10   Transforming queries into NFA.

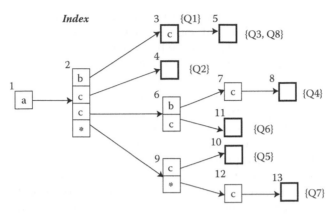

FIGURE 9.11 Indexed query structure from NFA.

our algorithm and conclude our semantic rules. We will introduce them in the rest of this section for the sake of their importance.

All known algorithms [24,25,29,36] for evaluating XPath and XQuery queries over XML streams suffer from excessive memory usage on certain queries and documents. The bulk of memory used is dedicated to two tasks: (1) storage of large transition tables and (2) buffering of document fragments. The former emanates from the standard methodology of evaluating queries by simulating finite-state automata. The latter is a result of the limitations of the data stream model. Finite-state automata or transducers are the most natural mechanisms for evaluating XQuery/XPath queries. However, algorithms that explicitly compute the states of these automata and the corresponding transition tables incur memory costs that are exponential in the size of the query in the worst case. The high cost is a result of the blow-up in the transformation of nondeterministic automata into deterministic ones. In papers such as [21], some work

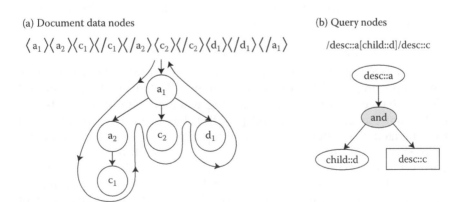

FIGURE 9.12 SPEX for query evaluation step by step.

investigated the space complexity of XPath evaluation on streams as a function of the query size, and showed that the exponential dependence is avoidable. An optimal algorithm whose memory depends only linearly on the query size (for some types of queries, the dependence is even logarithmic) has been exhibited. The other major source of memory consumption has also been studied: buffers of (representations of) document fragments. Algorithms that support advanced features of the XPath language, such as predicates, fully fledged evaluation (as opposed to filtering only), or closure axes, face the need to store fragments of the document stream during the evaluation. The buffering seems necessary because, in many cases, at the time the algorithm encounters certain XML elements in the stream, it does not have enough information to conclude whether these elements should be part of the output or not (the decision depends on unresolved predicates, whose final value is to be determined by subsequent elements in the stream). Indeed, all the advanced evaluation algorithms maintain some form of buffer (e.g., the stack of the XPush Machine [29], the BPDT buffers of the XSQ system [25,26], the predicate buffers of TurboXPath, and the buffer trees of the FluX query engine [35]). It has been noted anecdotally that for certain queries and documents, buffering seems unavoidable. However, to date, a formal and theoretical study that quantifies the amount of buffering needed to support advanced features of XPath has been demonstrated.

With respect to the advanced features of XPath, two major classes of XPath evaluation problems necessitate buffering. Space lower bounds quantify the amount of buffering required in terms of some document properties such as nonrecursive document, star-free, and so on.

Bar-Yosseff et al. [33,37] investigated the upper two types of space complexity of XPath evaluation on streams and proved, for any algorithm A that evaluates a star-free XPath query Q on an XML streaming document D, the minimum bits of space that A needs to use. These two types of space complexity are so-called theoretic lower bounds for the query processing over XML data streams. These two classes of evaluation problems are the following.

9.3.3.3.1 Full-fledged Evaluation of Queries with Predicates

There are two typical modes of query evaluation: fully fledged evaluation (i.e., outputting all the document elements selected by the query) and filtering (i.e., outputting a bit indicating whether the query selects any node from the document or not) [29,30]. It has been demonstrated that fully fledged evaluation of queries with predicates requires substantially more space than just filtering documents using such queries. This concurrency lower bound is based on fully fledged evaluation. The lower bound is stated in terms of a property of documents called "concurrency" denoted as $\Omega(\text{CONCUR}(D,Q))$. This lower bound is defined as the concept of a concurrency.

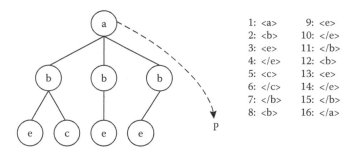

FIGURE 9.13 Concurrency of D with respect to $Q = a[p]/b[c]/e$.

As shown in Figure 9.13, document D is represented as a stream of 16 events called time steps. The concurrency of the document D with respect to query Q at step $t \in [1, m]$ is the number of content-distinct nodes in D that are alive at step t. As shown in Figure 9.13, let $Q = a[p]/b[c]/e$. At Step 14, two e elements are alive. The first is at Step 3 because whether it will be selected depends on whether its a grandparent will have a p child. The second is at Step 13, because whether it will be selected depends on whether its b parent will have a c child and its a grandparent will have a p child. So the concurrency at Step 14 is 2. The document concurrency of D with respect to Q, denoted as CONCUR(D,Q), is the maximum concurrency over all steps $t \in [1, m]$. For example, CONCUR(D,Q) in Figure 9.13 is 2. The concurrency lower bound is suitable for single-variable predicate queries. For queries with a multi-variable predicate, the dominance lower bound is defined in [29]. It is simple to verify that if Q is a nonpredicate query, CONCUR(D,Q) is 1. However, for queries with single predicate, it is easy to construct documents with arbitrarily large concurrency.

9.3.3.3.2 Multivariate Comparison Predicates

A predicate that consists of a comparison of two nodes (e.g., $a = b$ or $a > b$) is said to be a multivariate comparison predicate [29,30]. On the other hand, univariate comparison predicates are ones that compare a node with a constant (e.g., $a = 5$ or $b > 4$). It has been proved that evaluation (whether fully fledged or filtering) of queries that consist of multivariate comparison predicates may require substantially more space than evaluation of queries that have only univariate comparison predicates.

The existing semantics of XPath implies that a predicate of the form /$c[R(a, b)]$, where R is any comparison operator (e.g., $=$, $>$), is satisfied if and only if the document has a c node with at least one a child with a value x and one b child with a value y, so that $R(x; y) =$ true. Thus, if all the a children of the c node precede its b children, the evaluation algorithm may need to buffer the (distinct) values of the a children, until reaching the first b child. It is proved that such a buffering is indeed necessary when R is an equality operator (i.e., $=$, $!=$). It is not needed for inequality operators (i.e., <, <=, >, >=), because for them it suffice to buffer just the maximum or minimum value of the *a* children.

The lower bound is in fact stated with respect to any relational operator *R*, not just comparisons. The bound is given in terms of a graph-theoretic property of relations, which is called the "dominance cardinality." So this type of lower bound is named dominance lower bound.

Theorem 1. For the query $Q = /c[R(a, b)]$, for every integer k, and for every filtering algorithm A for Q on XML streams, there exists a document D of candidate cardinality k with respect to u on which A uses Ω (log DOM$k(R)$) bits of space.

The proof of Theorem 1 can be referred to in [20]. Theorem 1 clarifies the minimum space consumption clearly for any query with several predicates belonging to the same query node. Note that for the equality operators (=, !=), the above lower bound is $\Omega(\log(U/k)) = \Omega(\log(|U|))$ for sufficiently small k's. That means that if R is an equality operator, evaluation of Q on documents that have a node with k children matching u^a (or u^b) would require buffering the distinct data values of these children. In addition, when R is an inequality operator (<, <=, >, >=), evaluation of Q requires only $\Omega(\log|U|)$ bits of space, which is what is needed to buffer the maximum or minimum data value of the k children that match u^a (or u^b). The detailed definitions and proof of two space lower bounds can be found in [33,37].

In traditional workflow modeling, most attention has been paid to the definition of the activities. However, considering the scientific workflow requirements of data and computation intensity, we should focus on the calculation to coordinate different activities such as synchronization, parallelisation, and so on.

Considering the scientific workflow environments, we remove the concepts of predicate, and keep the previous concepts such as *"live"* to define the activated activities. Similarly, we can get the concept for the concurrency activity and the dominance activity. Using the upper lower bounds, we can get the minimum expense required to execute a given scientific workflow schema.

9.4 Scientific Workflow Runtime Scheduling and XML Stream Optimization

In this section, the comparison of DSMS and DSWMS is given to explore the similarity between them. At a conceptual level, query processing can also be treated as a workflow application, so the results of DSMS research will certainly enrich the research of DSWMS. One of them is *"optimization,"* which is indispensable in terms of any DSMS.

9.4.1 What is Optimization?

Imagine yourself standing in front of an exquisite buffet with numerous delicacies. Your goal is to try them all out, but you need to decide in what order. With the exchanging order and some trade-off, it can guarantee you achieve your goal with the maximum benefit. In other words, it will maximize the total enjoyment for your palate. This exchanging order or trade-off is a successful optimization.

In DSMS, to evaluate a query, there are many plans that can be adopted to produce the satisfactory answers. From a functional point of view, we can call these equivalent query plans. But the query processing performances vary from one to the other, such as their space cost or the amount of time that they need to run. What is the query plan solving the problem with the least time and the least system cost? To find out, we have to depend on optimization [36,38].

In the following query flow (shown in Figure 9.14), we can see that optimizations are widely involved in the query flow processing for DSMS. Specifically, a query process in the DSMS can be divided into five steps: *Query Parser, Query Optimizer, Code Interpreter, Query Processor,* and *Runtime Optimizer*.

- The *Query Parser* checks the validity of the query and then translates it into an internal presentation, usually a calculus expression or something equivalent.

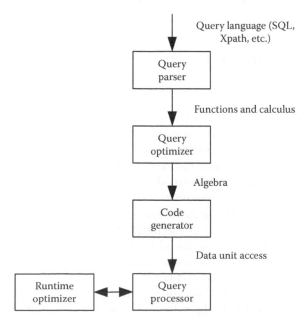

FIGURE 9.14 Query flow through a DSMS.

- The *Query Optimizer* examines all algebraic expressions that are equivalent to the given query and chooses the one that is estimated to be the cheapest.
- The *Code Interpreter* transforms the access plan generated by the optimizer into calls to the query processor.
- The *Query Processor* actually executes the query.
- The *Runtime Optimizer* modifies and adapts the query plans for the Query Processor during the query evaluation.

It is obvious that under a streams environment, optimization techniques are deployed before and after the execution of the query evaluation. The situation is quite similar in terms of the data-intensive scientific workflow. The optimization can be done both on the workflow schema level and on workflow execution instance. In terms of specific optimization techniques, there are many categorization methods. Here we will introduce two of them. Other methods can be referred to some other references such as [39].

9.4.1.1 Static Optimization and Runtime Optimization

When a scientific workflow schema is given, we can evaluate and adjust the execution plan, to make an optimized strategy to execute it. That is a static optimization. During the execution of a workflow, with the instance data coming, the WFMS can know more and more extra information to adapt its executing plan to gain a more optimized strategy. This is a runtime optimization.

9.4.1.2 Semantic Optimization

Semantic optimization is a comparatively recent approach for the transformation of given queries into equivalent alternative queries using matching rules in order to select an optimal execution plan based on the cost of executing alternatives [36,38]. To understand the semantic optimization, we have to introduce two important concepts of semantically equivalent and semantic transformation.

Two plans are semantically equivalent if they return the same answer or fulfill the same task. A semantic transformation transforms a given schema into a semantically equivalent one. Semantic optimization is the process of determining the set of semantic transformations that results in a semantically equivalent schema with a lower execution cost. The discussion of optimization is indispensable in terms of any processing algorithm and its corresponding execution plan.

XML stream-specific optimization techniques have been well developed [36,38]. For example, the schema-based optimization (SQO) works on one abstraction level, which uses schema knowledge to rewrite a query into a more efficient one. Most current literature on SQOs in XML focuses on techniques that are either (1) general regardless of persistent or streaming XML sources, or (2) specific to a persistent XML source. For example, query tree minimization is a general technique. It eliminates a pattern from the query if the pattern is known to always exist. Since the pruned query involves less computation than the original one, it is more efficient to evaluate regardless of the nature of data sources. As another example, the query rewriting using the state extents technique requires indices of the data. Applications on persistent XML can usually afford the preprocessing of building indices, while this is often not the case for XML stream applications due to the on-the-fly arriving nature of their data. Therefore, ,this technique is more suitable for persistent XML.

Being rich in semantic elements and expressions is an important feature of XML language itself, which means SQO can be used to optimize the query on the XML document, and consequently on XML data streams. The work in [36,38] focuses on SQOs specific to XML streams instead of XML documents. To our knowledge, their work has proposed a comprehensive solution for XML stream-specific SQO techniques. It handles only limited queries (i.e., Boolean XPath or XQuery match) with one type of constraint. In contrast, first, they handle more complex query types; that is, a subset of XQuery. Second, their system supports most commonly used constraints in XML schemas. However, the application of SQO on XML data streams is at its initial stage. Some progress can be seen in [36,38], but that is far from completion and soundness. According to the above discussion, the techniques of SQO over XML data streams are well developed, which can be applied to the semantic optimization over the scientific workflow.

9.4.2 Execution of Optimized Scientific Workflow

The high data and computation intensity is a fundamental feature of scientific workflow. Scheduling scientific workflow execution means to allocate the resource or service for the workflow instances effectively and efficiently. Whatever the environment is, the goal of the scheduling of the scientific workflow is lower expense and shorter time. Naturally, the stream-like flow makes it possible to deploy optimization techniques from DSMS to cut the system space and time cost. When defining a workflow, the generated scientific workflow schema will be refined by the schema optimizer. It is a static optimization before the execution of the workflow. During the workflow execution, the runtime optimizer as shown in Figure 9.15 will analyze the information of instances and exchange the information with SWFMS. Consequently, the runtime optimizer will help the SWFMS reschedule the execution plan to improve the system performance. The optimization procedure itself is a kind of scheduling for the scientific workflow execution. So the optimization research on XML data streams will greatly enhance the scheduling over the scientific workflow.

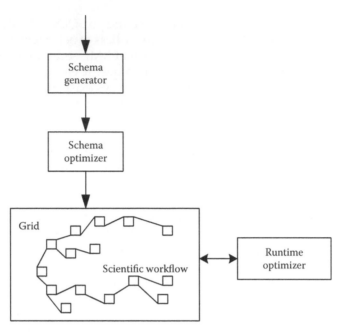

FIGURE 9.15   Execution of scientific workflow.

9.5 Conclusions and Future Work

In this chapter, we introduced the current technique: a database view of the scientific workflow. We have analyzed the scientific workflow from the point of view of a data stream. The advantages and the constraints were also given. From data model level, we explained why we chose the data stream view for the scientific workflow. We then compared similarities and differences between scientific workflow schemas and data streams. Based on XML technology, we proposed a new way to process scientific workflow based on automata-based filters. Finally, with inspiration from the XML data stream, we introduced the idea of using optimization techniques over XML data streams for scientific workflow scheduling. In future, we will further investigate data-stream-based techniques for scientific workflow execution.

References

1. SwinDeW-G Team. *System Architecture of SwinDeW-G*, 2007. Available at: http://www.swinflow.org/swindew/grid/, accessed on December 1, 2007.

2. Chen, J. and W.M.P. van der Aalst. *On Scientific Workflow*, 2007. Available at: http://www.ieeetcsc.org/newsletters/2007-01/scientificworkflow.html, accessed on December 1, 2007.
3. Gong, Y., M.E. Pierce, and G.C. Fox. "Matchmaking scientific workflows in grid environments," in *Proceedings of the IASTED International Conference on Parallel and Distributed Computing and Systems (PDCS 2007)*, Cambridge, MA, pp. 378–393, 2007.
4. Yu, J. and R. Buyya. "A taxonomy of scientific workflow systems for grid computing." *Journal of Grid Computing*, 2005, 34 (3), 171–200.
5. Cao, J., S.A. Jarvis, S. Saini, and G.R. Nudd. "Gridflow: Workflow management for grid computing," in *Proceedings of the 3rd International Symposium on Cluster Computing and the Grid (CCGrid)*, Tokyo, 2003.
6. Ailamaki, A., Y.E. Ioannidis, and M. Livny. "Scientific workflow management by database management," in *Proceedings of the 10th International Conference on Scientific and Statistical Database Management*, Capri, Italy, pp. 190–199, 1998.
7. Deelman, E. "Pegasus: Mapping scientific workflows onto the grid," in *Proceedings of the 2nd European Across Grids Conference*, Nicosia, Cyprus, 2004.
8. Deelman, E. and Y. Gil. "Managing large-scale scientific workflows in distributed environment: Experiences and challenges," in *Proceedings of the 2nd IEEE International Conference on e-Science and Grid Computing (e-Science'06)*, Amsterdam, Netherlands, pp. 144–152, 2006.
9. Yan, J., Y. Yang, and G.K. Raikundalia. "SwinDew—A peer to peer based decentralized workflow management system." *IEEE Transactions on Systems, Man and Cybernetics (Part A)*, 2006, 36 (5), 922–935.
10. Chimera Virtual Data System (VDS) part of the Grid Physics Network (GriPhyN). Available at: http://www.extreme.indiana.edu/swf-survey/Chimera.html, accessed December 1, 2007.
11. Isert, C. and K. Schwan. "Adapting computational data streams for high performance," in *Proceedings of IEEE International Parallel & Distributed Processing Symposium*, Cancun, Mexico, May 1–5, pp. 641–646, 2000.
12. Aeschlimann, M., P. Dinda, J. Lopez, B. Lowekamp, L. Kallivokas, and D. O'Hallaron. "Preliminary report on the design of a framework for distributed visualization," in *Proceedings of International Conference on Parallel & Distributed Processing*, Las Vegas, Nevada, pp. 1833–1839, 1999.
13. Kuo, D., et al. "A model for transactional workflows," in *Proceedings of the Australian Database Conference (ADC)*, 1996.
14. van der Aalst, W.M.P., Arthur H.M. ter Hofstede, B. Kiepuszewski, and A.P. Barros. "Workflow patterns." *Distributed and Parallel Databases*, 2003, 14 (1), 5–51.
15. World Wide Web Consortium. Available at: http://www.w3.org/, Addison-Wesley Professional, Old Tappan, NJ.
16. Bradley, N. *The XML Companion*, Addison-Wesley, Old Tappan, NJ, 2002.
17. Babcock, B., S. Babu, M. Datar, R. Motwani, and J. Widom. "Models and issues in data stream system," in *Proceedings of the 21st ACM SIGMOD-SIGACT-SIGART Symposium on Principles of Database Systems*, pp. 1–16, 2002.
18. Darabi, H. "Finite automata modelling and analysis of workflow management systems." *International Journal of Industrial and Systems Engineering*, 2006, 1 (3), 388–411.

19. Altinel, M. and M.J. Franklin. "Efficient filtering of XML documents for selective dissemination of information," in *Proceedings of the International Conference on Very Large Data Bases (VLDB)*, Cairo, pp. 53–64, 2000.
20. Yang, C., Liu, C., Li, J., Yu, J.X., and Wang, J., "Semantics based buffer reduction for queries over XML data streams," in *Proceedings of the Nineteenth Conference on Australasian Database Gold Coast*, Australia, pp. 145–153, 2008.
21. Josifovski, V., M. Fontoura, and A. Barta. "Querying XML streams." *VLDB Journal*, 2005, 14 (2), 197–210.
22. Diao, Y., P. Ficher, and M.J. Franklin. "YFilter: Efficient and scalable filtering of XML documents," in *Proceedings of International Conference on Data Engineering*, San Jose, CA, pp. 341–342, 2002.
23. World Wide Web Consortium XML Schema. Available at: http://www.w3.org/XML/Schema, accessed on December 1, 2007.
24. Peng, F. and S.S. Chawathe. "XPath queries on streaming data," in *Proceedings of ACM SIGMOD Conference*, San Diego, CA, pp. 431–442, 2003.
25. Peng, F. and S.S. Chawathe. "XSQ: Streaming XPath queries: A demonstration," in *Proceedings of International Conference on Data Engineering*, Bangalore, India, pp. 780–782, 2003.
26. Peng, F. and S.S. Chawathe. "XSQ: A streaming XPath engine." *ACM Transactions on Database Systems*, 2005, 30 (2), 577–623.
27. Ludascher, B., P. Mukhopadhyay, and Y. Papakonstantinou. "A transducer-based XML query processor," in *Proceedings of the International Conference on Very Large Data Bases (VLDB)*, Hong Kong, China, pp. 227–238, 2002.
28. Bry, F., F. Coskun, S. Durmaz, T. Furche, D. Olteanu, and M. Spannagel. "The XML stream query processor SPEX," in *Proceedings of the 21st International Conference on Data Engineering*, pp. 1120–1121, 2005.
29. Diao, Y., et al. "Path sharing and predicate evaluation for high performance XML filtering." *ACM Transaction on Database Systems*, 2003, 28, 467–516.
30. Gupta, A.K. and D. Suciu. "Stream processing of XPath queries with predicates," in *Proceedings of ACM Special Interest Group on Management of Data Conference (SIGMOD)*, San Diego, CA, pp. 419–430, 2003.
31. World Wide Web Consortium XPath1.0. Available at: http://www.w3.org/TR/xpath, accessed on December 1, 2007.
32. Koch, C., S. Scherzinger, N. Schweikardt, and B. Stegmaie. "Schema-based scheduling of event processors and buffer minimization for queries on structured data streams," in *Proceedings of the International Conference on Very Large Data Bases (VLDB)*, pp. 228–239, 2004.
33. Bar-Yossef, Z., M. Fontoura, and V. Josifovski. "Buffering in query evaluation over XML streams," in *Proceedings of International Conference on Principles of Database Systems*, Baltimore, MD, pp. 216–227, 2005.
34. Koch, C., S. Scherzinger, N. Schweikardt, and B. Stegmaie. "An optimizing XQuery processor for streaming XML data," in *Proceedings of the International Conference on Very Large Data Bases (VLDB)*, pp. 1309–1312, 2004.
35. Bose, S. and L. Fegaras. "Data stream management for historical XML data," in *Proceedings of ACM SIGMOD Conference*, Paris, France, pp. 239–250, 2004.
36. Su, H., E.A. Rundensteiner, and M. Mani. "Semantic query optimization for XQuery over XML stream," in *Proceedings of the International Conference on Very Large Data Bases (VLDB)*, Trondheim, Norway, pp. 277–288, 2005.

37. Bar-Yossef, Z., M. Fontoura, and V. Josifovski. "On the memory requirement of XPath evaluation over XML streams," in *Proceedings of International Conference on Principles of Database Systems*, Gold Coast, Australia, pp. 145–153, 2004.

38. Su, H., E.A. Rundensteiner, and M. Mani. "Semantic query optimization in an automata-algebra combined XQuery engine over XML streams," in *Proceedings of the International Conference on Very Large Data Bases (VLDB)*, Toronto, Canada, pp. 1293–1296, 2004.

39. Hellerstein, J.M. "Optimization techniques for queries with expensive methods." *ACM Transactions on Database Systems (TODS)*, 1998, 23 (2), 113–157.

10

Design of a Model-Based SLA Management Service and Its Applications over Distributed SOA Environments

Xiaorong Li, Subu Iyer, and Henry Palit

CONTENTS

10.1 Introduction

As a result of the advances in wide-area networks, distributed scalable systems such as grid computing [1], and the evolutions of Web technologies and service-oriented architecture (SOA), there is a lot of demand for providing efficient services over a shared infrastructure where heterogeneous resources can be aggregated and manipulated to improve the service capacities over the Internet (e.g., to provide high-performance computational services that are impossible to handle with a local PC or a local cluster). The cooperative infrastructures (e.g., LCG [2], DataGrid [3], PPDG [4], NGPP [5]) have the advantages of dramatically decreasing the processing time for high-performance computing applications, avoiding the capital investment on physical resources, enabling cross-domain cooperation, and allowing various service providers to provide versatile services by integrating a variety of resources, functions, applications, and so on.

The shared services platform (SSP) [6] is a cooperative infrastructure where distributed resources including applications (e.g., software, tools, functions, etc.) and physical resources (e.g., CPU, storage, data, virtual machines, etc.) are aggregated and shared by the users. It encapsulates the distributed resources as services and builds an intergraded platform [7] for the service providers to integrate their services to fulfill complex business functionalities. However, to ensure the desirable operations of the services, the service interactions need to be reliable and accurate so as to meet various constraints; for example, the deadline of data processing and the valid time period of resource rendering. Therefore, it is a crucial challenge to support efficient service interactions and ensure the smart and reliable accesses to the services in such a cooperative infrastructure.

The introduction of the service-level agreement (SLA) [8] provides a powerful instrument for service interaction and provisioning. An SLA is a bilateral service agreement that specifies the user requirements, the quality of service (QoS), and the responsibility of both the service provider and its customer in the form of a service contract [9]. Once an SLA is agreed by both the service provider and its customer, the service provider has to ensure that the service meets the QoS requirements in the SLA.

In a typical business scenario, customers specify their QoS requirements and negotiate with the service provider for the terms of a service. During the negotiation process, the service provider has the ability to provide offers and counter-offers for individual requests according to the resource availabilities and the economic considerations so as to maximize their business benefits [10].

The agreement-oriented service provisioning (which has been applied to many aspects in our daily life) has a big market in the computer world for providing network-based services with better quality and cooperation. To realize that, efficient SLA management services [11] are needed for automating the process and facilitating the communications between the service provider and its customer for SLA-relevant operations over distributed network environments. An SLA model is also required to describe the SLA specifications, such as QoS, cost, reliability, price, violation penalty, and payment terms, in a structured way.

In this chapter, we present a model-based design of an SLA manager which serves as a generic component to allow the client and service provider to establish the SLAs before the service is delivered. Such an SLA manager should be highly flexible and extensible to fulfill complex business requirements. Our SLA Manager is based on a lightweight core, referred to as an "SLA-based service." The service providers can easily modify and tailor the SLA-based service by editing the service models to meet their own needs. We adopt the common information model (CIM) [12] for describing the meta-model using object-oriented constructs and design. Most of the terminologies of our proposed SLA model follow the WS-agreement [9] specifications. We have tested our proposed SLA manager on a distributed service-oriented platform called the shared services platform [6]. The SLA manager can be easily incorporated to provide customized SLA management services for various applications within this platform. The SLA-based service allows the service provider to manage the copies of SLAs, retrieve the SLA templates, and facilitate the client and the service provider to propose and counter-propose SLAs. In the rest of this chapter, we will describe the model of our proposed SLA manager, the implementation of the SLA model and the SLA-based service, and the working scenarios that demonstrate service interactions. We use the fire dynamic simulator (FDS) as an example of high-performance application services over the shared services platform.

This chapter is organized as follows. The next section presents the related work. This is followed by an introduction to the framework of the model-based SLA manager and describes the creation of the model-based SLA management service over the SSP and the interactions with other components. We then present the CIM model of the SLA and show how it works in SLA negotiations. Finally, we describe the implementation of a prototype and its performance for providing a high-performance application service, FDS, over the SSP.

10.2 Related Work

Negotiation has been widely studied in recent decades, in particular human negotiation (e.g., negotiation between a seller and a buyer). SLA negotiation and specifications [8,10,13] for computation services has recently attracted much interest as it points out the value of economic-driven computer service provision. There are a number of related studies in SLA management that look at automation of service negotiation [8,9] to provide network-based services according to established service-level management. Literature [14,15] studies the efficiency of SLA negotiation processes and suggests ways of achieving mutually satisfactory agreement with shorter renegotiation rounds/times and higher probabilities of the acceptance. In [14], analytical approaches for negotiation support systems (NSS) are proposed to optimize the outcome for negotiation parties. WS-agreement was proposed by the GGF [16] to describe the SLA-relevant data in a structured way. WS-agreement [9] proposes an XML-based language to specify SLA templates that the service providers can issue to their clients for filling their requirements. It also suggests a generic way of constructing the SLAs, SLA proposals, and offers in XML. Methods in [15] suggest extending WS-agreement by specifying the conditions of the services in *guarantee terms* so as to improve the chances of reaching an agreement and reduce negotiation traffic.

In a distributed cooperative infrastructure, SLA management [17] is not only the gateway between the service provider and its customer for the service negotiation, but also the interface for services to interact with each other. For example, an application service provider may negotiate with a resource service provider to ask for the physical resources. SLA decomposition [18] is an important technique for translating the QoS requirements of the application into the requirements of the physical resources. However, the difficulty of accurately predicting the performance motivates the study of *adaptation and optimization* in SLA management. Literature [19] suggests establishing SLAs with various QoS levels where each QoS level is associated with service price and violation penalty. Heuristic algorithms are proposed for the service provider to admit the requests selectively and adapt the services according to the runtime resource availability so as to maximize its global profits.

The purpose of SLA-driven resource management [20] is to manage the distributed resources to ensure service quality and this has been studied in recent research work [21–24]. Literature (e.g., GRAM [21], GARA [22], SNAP [23], VAS [24]) considers ensuring the service quality by reserving resources in advance; for example, admitting/rejecting requests, choosing appropriate resources, or reserving and allocating the various resources for delivering the services over a distributed environment. In SNAP [23], a multiphase SLA-driven resource management protocol is presented to manage distributed resources for a task SLA, resource SLA, and bounding service level agreement to fulfill complex tasks. The Globus Architecture for Reservation and Allocation (GARA) has been proposed in [22] to reserve resources to ensure the end-to-end QoS in grids. The Virtual Application Service (VAS) [24] is an extension of GARA, which supports the QoS negotiation and the submission of jobs with deadline constraints. Adaptive resource management has been proposed (e.g., G-QoSm [25]) to cope with the dynamic feature of resource availability in cooperative environments.

As mentioned above, realizing SLA management has various aspects covering SLA negotiation, SLA decomposition, resource allocation, adaptation, and optimization. In real-life situations, not all services require the full functionality mentioned above. Hence, a generic SLA management service is required to facilitate SLA-related operations. In this chapter, we propose a model-based design for an SLA manager, which provides a generic SLA management service in distributed environments. It facilitates the SLA negotiation between the service provider and the client and manages the SLA copies for SLA-relevant process in distributed networked environments.

10.3 Architecture of an SLA Manager

10.3.1 SLA Manager and Its Environment

SLA Manager is a software module that provides APIs to manage the SLA copies and facilitate the communication between a service provider and its service clients for establishing the SLAs. It supports SLA-relevant operations such as retrieving the SLA templates, negotiating the services, settling the agreement, monitoring the SLA status during the service delivery, and so on.

Figure 10.1 shows the relationships between SLA Manager and its working environment. SLA Manager provides interfaces through which service clients can retrieve service templates, specify their requirements, and negotiates an SLA with the service provider regarding the service terms and conditions. In addition to this, SLA Manager closely interacts with the service manager to provide offers and counter-offers to the service clients. Overall, SLA Manager defines a standardized protocol and a set of methods that help in the SLA-relevant operations.

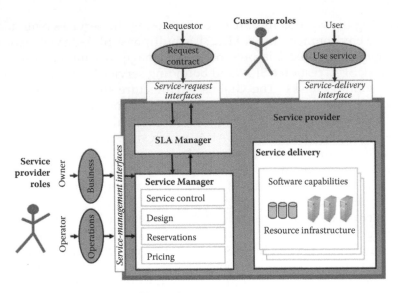

FIGURE 10.1   SLA Manager and its environments.

10.3.1.1 Definitions of the Roles

We define the roles of different entities that interact with SLA Manager as follows:

- Service client/customer: the service client may be an application or another service implementation with sufficient authority for service subscription and usage. It can be a software module that requests a service from a service provider. The service client assumes two roles. The first role is that of a service requestor when it queries the service provider regarding service offerings and negotiates an SLA. The role of the service client changes to a service user when the service is being delivered.

- Service provider: the service provider is a network-addressable software entity that executes requests and provides services to their clients. The service provider may publish the descriptions of the services and provide the SLA templates for its potential service clients to access.

- SLA Manager: this is a software module/component located inside a service entity to facilitate the interaction between a service provider and its service clients for establishing SLAs and completing SLA-relevant operations.

- Service manager: this is a software module/component within a service entity that manages the service. The service manager is the central management component module that manages service requests, pricing, resource allocations, and so on.

- Service operator: the service operator is responsible for day-to-day operation and supervision of the service's functioning. This role may be filled by a person or a piece of software, or both.
- Service owner: the service owner is responsible for setting business policies to be followed by the service provider. This role may be filled by a person or a piece of software, or both.

With the above definitions of the roles, we note that the SLA Manager is located inside a service entity as a software component to facilitate the establishment of SLAs for the service provider and its customers. We therefore consider designing a generic model-based design for SLA management, which can not only provide basic functionalities for the SLA management but also be customizable to meet the needs of various service providers in distributed environments.

10.3.1.2 SLA Manager

SLA Manager provides interfaces for the service client to query the service provider's offerings, interfaces to communicate with the service manager and operator-specific interfaces. The core of SLA Manager is a software module called the SLA-based service that facilitates the basic communications between the service provider and the service client. SLA Manager includes different submodules to support various SLA-relevant operations. The operations can be classified into four groups: (1) specification and negotiation, (2) validation and creation, (3) monitoring and archiving, and (4) violation and adaptation.

10.3.1.2.1 Specification and negotiation

When a client requests a service, SLA Manager will allow the client to retrieve the SLA templates offered by the service provider and specify their requirements in an SLA proposal. During the negotiation process, SLA Manager allows the clients to submit a proposal based on the template and enables both parties to go through a negotiation process that results in the proposal being either accepted or rejected. During the negotiation process, both parties can offer counter-proposals and maintain copies of the intermediate states and the final SLA. Some of the basic operations for the SLA-based service include getSLATemplate(), proposeSLA(), signSLA(), rejectSLA(), and cancelSLA().

10.3.1.2.2 Validation and creation

Before an SLA is created, a validation process is required to check whether the SLA proposal satisfies the SLA creation constraints [9]. The service provider may describe the creation constraints in the SLA templates to define the terms and conditions that a valid proposal should conform to. Once an SLA proposal has been submitted, SLA Manager will check whether it is a valid proposal. An SLA will be created only if the proposal conforms to the creation constraints and is agreed upon by both parties.

10.3.1.2.3 Monitoring and archiving

The purpose of SLA monitoring and archiving is to automatically measure and monitor the status of the SLAs during both SLA negotiation and service delivery stages. The status of an SLA may vary during the lifecycle of an SLA. For example, an SLA can be in "active," "expired," or "canceled" state, or even in "changed" or "violated" state. Operation getState() checks the SLA status. Each SLA status may be associated with a time stamp or certain conditions that are specified in the SLAs. Therefore, it requires efficient mechanisms to automatically monitor the status of SLA, maintain the different versions of SLAs, and notify the relevant parties when changes happen. Sometimes, SLA monitoring mechanisms are required to check the status of a job execution, predict the performance, and inform the service manger to adjust the resource allocation at runtime so as to prevent SLA violation or service suspension.

10.3.1.2.4 Violation and adaptation

An SLA violation may happen when some of the agreed terms in an SLA are violated or the terms cannot be successfully fulfilled by the service execution environment. Adaptation mechanisms are required to react to the violations and trigger the actions of automatic adaptation and adjustment. For example, it is hard to accurately predict the performance for many of the high-performance computing applications just by looking at the input parameters. In such cases, the SLA needs to be established based on prior estimation regarding resource requirements and time to completion. Therefore, efficient adaptation mechanisms are crucial to handling the uncertainties in performance and resource availabilities in dynamic environments.

In real-life situations, not all the SLA-relevant operations are required for every service provider. Therefore, we propose an SLA manager that is based on a generic core model with pluggable modules/components so as to allow the service providers to customize their implementations. Figure 10.2 shows the components of SLA Manager with the SLA-based service as its core to support various SLA-related operations. The functionalities of each group are governed by a pluggable component that connects to the SLA-based service.

10.3.1.3 Service Model Creation over the Shared Services Platform

Figure 10.3 demonstrates the process of creating a service model for SLA management over the shared service platform. To instantiate a service

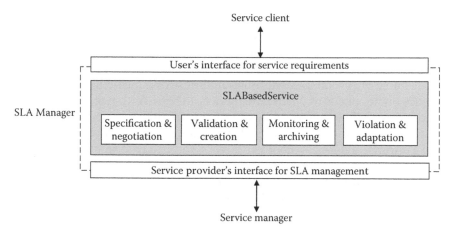

FIGURE 10.2 SLA Manager and its core SLA-based service.

instance, the service owner needs to create its service model, upload the model instance to a service access point (SAP), and register the service. The SAP is a service end-point that provides a standard interface for managing the service and supporting the interaction between services [7]. In a distributed service-oriented architecture, the SAP serves as a proxy for service clients to access the services.

In SSP, once a SAP receives the service model instance from the service owner, it creates a service instance with an entity reference and

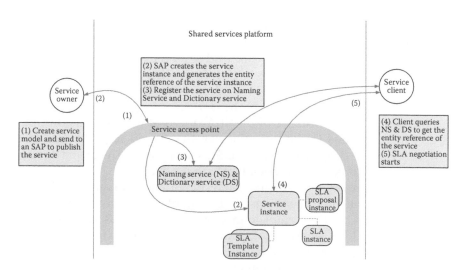

FIGURE 10.3 SLA management model creation over the shared services platform.

acknowledges the service owner that the service has been started. The entity reference of the newly created service instance will be published via a dictionary service and a naming service for customers to discover and access the service.

To enable SLA management functionalities for a particular service, the service owner needs to upload the model of the SLA Manager to SAP. The model may follow a generic SLA Manager model or a modified version based on a core model (e.g., an SLA-based service). Once the service is up and running, a service client can access the external interfaces of SLA Manager via the entity reference of the service. The service provider can also send requests to the SAP for modifying the existing service model or uploading a new model to the SAP. The model-based design gives the service provider the flexibility to edit or modify the service model for customization.

10.3.1.4 CIM Models of SLA Manager

In SSP, the models are described using the common information model (CIM) meta-model. The models contain meta-information of the services in a standardized construct format and are often described using MOF [12]. The MOF file describes the classes, associations, references, types, methods, instance declarations, comments, and so on in textual format. The CIM model of the SLA manager may include the operation models, the data models, and, sometimes, the instance models in one or multiple text-based MOF files. An operation model defines the service type and declares the service's external interfaces; a data model describes the data structures of certain entities; and an instance model creates the instance of certain entities and specifies the values of the instances.

The core of SLA Manager is the SLA-based service, which consists of the operation model that describes SLA negotiation operations, and the data model that defines the data structure, and several instance models for creating the SLA template instances. Once the models are uploaded to SAP, SLA template instances will be created so as to allow service clients to retrieve the SLA templates from the service provider. When a client proposes an SLA, an instance of the SLA proposal will be created in the SAP which contains the specifications of the client's requirements. We will describe the copy management of SLA proposals/agreements during the negotiation processes described later in the chapter.

Figure 10.4 provides an example MOF file of an SLA-based service for basic SLA negotiation operations. For example, operation getSLATemplates() retrieves the SLA template instances from the SAP. The operation proposeSLA() allows service client/provider to propose or counter-propose an SLA by filling in the detailed requirements in an SLA proposal instance.

As mentioned before, SLA Manager is one of the software modules within a service entity. Besides SLA Manager, the service provider may

```
[Description("The model for the SLA Based Service"), Version("1.0"), Author("WP5")]
class SSP_SLABasedService_negotiation : SSP_Service
{
        [Description("Client retrieves one or more SLA templates")]
        string getSLATemplates([IN] string selector,[OUT] SSP_SLATemplate REF slaTemplate[]);

        [Description("Client proposes an SLA and the return value is either an accepted
        (with Service Provider signature) or modified (counter-propsal) SLAProposal from the Service Provider")]
        boolean proposeSLA([IN] SSP_SLAProposal REF proposedSLA,
                                           [OUT] SSP_SLAProposal REF slaProposal);

        [Description("Client rejects the SLA proposal identified by slaID")]
        boolean rejectSLA([IN] string slaID, [IN] Boolean byclient);

        [Description("Client ends the SLA identified by slaID")]
        boolean cancelSLA([IN] string slaID, [IN] Boolean byclient);
```

FIGURE 10.4 Example of a MOF file of SLA-based service for SLA negotiation operations.

declare a number of other modules by uploading the respective models to the SAP. It is possible for the service provider to modify the meta-models of certain modules without affecting the others. The SAP maintains the model validation mechanisms to check and validate the modification of the models at runtime to ensure that violations or conflicts do not occur.

10.4 SLA Model and Its Implementation

In this section, we will describe the data model in SLA Manager, referred to as an SLA model, and present its UML design. The SLA model defines the data structure of the SLA templates and SLAs which are used to create respective SLA templates and SLA proposals/agreements during the SLA negotiation process. We will discuss the lifecycle of an SLA and the implementation details to realize SLA negotiations in this section.

10.4.1 SLA Model

The SLA model contains complete information regarding the service type, user requirements, business values, and so on. It takes advantage of the standard object-oriented paradigms such as inheritance and class hierarchy. We define the SLACommon class as the core of the SLA model. This allows us to group common functionalities used by both the SLA class and the SLATemplate class. Figure 10.5 shows the UML diagram of the SLACommon class and its associations with the SLA class and the SLATemplate class. The SLACommon class consists of four major data structures: context information, service terms, guarantee terms, and service-level

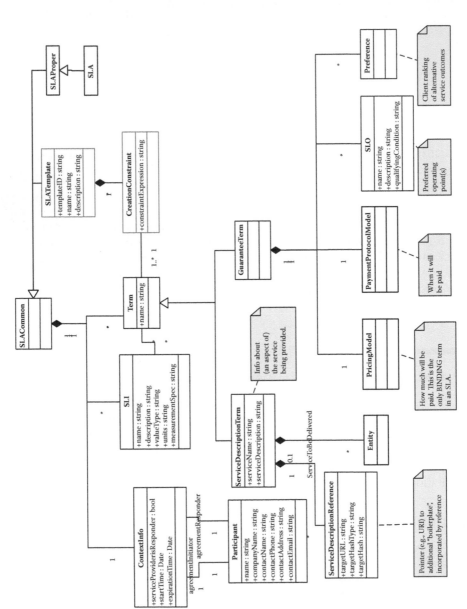

FIGURE 10.5 UML design of the SLA model.

indicators (SLIs). The SLA class inherits from the SLACommon class and has the same data structures as the SLACommon class. The SLATemplate class is derived from the SLACommon class. Most of the terminology described in our proposed SLA model follows the WS-agreement specification draft version 1.1 [9].

As shown in Figure 10.5, the SLACommon class consists of the following classes: (1) ContextInfo, which includes the information of the participants, (e.g., who are agreeing to the SLA, and how long is the SLA valid); (2) ServiceDescriptionTerm, which describes the service properties (e.g., service name, type); (3) GuaranteeTerm, which expresses the desired or required service properties, behaviors, consequences, and so on; (4) the SLI, which represents the name and details of the terms used, how they are measured, and so on. GuaranteeTerm consists of a PricingModel that describes the service charges, PaymentProtocolModel that describes how and when the payment will be made, and ServiceLevelObjectives (SLO) that specify the measurable characteristics of the service such as response time, reliability, availability, throughput, and so on. With the structured information in the GuaranteeTerms, the proposed SLA model allows the service provider and its clients to express the business values associated with the service qualities in the SLA.

10.4.2 SLA Negotiation and the Implementations

Based on the SLA model, a service client and the service provider can negotiate with each other via the SLA-based service. A basic SLA negotiation process may involve three phases: (1) retrieving SLA templates, (2) proposing SLAs, and (3) agreeing/disagreeing to an SLA.

10.4.2.1 Retrieving SLA Templates

As explained before, SLA templates hold the default values and creation constraints for the service. A service customer uses the SLA template to put together an SLA proposal. To subscribe a service, the client may first request one or more SLA templates from the service provider. As a result, a set of entity references of the SLATemplate instances will be sent to the client. The client can then access the SLATemplate instances using these entity references.

10.4.2.2 Proposing SLAs

Based on the SLATemplate obtained from the service provider, the client can create an SLA proposal object by adding the user requirements. To propose an SLA, the client invokes the proposeSLA() method on the service provider interface. Once an SLA is proposed, the service provider receives a copy of the SLA proposal. After receiving the SLA proposal, the service provider may reject the SLA proposal, accept the proposal, or offer a counter-proposal.

- *Reject proposal:* The service provider rejects the request by returning an empty SLA proposal object to the client.
- *Accept proposal:* The service provider accepts the offer by providing additional information regarding the service such as service price and end-point details. The service provider may attach its digital signature agreeing to the proposal. At this stage, this SLA becomes binding, once the client signs the SLA.
- *Offer counter-proposal:* Sometimes, the service provider may not agree with the terms offered by the customer. In this case, the service provider may return a counter-offer that is different from the original offer.

10.4.2.3 Agreeing/Disagreeing to SLA

The negotiation of an SLA may be terminated when any one of the following scenarios happens:

- *Both parties have signed the SLA proposal.* When both the service provider and the client add their digital signatures in the SLA proposal, an SLA is considered to have been established. Hence, the negotiation terminates.
- *Both or either one of the parties reject(s) the proposal.* Both the parties have the right to reject a proposal from its counter-part. In our proposed SLA manager, the service client can reject the service provider's SLA proposal by invoking the rejectSLA() function, while the service provider can decline the client request by returning an empty SLA proposal.
- *Negotiation time-out.* Either side can abandon the negotiation process if the other party does not respond in time. The SLA negotiation will automatically stop if there is no reply from the other party till the required response deadline.

Once an SLA has been established, the service provider will provide the service to the client under the terms of the SLA.

10.4.2.4 SLA Lifecycle

Figure 10.6 describes the lifecycle of an SLA. When an SLA is first created based on an SLATemplate, the SLA proposal is in *pending* state. At this state, the proposal is ready to be submitted to the service provider. Once the proposal is submitted, it enters the *proposed* state. When both the service provider and the client agree on the SLA proposal by adding their

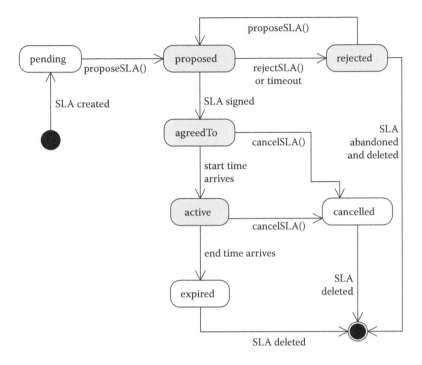

FIGURE 10.6 The lifecycle of an SLA.

digital signatures, the SLA proposal enters the *agreedTo* state. Typically, every SLA will have a designated start and end time when the SLA is active. Once the start time recorded in the SLA arrives (which may or may not be immediately), the status of the SLA becomes *active*.

If an agreement is never reached, the SLA proposals will eventually be abandoned, and deleted. The active state is exited normally when the SLA expiration time is reached (the *expired* state), or abnormally if the SLA is cancelled by either side (the *cancelled* state). Eventually, the SLA will be deleted.

There may be additional substates of *active*. For example, the service provider may refuse to provide service until certain conditions are met—for example, payment up front for its services, or if interim payments are late. Such substates will need to be handled by the service provider on a case-by-case basis.

10.4.2.5 Management of SLA Copies

During the SLA negotiation process, multiple versions of the SLA proposals may be created before reaching an agreement. Efficient means of SLA copy management is required to ensure that different versions of SLA proposals are preserved until final agreement is reached.

Figure 10.7 shows a possible sequence of the operations where an SLA proposal is created, modified, and agreed upon during the SLA negotiation. In this case, we assume that the service client and the service provider are located at different service access points. Figure 10.7 demonstrates how the SLA copy management is realized by the SLA Manager over the SSP. As shown in Figure 10.7, there would be multiple copies of SLA/proposals in various stages of a simple client-initiated SLA negotiation.

1. The service client retrieves the SLA templates from the service provider.
2. The service client creates an SLA proposal object (i.e., SLA proposal1) according to an SLA template and caches it in its local memory.
3. The created SLA proposal1 is pushed from the client's local memory to the SAP1, the service access point at the client side.

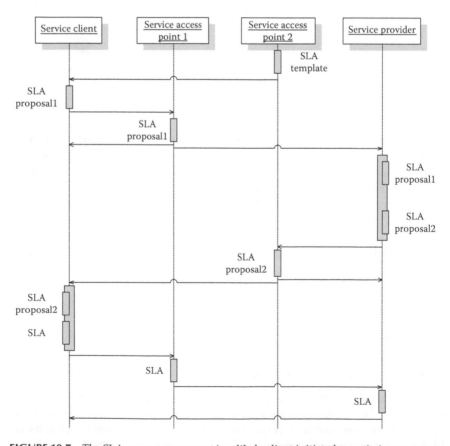

FIGURE 10.7   The SLA copy management in a likely client-initiated negotiation sequence.

4. SLA proposal1 is transmitted from SAP1 to the service provider's local memory. If the service provider is willing to give an offer based on SLA proposal1, an SLA proposal2 is created by the service provider by filling the other details and attaching its signature.

5. The SLA proposal2 is propagated from the service provider's local memory to the SAP2, which locates at the service provider's side.

6. The SLA proposal2 is transmitted from SAP2 to the client, and cached in the client's local memory.

7. An SLA will be established if the service client adds its digital signature on SLA proposal2 to create an SLA.

8. The SLA copy is pushed from the client's local memory to SAP1.

9. The SLA is delivered to the service provider for him/her to keep locally. Once the service provider receives the SLA, an ACK signal will be returned to the client to acknowledge receiving the SLA.

Step 0 is executed by operation getSLATemplates() in the SLA-based service. Steps 1, 2, 6, and 7 are done by the respective operations in a client library for operating the SLA proposal object and pushing the objects to the client's local SAP. Steps 3, 4, and 5 are completed by invoking propose-SLA() in the SLA-based service. In Step 8, the client calls an explicit method (e.g., addMySignature) to add its signature in the SLA.

The mechanisms of managing SLA copies via SLA Manager, as shown in Figure 10.7, make sure that both the service provider and its client have all the versions of the SLA proposals and the final SLA, and what's more, they are synced up with each other during the SLA negotiations.

10.4.2.6 Signing the SLA Proposals

We define an SLA as being active if and only if both the service provider and its client have signed an SLA proposal. However, the sequence regarding which party signs first is flexible. Depending on who signs first, the negotiation sequences may change. For example, in a client-initiated SLA negotiation, the client may sign the original SLA proposal1 when it creates SLA proposal1. Hence, an agreement will be reached once the service provider agrees to SLA proposal1. The established SLA therefore has signatures from both the service provider and the client.

For some circumstances, the client may not sign the original SLA proposal1 but may be waiting for the service provider to fill in the next level details before signing the agreement, as shown in Figure 10.7. If the service provider is the first to sign, an additional interaction is required for the service client to sign. The client will first review the SLA proposal2 created by the service provider before signing it. The SLA proposal2 will finally become an SLA as the signatures from both the service provider and the client are present.

In real-life situations, there may be other possible SLA negotiation sequences determined by who initiates the SLA negotiations, who signs the SLA proposal first, whether the service provider or/and the client will give a counter-proposal by modifying the previous SLA proposal, and so on. The SLA copy management in our proposed SLA-based service is flexible and extensible to deal with various scenarios of SLA negotiation in distributed environments.

10.5 Applications of Fire Dynamic Simulation Services

In this section, we will examine the performance of our proposed SLA manager by incorporating it with Fire Dynamic Simulation (FDS) for providing high-performance service over the shared services platform.

10.5.1 Deployment for FDS Services with an SLA Manager

FDS is a computer program based on computational fluid dynamics (CFD) models to simulate fire growth and spread. "The software solves numerically a form of the Navier–Stokes equations appropriate for low-speed, thermally-driven flow, with an emphasis on smoke and heat transport from fires" [26]. FDS was developed and maintained by the National Institute of Standards and Technology (NIST) from 2000, and it was widely used to simulate fireworks and predict the temperature and radiation of heat and smoke. The FDS works with Smoke View, a visualization program, to demonstrate the propagation of the fire and estimate the damages it may cause within a certain amount of time.

As a high-performance application, FDS generates a large amount of data and consumes significant computation resources. It may take extraordinary processing time to run a simulation test especially when there are no sufficient resources available. In general the processing time of FDS will be affected by resource availability, environmental data setting, and simulation parameters such as simulation duration, number of CPUs, and so on. For example, one FDS simulation test (e.g., 3st-5grid) with 1-minute simulation duration on five CPUs takes about 30 minutes to complete the job running. FDS is programmed into two different modes: sequential and parallel. In a sequential mode, the simulation is done on one processor only. In a parallel mode, the simulations are divided into multiple meshes that can run in parallel on multiple processors using MPI. Users can specify input parameters such as the number of processors, the value of IBAR, JBAR, KBAR (i.e., x-, y-, z-dimensional information) of a mesh, and so on in an input data file. If there are enough resources, all the meshes can run in parallel; however, if the number of available processors is less than the number of meshes, it will take a longer time to process.

We have implemented FDS as a service in SSP with SLA Manager as one of the modules, as shown in Figure 10.8. Clients can subscribe to FDS services via SLA Manager by specifying their job requirements. In the meantime, the FDS service provider can negotiate with its potential clients on the service terms and conditions based on client input parameters, estimated simulation response time, resource availability, and so on.

The FDS service provider can not only establish SLAs with its service clients, but also establish SLAs with other service providers to provide comprehensive services for its clients. For example, the FDS service could establish SLAs with other service providers that offer compute and storage resources. As shown in Figure 10.8, the FDS establishes an SLA1, with a physical resource service provider (PRsp) to obtain the computation resources; for example, PC cluster, CPUs, and so on. In the meantime, an SLA2, between the FDS service provider and the visualization service

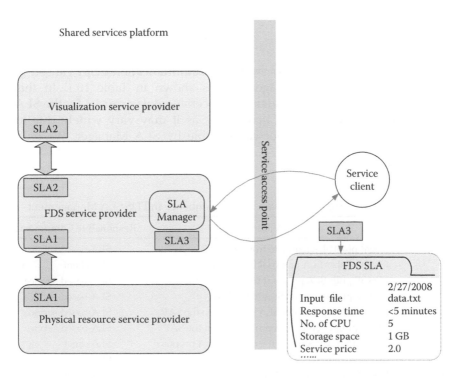

FIGURE 10.8   SLA management of FDS service over SSP.

provider (VSP), for example, PyroSim, can be established for visualizing the simulation results.

In our experiments, the FDS establishes an SLA with PRsp first before agreeing to SLAs with its clients. In this way, the FDS service provider can allocate resources to the individual FDS jobs without asking PRsp for each request. This method avoids the delay caused by the negotiation "on the fly" where the FDS service provider negotiates with the PRS service provider on resource allocation each time an FDS request arrives.

The entire platform is developed in Java and we used *eclipse* [27] as the IDE for program development. The SLA templates, the SLA proposals, and the SLAs are Java instances associated with a certain lifetime. The service models are stored in MOF files, which are in turn uploaded to the SAPs for activating the service. FDS clients have the ability to query the FDS service for SLA templates, specify their requests, and submit their proposal to the FDS service provider. Another important step in the process is to accurately predict resource requirements for running the job when an FDS client submits a request. We call this process SLA decomposition. We perform SLA decomposition by combining performance modeling and profiling. We use a constraint satisfaction solver called "Cauldron" [28] for deriving resources requirements for executing the FDS jobs. If the FDS service provider accepts a request, the client will receive an offer with specifications of response time, service price, penalty, deadline to reply, and so on.

10.5.2 Performance Study

We conducted experiments to study the communication delay caused in various stages of SLA management, as shown in Table 10.1. In this experiment, we did not consider the processing delay caused by the SLA decomposition and pricing calculations, as it may vary widely due to other service components not controllable by SLA Manager. We only

TABLE 10.1

Various Stages of a Simple SLA Negotiation Sequence in FDS SLA Management

Index	Action	Responsible Component
0	Client pulls SLA template from the SAP	SLA-based service
1	Client pushes SLA proposal to the SAP	SLA client
2	FDS SP pulls SLA proposal from the SAP	SLA-based service
3	FDS SP pushes signed SLA to the SAP	SLA-based service
4	FDS SP pushes counter SLA proposal to the SAP	SLA-based service
5	Client pulls counter SLA proposal from the SAP	SLA-based service
6	Client adds signature to the counter SLA proposal	SLA client
7	Client pushes SLA proposal to the SAP	SLA client
8	FDS SP pulls signed SLA from the SAP	SLA client

consider the communication delays caused by propagating the SLA objects among the relevant parties in the distributed environments. Figure 10.9a shows the absolute values of the average communication delay of different stages, and Figure 10.9b shows the percentage value of the overall communication delays. As we can see the communication delays of stages 1, 3, 4, 6, and 7 are very short (within just a few seconds), while stages 0, 2, 5, and 8 consume longer time than the previous ones and take up to 28% of the communication delay of overall communication delay in SLA negotiations. The communication time for retrieving data objects remotely from either client or service provider to its remote SAP takes a much longer time compared with the other stages in the SLA negotiations. This is

(a)

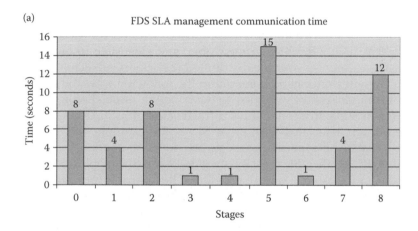

(b)

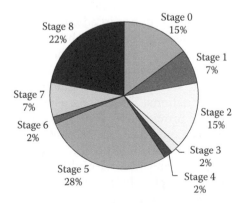

FIGURE 10.9 FDS SLA manager's communication delay over SSP: (a) communication time of different stages; (b) pie chart of communication time of different stages.

because the operations in those stages involve data transmission and retrieving a data entity from remote SAP repositories.

10.6 Conclusions

In this chapter, we proposed a CIM model-based design of an SLA manager which allows the service provider and its customers to specify requirements regarding SLAs and customize SLA management. We presented an SLA model where SLA template class and SLA class are derived from a common data structure. A model-based SLA manager and its core, called SLA-based services, were introduced to facilitate the communication between the service provider and the client and to manage the SLA copies in a distributed environment. The proposed model-based design of SLA Manager is highly flexible, customizable, and extensible to provide SLA management services over the distributed service-oriented environments.

Our present work can be extended in the following ways. First, the current SLA model can be extended to describe the association information of different SLAs. In a cooperative infrastructure, the SLAs may not stand alone but are associated with other SLAs. For example, the SLA established between an application service provider and its client may be associated with another SLA established between the application service provider and a resource service provider. The failure of fulfilling one SLA may result in the violation of other SLAs or the suspension of the other services. Therefore, the status of one SLA may affect the execution of other SLAs. Efficient mechanisms to specify and handle the associations of the SLAs help the service provider to fulfill the SLAs and eventually reduce the loss.

Second, the existing SLAs may be violated due to the difficulties of accurately predicting the performance in advance. Especially in the case where one single SLA consists of multiple jobs, the failure of executing one job may affect the execution of the others. Therefore, the mechanisms for adaptive SLA management are important. Some may consider the possibility of renegotiating the services and changing the established SLA to cope with the dynamic situations. The mechanisms of adapting existing SLAs may help the service provider to avoid the failures caused by inaccurate predictions and overcome the uncertainties of the decompositions and resource availabilities in distributed environments.

In addition, SLA agreement may involve multiple parties and each party will share the responsibilities of fulfilling the SLA. These scenarios may be common in computer services (e.g., service orchestration/compositions) where multiple parties are involved in completing a business process. As there might be multiple parties involved in the service negotiations, the specifications of the requirements and responsibilities of different parties need to be included in the SLA. It would be interesting to extend the current SLA management model for the multiparty negotiation over distributed environments.

Acknowledgments

The authors would like to thank those who participated in the design and implementation of the SLA-based service and/or the SSP platform: Sharad Singhal, John Wilkes, Akhil Sahai, Jim Prym, Andre Lopes, Karin Becker, Rully Adrian Santosa, Yong Siang Foo, Lester Leong, Elaine Wong, Chen Yuan, Pranavi Reddy, and many others. They also thank Sharad Singhal, Terence Hung, John Wilkes, Dejan Milojicic, and Hoong Maeng Chan for the pleasant discussions and the insightful comments on improving the quality of the chapter. The work is co-sponsored by the Singapore A-star Research Organization and HP labs.

References

1. Foster, I. and C. Kesselman. *The Grid: Blueprint for a New Computing Infrastructure.* Morgan Kaufmann, San Francisco, CA, 1998.
2. Large Hadron Collider Computing Grid (LCG). Available at: http://lcg.web.cern.ch/LCG/overview.html.
3. Data Grid. Available at: http://eu-datagrid.web.cern.ch/eu%2Ddatagrid/Intranet_Home.htm.
4. Particle Physics Data Grid. Available at: http://www.ppdg.net/.
5. Singapore National Grid Pilot Platform (NGPP). Available at: http://www.a-star.edu.sg/astar/about/action/pressrelease_details.do? id=0f8c16f87465.
6. Shared Services Platform. Available at: http://www.hp.com/hpinfo/newsroom/press/2006/060614a.html.
7. Singhal, S., J. Pruyne, and V. Machiraju. *Picasso: A Service Oriented Architecture for Model-Based Automation.* HP Technical Report, HPL-2007-50R1, Palo Alto, CA, 2007.
8. Dan, A., D. Davis, R. Kearney, R. King, A. Keller, D. Kuebler, H. Ludwig, M. Polan, M. Spreitzer, and A. Youssef. "Web services on demand: WSLA-driven automated management." *IBM Systems Journal,* 2004, 43 (1), 136–158.
9. Web Services Agreement Specification (WS-Agreement) V1.0. Available at: http://www.gridforum.org/documents/GFD.107.pdf, 2007.
10. Hellerstein, J.L., K. Katircioglu, and M. Surendra. "An on-line, business-oriented optimization of performance and availability for utility computing." *IEEE Journal on Selected Areas in Communications,* 2005, 23 (10), 2013–2021.

11. Debusmann, M. and A. Keller. "SLA-Driven management of distributed systems using the common information model," in *Proceedings of the 8th IFIP/ IEEE International Symposium on Integrated Network Management*, Colorado Springs, 2003, pp. 563–576.
12. Common Information Model (CIM) specification, Version 2.2, Distributed Management Task Force, June 14, 1999. Available at: http://www.dmtf.org/ standards/cim/cim_spec_v22.
13. Sahai, A., S. Graupner, V. Machiraju, and A. van Moorsel. "Specifying and monitoring guarantees in commercial grids through SLA," in *Proceedings of the 3rd IEEE/ACM International Symposium on Cluster Computing and the Grid*, Tokyo, Japan, May 12–15, 2003, pp. 292–299.
14. Foroughi, A. "Minimizing negotiation process losses with computerized negotiation support." *Journal of Applied Business Research*, 1998, 14 (4), 15–26.
15. Sakellariou, R. and V. Yarmolenko. "On the flexibility of WS-agreement for job submission," in *Proceedings of the 3rd International Workshop on Middleware for Grid Computing*, Grenoble, France, 2005, pp. 1–6.
16. Global Grid Forum (GGF). Available at: http://www.gridforum.org/ L_About/about.htm.
17. Schmietendorf, A., R. Dumke, and D. Reitz. "SLA management-challenges in the context of Web-service-based infrastructures," in *Proceedings of the IEEE International Conference on Web Services*, San Diego, CA, July 6–9, 2004, pp. 606–613.
18. Chen, Y., S. Iyer, X. Liu, D. Milojicic, and A. Sahai. *SLA Decomposition: Translating Service Level Objectives to System Level Thresholds*. HP Technical Report, HPL-2007-17, Palo Alto, CA, 2007.
19. Atdelzater, T.F., E.M. Atkins, and K.G. Shin. "QoS negotiation in real-time system and its application to automated flight control." *IEEE Transaction on Computers*, 2000, 49 (11), 1170–1183.
20. Hasselmeyer, P., B. Koller, L. Schubert, and P. Wieder. "Toward SLA-supported resource management," in *Proceedings of the International Conference on High Performance Computing and Communications (HPCC-06)*, Munich, Germany, 2006, pp. 743–752.
21. Foster, I., C. Kesselman, C. Lee, R. Lindell, K. Nahrstedt, and A. Roy. "A distributed resource management architecture that supports advance reservations and co-allocation," in *Proceedings of the International Workshop on Quality of Service*, London, UK, 1999.
22. Foster, I., A. Roy, and V. Sander. "A quality of service architecture that combines resource reservation and application adaptation," in *Proceedings of the 8th International Workshop on Quality of Service (IWQOS)*, Pittsburgh, PA, June 2000, pp. 181–188.
23. Czajkowski, K., I. Foster, C. Kesselman, V. Sander, and S. Trecke. "SNAP: A protocol for negotiating service level agreements and coordinated resource management in distributed systems," LNCS 2537, Springer, 2002, pp. 153–183.
24. Keahey, K. and K. Motawi. *Taming of the Grid: Virtual Application Services*. Argonne National Laboratory, Mathematics and Computer Science Division Technical Memorandum ANL/MCS-TM-262, 2003.
25. Al-Ali, R.J., O.F. Rana, D.W. Walker, S. Jha, and S. Sohail. "G-QoSM: Grid service discovery using QoS properties." *Journal of Computing and Informatics*, 2002, 21 (4), 363–382.

26. McGrattan, K.B. "Verification and validation of selected fire models for nuclear power plant applications," Fire Dynamics Simulator (FDS), NUREG-1824, EPRI 1011999, Volume 7, May 2007, pp. 205.
27. Eclipse. Available at: http://www.eclipse.org/.
28. Ramshaw, L., A. Sahai, J. Saxe, and S. Singhal. "Cauldron: A policy-based design tool," in *Proceedings of the Seventh IEEE International Workshop on Policies for Distributed Systems and Networks*, June 2006.

11

Portal and Workflow in Grid Computing: From Application Integration to Service Integration

Xiaoyu Yang

CONTENTS

11.1 Introduction

Grid computing involves access to geographically distributed computing and data resources, and a grid-based computation usually requires defining a sequence of activities such as metascheduling, job submission and monitoring, file transfer, analysis and simulation, and data harvesting. These determine a need for an integrated platform that can provide uniform access to a variety of grid applications and services, and a need for process automation that can manage and coordinate the activities required for grid-based computation. Portal and workflow technologies can be employed to accommodate these needs in grid computing. A grid-enabled portal can provide an integrated access to applications and services that are located throughout the grid. Workflow is the automation of a process where documents, information, or tasks are passed from one participant to another to be processed, according to a set of procedural rules. In the context of service-oriented architecture, a service-based approach is often adopted to build grid workflow where core grid-computation-related tasks are deployed as Web services that are then integrated as workflow processes.

This chapter discusses portal and workflow technologies used in grid computing for application integration and service integration. First, portals and grid-enabled portals are introduced, which include basics of portals, the portal inherent feature single sign-on (SSO), using a portal as a front end for a grid accounting system, and using AJAX in portal applications. Various portal frameworks that can be employed in building grid-enabled portals are reviewed. Second, workflow technologies and the state-of-the-art service component architecture (SCA) are discussed. Finally, in an actual case study within the UK e-Science program, a grid-enabled science portal for running quantum mechanical simulation for material properties over grids is developed, and a service-oriented workflow used for orchestrating services required for running quantum mechanical simulation is created.

11.2 Portals for Grid Application Integration

This section covers the following topics in relation to grid portals: (1) the basics of portals; (2) SSO, one major inherent feature of portal that is often used in the grid environment; (3) portals in grid accounting; (4) integration of AJAX into the portal; and (5) a survey on major grid portal development tools and technologies that can be employed to facilitate grid portal development.

11.2.1 Portals and Portlets

11.2.1.1 Portals and Grid-Enabled Portals

Web portals are sites on the World Wide Web that typically provide personalized capabilities to their visitors. They are Web-based applications that commonly provide personalization, SSO, content aggregation from different sources and hosts the presentation layer of information systems. They are designed to integrate distributed applications, different numbers and types of middleware, and hardware to provide services from a number of different sources.

A grid portal is "a Web-based application server enhanced with necessary software to communicate to grid services and resources" [2]. It provides access to grid technologies through shareable and reusable components for Web-based access to domain-specific applications. The grid portal is built upon the familiar Web portal model, such as Yahoo or Amazon, to deliver the benefits of grid computing to virtual communities of users, providing a single access point to grid services and resources.

11.2.1.2 Grid Portal: Nonportlet Based versus Portlet Based

According to the way of building portals, grid portals can be classified into nonportlet based and portlet based [3]. Many early grid portals or early version of existing grid portals are nonportlet based; for example, the Astrophysics Simulation Collaboratory (ASC) portal [4] and UNICORE [1,5]. These grid portals can provide uniform access to the grid resources. Usually these portals were built based on typical three-tier Web architecture: (1) Web browser, (2) application server/Web server which can handle HTTP requests from the client browser, and (3) back-end resources that include computing resources, databases, and so on.

Portlet-based portals have become one of the most exciting areas for the portal server platform in recent years [6]. A portlet is a Web component that generates fragments—pieces of markup (e.g., HTML, XML) adhering to certain specifications. Fragments are aggregated to form a complete Web page. The portlet-based portal is based on component-based development, and portlets in the portal are engineered independently from each other. A portlet is an individual class that processes the user request and returns the content for display within a portal. It is contained in portlet container, which is part of portal and instantiates and executes the portlet classes.

Developing portlet-based portals can bring many benefits to both end-users and developers, which now gets more recognition [8]. This can be reflected through evolution of some grid portal projects. For example, although the ASC portal [4] did provide functionalities for the astrophysics community to remotely compile and execute applications, it was difficult to maintain when the underlying supporting infrastructure evolved. Eventually, the ASC portal was retired and its functionality moved into the Cactus portal developed by adopting GridSphere [2]. Another example is the GridPort portal [4]. The early GridPort was implemented in Perl and made use of HotPage [9] technology for providing access to grid access. Now the GridPort adopts GridSphere.

11.2.1.3 Portlet Standards

One of the main advantages of using a portlet-based portal is that there are two standards for portlet development, namely JSR-168/JSR-286* and WSRP (Web Service for Remote Portlets). JSR-168 establishes a standard API for

---

* JSR 286 is an extension of JSR-168. In this chapter, the JSR-168 is used for the standard portlet.

creating portlets [11], while WSRP is a standard for Web portals to access and display portlets on a remote server [1]. JSR-168 and WSRP work at different levels. JSR-168 specifies the interfaces for local portlets, while WSRP specifies the interfaces for accessing portlets across portal frameworks. Using standard portlets can ensure they can plug-and-play in any standard-compliant portlet containers (e.g., WebSphere Portal, GridSphere).

11.2.2 Single Sign-On in Grid Portals

SSO is one of the important requirements for grid portals. Grid portals aim to provide an integrated access to geographically distributed, cross-organization computing resource and services, each of which has its own user name and password. It is not a user-friendly approach for a user to enter several different passwords to gain the access to distributed resources; also, users are not good at remembering many passwords. SSO provides an access control mechanism that enables a user to be authenticated once and gain access to the resources of multiple applications. By using SSO, a user needs just a single user name and password to be authenticated by the portal, which can then access the different applications and services on behalf of the user without entering username and password again. This section introduces two important topics relating to SSO in grid portals: (1) Globus grid security infrastructure and (2) credential management.

11.2.2.1 Globus Grid Security Infrastructure

Cryptographic techniques are heavily used in grid environments to meet the strong security requirements, which are crucial for enforcing resources sharing policies in a virtual organization (VO); for example, granting access to VO members only [13]. Grids require all communication between a user (or an entity acting on behalf of a user) and resources in the grid to be authenticated and authorized. Grid users should authenticate themselves once and from that point on, all further security interactions are carried out transparently by the grid.

The Grid Security Infrastructure (GSI) [1], developed by the Globus Alliance [16], provides security mechanisms that are commonly used in grid environments. The GSI uses public key cryptography (also known as asymmetric cryptography) as the basis for its functionality, where the certificate is a central concept in GSI authentication [1]. Every user and service in the grid are identified via a certificate, which contains information vital to identify and authenticate the user or service. A GSI certificate includes four primary pieces of information: (1) a subject name, which identifies the person or object that the certificate represents; (2) the public key belonging to the subject; (3) the identity of a Certificate Authority (CA) that signed the certificate to certify that the public key and the identity both belong to the subject; and (4) the digital signature of the named CA. If two parties have certificates, and if both parties trust the CAs that signed each other's certificates, then the two parties can prove to each other that they are who they claim to be. This is known as mutual authentication [1].

One of the core concepts in GSI is the delegation capability, which can be used to avoid re-entering the user's password if a grid computation requires several grid resources to be used (each requiring mutual authentication), or if there is a need to have agents (local or remote) requesting services on behalf of a user. In order to accommodate this need, a proxy is used in GSI. A proxy consists of a new certificate (with a new public key in it) and a new private key. The new certificate contains the owner's identity, modified slightly to indicate that it is a proxy. It is signed by the owner, rather than a CA. The proxy certificate also includes a life span after which the proxy should no longer be accepted by others.

11.2.2.2 Credential Management: MyProxy for Single Sign-On in Grid Portals

MyProxy is open source software widely used in grid portals for managing X.509 public key infrastructure (PKI) security credentials (certificates and private keys) to help realize the SSO [16]. It provides a solution for delegating credentials to grid portal to allow the portal to authenticate to grid services on the user's behalf [18].

The work mechanism of MyProxy with a grid portal can be described as follows. First, users need to store their grid credentials on a MyProxy server that the portal can use. Usually this can be achieved by running the *myproxy-init* command on a user's computer where grid credentials are located to upload the credential to the MyProxy server. This approach is simple but requires some MyProxy commands to be installed. There is a GUI-based open-source tool, MyProxy Upload Tool [18], that can be used to help upload credentials to the MyProxy server via a Web browser without any software installed. Second, once the user grid credentials are stored in the MyProxy server, the grid portal can retrieve a short-term proxy credentials for a grid portal to access the grid services on behalf of the user. As a commitment of SSO, the grid portal can be developed to authenticate a user by using the user's MyProxy username and password rather than using a separate username and password.

11.2.3 Portal in Grid Accounting

Resources sharing in grid computing are not always free. Many organizations do not want their resources to be shared unless they get paid for the resources provided. Grid computing will be widely available for commercial use once the issue of charging for resource usage is addressed. This has resulted in the issue of grid accounting, where the grid portal plays a critical role as a front end. In this section, related work to grid accounting is selectively reviewed, and the requirements specification for grid portal serving as a frond-end of grid accounting system has been identified.

11.2.3.1 Related Work to Grid Accounting

11.2.3.1.1 Usage Record and Resource Usage Service

The Open Grid Service Architecture (OGSA) compatible Resource Usage Service (RUS) was proposed by the Open Grid Forum (OGF) to enable the recording and retrieval of consumed resource information [19]. The RUS has defined two primary functions: the upload of resource usage information, and the extraction of resource usage data from the service. The Usage Record (UR) XML schema has been employed in RUS, which provides a common format for exchanging basic accounting and usage data over the grid. UR contains usage information that could potentially be collected at grid sites, such as CPU-time, memory, and network usage. The UR and RUS were mainly defined to support the Grid Economic Services Architecture (GESA) [20], which outlines service interfaces for enabling charging of grid services in OGSA.

11.2.3.1.2 Distributed Grid Accounting System

The Distributed Grid Accounting System (DGAS) is a grid accounting effort [22] that is now maintained and re-engineered within the EU-funded Enabling Grids for E-sciencE (EGEE) project [22]. The system can be divided into three layers: (1) resource usage metering, (2) accounting, and (3) account balancing (through resource pricing) in a fully distributed grid environment. The usage metering is done by lightweight sensors (i.e., DGAS Gianduia) installed on the computing elements by parsing PBS/LSF/Torque event logs to build a UR that can be passed to the accounting layer. The data collected are associated with user ID, resource ID, and job ID. The DGAS accounting layer provides home location registers (HLRs) that manage both user and resource accounts. Each HLR keeps the record of all grid jobs submitted or executed by each of its registered users or resources. The accounting balancing exchanges virtual credits between the user HLR and the resource HLR, and the resource pricing is done by dedicated Price Authorities (PAs) that may use different pricing algorithms such as manual setting of fixed prices or dynamic determination of prices according to the state of a resource.

11.2.3.1.3 Grid Bank

The Grid Bank [23] is another grid accounting effort. It presents a series of grid components that enable global computational economy in the context of the grid banking and accounting service. The workflow of components interaction can be briefly described as follows: (1) the *Grid*

Resource Meter extracts resource usage information from the operating system and presents it in a grid standard form (i.e., UR); (2) the *Grid Bank Charging Module* (GBCM) is responsible for determining payment instruments, setting up and removing temporary local accounts, calculating total charge, and so on; (3) the *Grid Bank Payment Module* (GBPM) receives requests for job execution from the *Grid Resource Broker* (GRB), obtains a payment instrument from the Grid Bank server, forwards the payment to GBCM and submits the job; (4) the *Grid Trade Server* negotiates service cost/rates with GRB and provides an interface for GBCM to obtain the information.

11.2.3.1.4   *SweGrid Accounting System*

A Swedish National Grid (SweGrid) grid accounting system uses the Bank Service and Accounting Manager [24] for grid accounting. The Accounting Manager controls the Job Manager, and the accounting mechanism can be described as follows: After a user submits a job request to resource, the job request is intercepted by the Accounting Manager, which then queries the Bank Service to check the credit balance. If the credit balance is allowed for a job run within the grid, the Accounting Manager forwards the request to the Job Manager, which submit the job to the grid. After the job has been completed (or canceled), the Accounting Manager gathers information about the resources consumed by the job and records this in a UR.

11.2.3.2 Portal: Front-End of a Grid Accounting System

The grid accounting system usually involves users checking their credit balance, resources used, bills, and so on. For example, a user may need to specify which account should be charged for resource utilization, to obtain the usage rates associated with a particular resource, to view the currently available quota (the account balance), to collect usage information from previous service interactions, and so on. Some accounting systems even provide functionality for online transactions. Hence, an accounting system should provide a unified front end with a user-friendly interface that integrates these operations. Using a portal as a front end to the accounting system is an ideal solution.

Generally, a portal as a front end for the grid accounting system should provide the following key functionalities:

- provide users with usage rates; that is, the cost associated with using a particular resource;
- grant/deny user access based on their previous resource utilization; for example, denying service request if exceeding the allocated quota;

- track resource usage;
- charge accounts for resource consumption;
- provide resource usage information.

11.2.4 Integration of AJAX into Portal Application

A single portal page may contain more than one portlet, which supply content to the portal page. A disadvantage of portal is that portal page refresh is an expensive operation as one portlet refresh can result in other portlets refreshing at the same time. In order to tackle this problem, one solution is to incorporate AJAX (Asynchronous JavaScript and XML) technology into the portlet. AJAX is a Web development technique for creating interactive Web applications, and is one of the major enabling techniques for Web 2.0 [25]. AJAX can improve the user experience and make Web pages feel more responsive by exchanging small amounts of data with the server behind the scene, so that the entire Web page does not have to be reloaded each time when a user makes a request. This is meant to increase the Web page's interactivity, speed, and usability.

AJAX is a combination of techniques such as JavaScript, DOM, XML, and HTML/DHTML, which allows a Web browser to update parts of a Web page asynchronously by communicating with a Web server using JavaScript through an *XMLHttpRequest* component. By using AJAX, only the data or the content is transferred over the networks where the data/content is marshaled in XML format. The HTML structure and frame are created locally within the Web browser using JavaScript. As each time only the data are delivered rather than the whole Web page, users do not have to reload the whole page to get a different dataset.

There are a number of open-source AJAX toolkits available such as Direct Web Remoting (DWR) [1], Dojo [27], Google Web Toolkit [28], Microsoft Atlas [29], Open Rico [30], Yahoo AJAX Library [31], and Zimbra's Kabuki AJAX Toolkit [32]. These tools offer a number of extremely useful user interface widgets and background tools for simplifying the process of building an AJAX application. However, in terms of a J2EE environment, most documentation and tutorials of these toolkits focus on building AJAX applications using servlets rather than portlets. Therefore, using these toolkits in portlets may cause some unexpected hassles as the differences between servlets and portlets may result in similar codes that are based on these toolkits or libraries working fine with servlets but not working properly with portlets. In order to facilitate the use of AJAX in a portlet application without depending on any AJAX toolkits/libraries to avoid any unpredicted hassles, a development model for the use of AJAX in JSR-168 portlets has been proposed [33], as shown in Figure 11.1. This model gives a clear view of how AJAX works, which is not always explicitly clear in AJAX toolkits, and makes

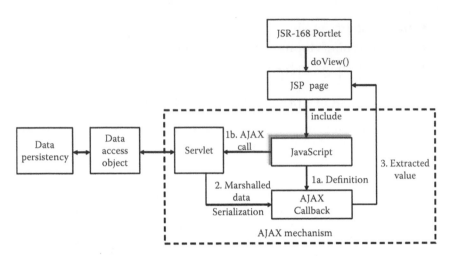

FIGURE 11.1   Development model of using AJAX in portal application.

it easy for new AJAX developers to start. The model indicates that in order to incorporate AJAX into the portlet the action request, which should be handled by the *processAction( )* method in JSR-168 portlet, is now handled by a servlet. The development model is illustrated as follows:

1. Create a JSP page that is rendered through the doView( ) method of *a* JSR-168 portlet.
2. Write AJAX code in JavaScript, and include this JavaScript in the JSP page. The procedures of writing AJAX code within JavaScript is summarized as follows:
   (a) initialize the XMLHttpRequest;
   (b) define a call-back function that processes the server-side response;
   (c) make AJAX call. This actually makes an HTTP request to the server by calling the open( ) and send( ) methods of the XMLHttpRequest object.
3. Develop a server-side component (i.e., the servlet) to handle the request from the AJAX call. Usually this involves the operation of retrieving data from a back-end database. The retrieved data are marshaled in XML and sent back to the defined AJAX call-back function.
4. Parse the XML within the call-back function using DOM and DHTML technologies.

However, using this model for AJAX-based portlet development requires a lot of client-side JavaScript coding, especially using DOM to interact with DHTML, which can be quite tedious and may cause JavaScript cross-browser problems. Hence, for experienced AJAX developers, the integration of AJAX into the portal can partially or wholly use some third-party AJAX toolkits or libraries. For example, it has been demonstrated that Direct Web Remoting (DWR) can be used in AJAX-based portlet development although it involves extra configuration and usage learning.

11.2.5 Survey of Tools and Technologies for Grid-Enabled Portal Development

Currently, there are a variety of development tools, frameworks and components that can support grid portal development. In order to provide a guideline for grid portal developers to choose an appropriate toolkit, a survey on major grid portal development tools and technologies that can be employed to facilitate grid portal development has been conducted.

11.2.5.1 GridSphere

The GridSphere Portal Framework [2] is developed as a key part of the European project GridLab [35]. The development of the GridSphere has combined the lessons learned in the development of the ASC portal and the Grid Portal Development Kit (GPDK), which was a widely used toolkit for building nonportlet-based portals. The GridSphere provides an open-source portlet-based Web portal, and can enable developers to quickly develop and package third-party portlet Web applications that can be run and administered within the GridSphere portlet container. One of the key elements in the GridSphere is that it supports administrators and individual users to dynamically configure the content based on their requirements [8]. Another distinct feature is that the GridSphere itself provides Grid-specific portlets and APIs for grid-enabled portal development. The main disadvantage of the current version of the GridSphere (i.e., GridSphere 3) is that it does not support WSRP.

11.2.5.2 eXo Platform

The eXo platform [37] can be regarded as a portal framework and Content Management System [36]. The eXo platform 1 was more like a portal framework. The eXo platform 2 proposed a Product Line Strategy [38] as it is realized that end-user customers need to get ready to use packaged solutions instead of monolithic products. The eXo platform 2 is now a core part on which an extensive product line can be built [38]. It features that it adopts Java Server Faces and the released Java Content Repository (JCR–JSR 170) specification. The eXo platform 2 adopts JSR-168 and supports WSRP.

11.2.5.3 Liferay Portal

The Liferay portal [39] is more than just a portal container [36]. It comes with helpful features such as a content management system (CMS), WSRP, and SSO. It is open-source, 100% JSR portlet API and WSRP compliant. Liferay is suitable for enterprise portal development. Institutions and companies that adopted Liferay to create their portals include Educa-Madrid, Goodwill, Jason's Deli, Oakwood, and Walden Media [36].

11.2.5.4 Stringbeans

Stringbeans [40] is a platform for building enterprise information portals. The platform is composed of three components: (1) a portal container/server, (2) a Web services platform, and (3) a process automation engine. Currently, the portal server and Web services platform have been released. The process automation engine will be released in the near future [40]. It is JSR-168 compliant and supports WSRP. Stringbeans was used in the UK National Grid Service (NGS) portal [41].

11.2.5.5 uPortal

uPortal [42] is a framework that aims to produce a campus portal. It is now being developed by the JA-SIG (Java Special Interest Group). It is JSR-168 compliant and supports WSRP. uPortal is widely used in creating university portals (e.g., the Bristol University staff portal) due to its built-in support. It is the only open-source portal framework that supports maximum types of portals ranging from Java portals to HTML portals, text portals to XML portals [36]. However, documentation of uPortal is not considered to be good and is not up to date. Most of the tutorials for uPortal are written by students doing their theses. Documentation related to uPortal is scattered about in several places such as the uPortal Website, a confluence-based wiki, email lists, the uPortal issue management system (using Jira), and external sources.

11.2.5.6 Pluto

Pluto is a subproject of the Apache Portal project. It is the reference implementation of the JSR 168 [1]. Pluto simply serves as a portlet container that implements the Portlet API and offers developers a working example platform to test their portlets. However, it is cumbersome to execute and test within the portlet container without a driver (i.e., portal). Pluto's simple portal component is built only on the requirements of the portlet container and the JSR 168 specification.

11.2.5.7 Jetspeed

Jetspeed is another Aapche Portal project that includes Jetspeed-1 and Jetspeed-2. Jetspeed-1 "provides an open source implementation of an Enterprise Information Portal, using Java and XML" [43]. Jetspeed-2 is the next-generation enterprise portal, and offers several architectural enhancements and improvements over Jetspeed-1 [44]. Jetspeed is more sophisticated than Pluto. It is concentrated on portal itself rather than just a portlet container.

11.2.5.8 Open Grid Computing Environment Toolkit

The Open Grid Computing Environment (OGCE) project was established to foster collaborations and sharable components with portal developers [46]. The OGCE portal toolkit consists of a core set of grid portlets that are JSR 168 compatible. Currently, OGCE 2.2 supports GridSphere and uPortal portlet containers. The portlets provided by OGCE include (1) grid credential proxy management portlets, (2) portlets for managing remote files through GridFTP and other file management systems, (3) portlets for running jobs through Globus Toolkit versions 2.4 and 4.0 (both Web service and pre-Web service versions), (4) storage resource broker (SRB) portlets, (5) grid portals information repository (GPIR) portlets, and (6) Condor job submission portlets.

11.2.5.9 GridPort Toolkit

The GridPort toolkit [1] enables the rapid development of highly functional grid portals. It comprises a set of portlet interfaces and services that provide access to a wide range of backend grid and information services. GridPort 4.0.1 was developed by employing GridSphere. GridPort 4.0.1 is the last release as the GridPort team has decided to shift the focus from developing a portal toolkit to developing production portals [1]. The GridPort toolkit has recently become part of the OGCE toolkit.

11.2.5.10 IBM WebSphere Portal

IBM's WebSphere portal [47] is a framework that includes a runtime server, services, tools, and many other features that can help integrate an enterprise into a single, customizable interface portal. It implements the JSR 168 Portlet API and WSRP [47]. The WebSphere portal is a powerful tool and is widely used in many business companies and enterprises. It provides a composite application or business mashup framework and the advanced tooling needed to build flexible, SOA-based solutions. IBM also offers a scheme, namely the IBM Academic Initiative [49], for academic and research institutes. Members of the IBM Academic Initiative can have the latest technology and majority of IBM software to use for free, which includes the IBM WebSphere portal.

11.2.5.11 A Comparison Matrix of Grid Portal Tools and Technologies

Having surveyed the major grid portal development tools, frameworks, and components, a comparison matrix table is produced as shown in Table 11.1. The evaluation criteria include: (1) JSR-168 compliant, (2) WSRP compliant, (3) provision of grid-specific portlets, (4) open source, (5) usage status, (6) performance, and (7) supported services. The table has revealed that

1. GridSphere has been widely used for grid-enabled portal development. GridSphere itself provides grid-specific portlets. The main disadvantage of the current version is that it does not support WSRP.

2. Other open-source portal frameworks and portlet containers, such as uPortal, liferay, eXo, and Jetspeed-2, are also appropriate for grid-enabled portal development. Although they do not directly provide grid-specific portlets, the existing open-source JSR-168 grid portlets (e.g., OGCE, GridPort) can be reused or new grid portlets need to be developed.

3. For developing commercial grid portals, the toolkit of the IBM WebSphere portal would be a good choice. Likewise, although the WebSphere portal itself does not provide grid-specific portlets, the existing JSR-168 grid portlets can be reused or new grid portlets need to be developed. Under the IBM Academic Initiative, academic and institutional researchers can use the IBM WebSphere portal for free.

4. OGCE provides a set of grid-enabled portlets that can be used to create grid portals. GridPort toolkit has now become part of OGCE.

11.3 Workflow to Integrate Services for Process Automation

Grid-based computation usually requires defining sequences of activities such as data gathering, metascheduling, job submission, file transfer, analysis and simulation, and data harvesting. This determines a need for process automation, which can integrate the activities required for grid-based computation without human interaction. Workflow technology

TABLE 11.1

A Comparison Matrix of Grid Portal Tools and Technologies

	JSR-168 Compliance	WSRP Compliance	Grid-Specific Portlets	Open Source	Usage Status	Performance	Supported Services	Total
GridSphere 3	5	0	5	5	4	4	4	27
GridPort 4.01	5	0	5	5	3	3	3	24
Liferay 4.3	5	5	0	5	3.5	3.5	3.5	25.5
eXo2	5	5	0	5	3	3.5	3.5	25
Stringbeans 3.2	5	5	0	5	2.5	3.5	3.5	24.5
uPortal 2.6	5	5	0	5	3	3.5	3.5	25
OGCE 3.2	5	0	5	5	2.5	3	3	23.5
Pluto 3.2	5	5	0	5	2	2.5	2.5	22
JetSpeed 2	5	5	0	5	2.5	3	3	23.5
IBM WebSphere Portal 6.0	5	5	0	4	5	5	5	29

Source: Adapted from X. Yang, M. Dove, M. Hayes, M. Calleja, L. He, and P. Murray-Rust, "Survey of tools and technologies for grid-enabled portals," in *Proceedings of the UK e-Science All Hands On Conference*, 2006.

can be used for the automation of a process where documents, information, or tasks are passed from one participant to another to be processed, according to a set of procedural rules. In the context of service-oriented architecture, service-based approach is often employed to build grid workflow where legacy code, business logic, and grid-computation- related tasks are wrapped or deployed as Web services that can then be integrated into workflow process. In this section, the basics of workflow technology and workflow used in the grid environment are introduced. The start-of-the-art service component architecture (SCA) that can be used to create a workflow is discussed.

11.3.1 Introduction to Workflow Technology

The workflow technology will be introduced from the following aspects: (1) service-oriented workflow, (2) workflow language, (3) workflow engine, and (4) workflow system.

11.3.1.1 Service-Oriented Workflow

According to the Workflow Management Coalition (WfMC), an international organization for workflow vendors, users, and research, the workflow can be defined as "the automation of a business process, in whole or parts, where documents, information or tasks are passed from one participant to another to be processed, according to a set of procedural rules" [49]. This definition can be extended to scientific workflow, which is a set of components and relations between them used to define a complex process from simple building blocks [50]. Relations may be in the form of data links, transferring information from the output of one component to the input of another, or in the form of control links that state some conditions on the execution of a component [50]. The component is a reusable building block that performs some well-defined function within a process, and may consume information and may produce output to provide information and knowledge. These components can be implemented as Web services and hosted by computing resources where external users can access.

Two approaches to compose the workflow have been identified: (1) job-based approach, and (2) service-based approach [51,53]. In the job-based approach, the workflow manager is responsible for the actual processing of data by programs on physical resources. As this approach is much close to the grid infrastructure, it allows the optimization of submission rate, the dispatch rate, the scheduling rate, and so on [51, 54]. For example, Condor DAGMan [54] is a workflow manager that adopts this job-based approach where a child node will not start until its parents have successfully completed. In the service-based approach, the workflow manager is responsible for the transmission of data to remote services and for the collection of the results. The workflow manager is only aware of the interface of the programs and does not access the actual binary files. Jobs are submitted to the grid by the services. The WSDL is used to describe the services. In the service-oriented approach, the workflow description stipulates that a particular output of a program is linked to a particular input of another. Hence, the iteration of such workflows on a number of input datasets is straightforward and does not require and rewriting of the workflow [51]. In the job-based approach, data dependencies between programs are not explicitly described. Iterating a single workflow on many datasets requires writing specific jobs. The service-based approach is more independent from the infrastructure than the job-based one. Services themselves are responsible for the submission of jobs but in the job-based approach it is the workflow manager that submits jobs to the grid.

11.3.1.2 Workflow Languages

The workflow language is a language that specifies the rules for connecting components to produce workflows. It uses XML notation to represent the intercomponent dependencies. In the workflow, each component has an input and outputs. The input receives tokens from a predecessor component, and the outputs send tokens to successor components. Receipt of a token from a predecessor triggers the component to perform its task. When the tasks completed, that component generates a token at each output, along with a result from the task. The workflow language can be used to coordinate the operation of the component within the workflow.

There are many existing standards for workflow such as XLANG, WSFL, BREL, ebXML, XPDL, and so on. Among these workflow languages, business process execution language (BPEL) is a leading candidate for the standard workflow language for business-oriented Web services. BPEL replaced IBM's WSFL and Microsoft's XLANG languages and is currently in the process of standardization in OASIS [55]. It is a combination of graph-oriented (WSFL) and procedural (XLANG) workflow languages and provides a convenient language to express the majority of workflow processes.

BPEL has become a very attractive workflow language to use in grid environments due to its strong support for Web services [55]. A BPEL workflow is designed to present as a Web service. It can easily interact with existing Web services by combining them into a new service. The BPEL workflow engine can receive messages, determine their destination, and send messages to other Web services depending on the workflow's execution state. BPEL uses the WS-Addressing Endpoint References to simplify passing addresses of Web services participating in a workflow. One very important advantage of using BPEL in grids is that it can easily and seamlessly interact with standard Web services (that may not be part of grids) and grid services provided by OGSI, WSRF, and so on.

11.3.1.3 Workflow Engines

A workflow typically needs to be enacted by a workflow engine. A workflow engine is defined by WfMC as "A software service or 'engine' that provides the runtime execution environment for a process instance." It coordinates the invocation of components within workflows, and manages the entire invocation process including progress reporting, data transfer between components and any other housekeeping required. As the BPEL is now a *de facto* standard workflow language, currently there are many workflow engines that support BPEL standard; for example, Oracle BPEL Process Manager, Microsoft BiTalk, IBM WebSphere Business Integration Server Foundation, IBM BPWS4J, ActiveBPEL Engine and ActiveWebflow, and OpenStorm Service Orchestrator.

11.3.1.4 Workflow Systems

A workflow system is defined by WfMC formally as "a system that defines, creates and manages the execution of workflows through the use of software, running on one or more workflow engines, which is able to interpret the process definition, interact with workflow participants and, where required, invoke the use of IT tools and applications." The widely used, open-source, and service-based workflow management systems are Kepler, Taverna, and Triana.

The Kepler system [1] focuses on many application areas from gene promoter to mineral classification. It can orchestrate standard Web services linked with both data and control dependencies, and implement various execution strategies. To support the interaction with Web services, Kepler uses a form of actor proxy for each Web service invoked. In addition, they have created a set of grid actors for doing grid FTP management and Globus GRAM execution [1]. Triana [57], from the GridLab project, provides composition and a large toolbox of ready-to-use components. Triana implements parallel and peer-to-peer policies and has been applied to various scientific fields. For grid applications, Triana provides a mechanism to move subgraph to remote grid resources for execution. It also serves as a programming environment, providing a user interface to enable the composition of scientific workflow applications. The Taverna from myGrid project [1] targets bioinformatics applications and is able to enact Web services and other components. Taverna creates a workflow based on service composition and provides over 1000 services that can be used as components in workflows [1].

11.3.2 Workflows in Grid Computing

Workflows in grid computing have some new issues [59]. First, on computational grids, workflow applications are not only from a business process but also from scientific and engineering computing; for example, bioinformatics, high energy physics, and astronomy. This requires the grid workflow systems to support new types of application. Furthermore, multiple grid middleware needs to be integrated into the workflow system, which is supposed to run on heterogeneous grid infrastructures. The research challenges and issues of grid workflow that have emerged in comparison with traditional business workflow environment are discussed as follows [59]:

- Workflow applications running on computational grids usually manipulate large amounts of data, which are of various types and from heterogeneous data sources. The grid workflow should support the efficient transfer of large amounts of data and resource discovery for heterogeneous data sources.

- Lifecycles of grid workflow applications may have a wide range; for example, from one minute to dozens of days. This brings challenges for fault tolerance, application monitoring, and steering, for example.

- The grid infrastructure and environment, including APIs, interfaces and protocols, are different from traditional workflow environment. A number of legacy codes and applications are not suitable for running on computational grids. Usually, these legacy codes need to be wrapped as Web services before integrating into the workflow.

- Some applications [16] running on computational grids need a real-time support environment; for example, real-time resource allocation and reservation. There are few real-time requirements for traditional business workflows. The workflow management system should be designed to meet the requirements of real-time support.

- The computational grid is a highly dynamic environment; for example, CPU load, network bandwidth, and resource availability. The dynamic feature brings difficulties for resource allocation and application steering for workflow system applications.

- The computational grid covers multiple administrative domains and lacks central control and complete information, which is not needed in traditional workflow enactment.

11.3.3 Start-of-the-Art: Service Component Architecture

The service-oriented architecture (SOA) approach promotes building systems or workflows from distinct and autonomous Web services. By using SOA to create the service-oriented workflow, the autonomous business logic or legacy scientific code can be presented as a service; hence, the development and maintenance of a workflow can be simplified. However, the understanding of SOA still remains at a conceptual level to many users. In order to help realize the SOA, IBM has proposed a new programming model, namely, service component architecture (SCA). SCA is a set of specifications that describe a model for building workflows, applications, and systems using SOA [1]. It now gets supported by many companies such as Oracle, Sun Microsystems, and BEA. SCA is based on the standard Web services and provides a model for service integration and creation of service components, including the reuse of existing application functions within SCA compositions. The IBM workflow engine WebSphere Process Server, and the associated workflow development toolkit WebSphere Integration Developer use the SCA programming model.

The SCA programming model is briefly described as follows [63]:

- *Module*—the module is a container for service components. It is a deployment unit and packaged as an EAR (Enterprise ARchive).
- *Component*—the service component is a basic building block in SCA, as shown in Figure 11.2. Each component has got interface, reference, and implementation.
- *Interface*—the interface is a specification of component's operations, which contains input, output, and faults. It can be defined using Java or WSDL.
- *Reference*—when implementing a service component, another service component may need to be invoked. This is realized through reference. A reference indicates what interface the called component should ultimately have. The calling component can have one or more reference. The reference can be expressed using Java or WSDL.
- *Wire*—a wire is simply a connection that indicates which reference is connected to which component.

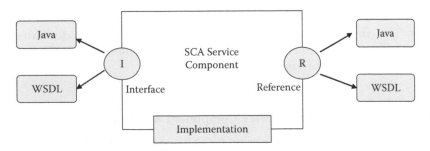

FIGURE 11.2 SCA service component model. (Adapted from S. Penugonda and P.K. Dash, "SCA application development, Part 1: An overview of service component architecture," *IBM developerWorks*, August, 2006.)

- *Import and export*—the components can only be wired within the same module. If a component outside the module needs to be called, an import has to be used. An import is the same as a service component, but without implementation. Instead, the import uses binding to define how to connect to the actual service. The binding usually includes (1) SCA binding, (2) Java message service (JMS) binding, (3) stateless session bean, and (4) Web service. The export is used to expose any module's components as services externally or to other modules.

- *Implementation*—the implementation type of a service component includes Java, BPEL, state machine, business rules, human tasks, selector, and interface map.

11.4 A Case Study within the UK e-Science Program

A case study within the UK e-Science program, namely MaterialsGrid [64], will be introduced, and how portal and workflow technologies are employed in the MaterialsGrid is briefly discussed in this section.

11.4.1 UK e-Science Program

The term e-Science is used to describe computationally intensive science that is carried out in highly distributed network environments, or science that uses immense datasets that require grid computing. According to John Taylor, a Director of the UK Government's Office of Science and Technology who initially created this term, "e-Science is about global collaboration in key areas of science, and the next generation computing infrastructure will enable it. e-Science will change the dynamic of way the science is undertaken" [65]. Another major aim of e-Science is to bring together researchers from science and IT/computing backgrounds, allowing them to use each other's experience and skills. Examples of this kind of science include particle physics, earth sciences, and bioinformatics. e-Science is different from traditional science and computer science. In e-Science, grid computing is one of the major enabling technologies used to develop an infrastructure where global research collaboration can be conducted.

e-Science in the United Kingdom has received significant funding from government and research councils. The main UK government support for grid computing has been through its £230 million e-Science program 0. The UK e-Science program comprises a wide range of resources, centers, and people. Two of the largest UK science grids are GridPP [67] and the National Grid Service (NGS) [68]. GridPP has £65 million funding from the Science and Technology Facilities Council to run a UK grid for particle physics. Its 17 sites and 5000 computers across the United Kingdom are part of an even larger international grid that will analyze data from Europe's next particle accelerator, the Large Hadron Collider [66]. NGS aims to provide the United Kingdom researchers with computation and data resources and facilities to help carry out their research, regardless of actual locations of resources and researchers. It currently pools resources from nine locations across the United Kingdom. Examples of the UK e-Science project in molecular simulation include e-Minerals [69] and the MaterialsGrid [70]. The vision of the e-Minerals project aims to combine developments in atomistic simulation tools with emerging grid computing technologies in order to stretch the potential for undertaking simulation studies under increasingly realistic conditions, and which can scan across a wide range of physical and chemical parameters [69]. The aim of the MaterialsGrid project is to create a pilot dynamic database of materials properties (such as elastic stiffness, dielectric constants, optical properties, heat capacity, and electronic band gap) based on quantum mechanical simulations run within Grid computing environments [70]. More about MaterialsGrid will be discussed in the next two sections.

11.4.2 MaterialsGrid: Large-Scale Simulation of Physical Properties of Materials

MaterialsGrid is a UK government DTI-funded e-Science research project ($5.4 million) that aims to create a unique database of critical electronic and physical materials properties based on highly accurate quantum mechanical simulation methods using the power of grid computing. Scientists can search the database, retrieve precomputed properties where available, or trigger new calculations that fill in missing data. Users will be able to donate data and grid resources in exchange for access, ensuring the continued growth of the service.

Running a quantum mechanical simulation for material properties (e.g., CASTEP, SIESTA) on the grid typically involves the following steps: (1) creating the simulation input files, (2) copying the input files and simulation code to remote computational resources, (3) logging into the computational resource and submitting the simulation job, (4) waiting for the job to finish, and once the job finishes copying back the simulation output files to the local machines, and (5) harvesting the data or metadata and storing them to the material property database. This approach works successfully but it does have disadvantages: first, the approach involves a lot of human interaction; second, in order to submit job(s) to remote resources, some grid software has to be installed (e.g., Globus) in a local machine, or logging into a machine where such grid software has been installed is necessary; third, it will be

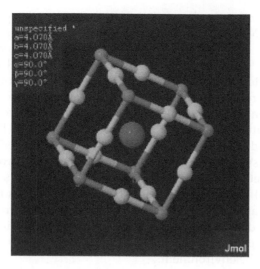

FIGURE 11.3 Integration of visualizing cell structure in Jmol in a portal.

ideal if we can have an integrated environment where operations such as setting-up, submitting, managing, and tracking of grid-enabled simulation jobs can be carried out.

In order to address these problems, a workflow-based reference architecture for running quantum mechanical simulations for material properties over grids is proposed. There are mainly four parts in the architecture: (1) a grid portal that can initiate a quantum mechanical simulation run in grid from just a Web browser without any grid software downloaded and installed, and provide a unified interface for end users to access distributed resources through SSO; (2) a workflow system that can facilitate the modeling of the workflow and automate the simulation process without human interaction; (3) Web services that wrap various scientific code and operations required for running a simulation such as job submission/monitoring, property calculation, and so on; and (4) data repositories that include a file system, and a material property database.

11.4.3 MaterialsGrid Portal and Workflow System

A prototype platform for running quantum simulation for material properties over the Grids has been developed. First, the IBM WebSphere Portal framework has been employed for building the grid portal [3,47] and AJAX technology has been employed to access the remote database for material properties. The portal provides an integrated platform for the material property query, creation of simulation input files in various approach (e.g., by uploading existing files, by the user providing data, by using the CML dictionary), access to the WebDav-based file system, visualizing the cell structure using Jmol, enacting the workflow, and monitoring the workflow. Second, the PipelinePilot [71] from Accelrys has been adopted as a workflow system, and a workflow that orchestrates services required for running a simulation in grid has been created. An associated monitoring system has also been developed that can monitor the status of both workflow and simulation job. Third, Web services integrated into the workflow include a metascheduling service, grid job submission and monitoring service, material property calculation service, metadata collection service and notification service. Fourth, the Oracle Real Application Clusters has been used as a material property database, and a WebDav-based file system has been used to stage data and provide an archive of all created data files. Finally, chemical markup language (CML) [73] has been incorporated in running the simulation within the environment.

The CASTEP [75] quantum mechanical simulation has been used to test the prototype. The simulation job is submitted to the NW-Grid [74] or the NGS Grid, and the whole job submission and simulation process are automated by the workflow. It has been demonstrated that the platform provides an integrated environment where the grid-based CASTEP simulation job can be initiated from a Web browser and submitted to any grid, the simulation output can be staged out to the file system, and the calculated material properties from the simulation can be stored into the database. All these operations are orchestrated through the workflow without any human interactions.

11.5 Summary

This chapter introduces the portal and workflow technologies that can be employed in grid computing for application integration and service integration. The following topics in relation to grid-enabled portal/portal are discussed: (1) the grid-enabled portal can take advantage of SSO, one inherent feature of portals, to provide an integrated access platform to resources and applications; (2) the portals is an ideal front end in grid accounting systems; (3) a development model of using AJAX in portal application is proposed to help address the issue that one portlet refresh can result in other portlets refreshing at the same time; (4) in order to provide a guideline for grid portal developers to choose an appropriate toolkit, a survey on major grid portal development tools and technologies that can be employed to facilitate grid portal development has been conducted.

In the section of service integration by workflow, the following topics are covered: (1) workflow technologies were introduced, which include service-oriented workflow, workflow languages, workflow engines, and workflow systems; (2) features of scientific workflow in grid computing were identified; and (3) the state-of-art service component architecture (SCA) that can be employed to build a service-oriented workflow was described.

In the case study, the literature related to the UK e-Science program was reviewed, and a MaterialsGrid project that employs both portal and workflow technologies to provide an integrated platform and to automate the operations required for running CASTEP quantum mechanical simulation over grids was introduced.

Acknowledgments

The authors would like to acknowledge the UK government DTI-funded MaterialsGrid project, and contributions from MaterialsGrid partners.

References

1. J. Novotny, "The Grid Portal Development Kit," *Concurrency and Computation: Practice and Experience*, 14 (13–15), 1129–1144, 2002.
2. GridSphere. Available at: http://www.gridsphere.org.
3. X. Yang, M. Dove, M. Hayes, M. Calleja, L. He, and P. Murray-Rust, "Survey of tools and technologies for grid-enabled portals," in *Proceedings of the UK e-Science All Hands On Conference*, pp. 353–356, Nottingham, UK, 2006.
4. M. Russell, G. Allen, G., Daues, I, Foster, E. Seidel, J. Novotny, J. Shalf, and G. von Laszewski, "The astrophysics simulation collaboratory portal: A science portal enabling community software development," *Cluster Computing*, 5 (2), 297–304, 2002.
5. UNICORE. Available at: http://unicore.sourceforge.net.
6. M. Romberg, "The UNICORE architecture: Seamless access to distributed resources," in *Proceedings of the Eighth IEEE International Symposium on High performance Distributed Computing*, pp. 287–293, Redondo Beach, CA, 1999.
7. Novotny, "Developing grid portlets using the GridSphere portal framework," IBM developerWorks: IBM Resource for Developers and IT professionals, 2004. Available: http://www.ibm.com/developerworks/java/.
8. I. Kelley, O. Wehrens, M. Russel, and J. Novotny. "The Cactus Portal," in *Proceedings of Third International Conference on Asian and Pacific Coasts (APAC) 2005: Advanced Computing, Grid Applications and eResearch*, 4–8 September, Seogwipo KAL Hotel, Jeju, Korea, 2005.

9. GridPort. Available at: http://gridport.net/main.
10. J. Boisseau, S. Mock, and M. Thomas, "Development of Web toolkits for computational science portals: The NPACI HotPage," in *Proceeding of the Ninth IEEE International Symposium on High Performance Distributed Computing*, pp. 308–309, Pittsburgh, PA, 2000.
11. JSR 168 and JSR 286. Available at: http://developers.sun.com/portalserver/reference/techart/jsr168/.
12 WSRP. Available at: http://www.oasis-open.org/committees/download.php/18617/wsrp-2.0-spec-pr-01.html.
13. WSRP4J. Available at: http://portals.apache.org/wsrp4j.
14. R. Butler Von Welch, D. Engbert, I. Foster, S. Tuecke, J. Volmer, and C. Kesselman, "A national-scale authentication infrastructure," *Computer*, 33 (12), 60–66, 2000.
15. Globus Grid Security Infrastructure. Available at: http://www.globus.org/toolkit/docs/4.0/security/key-index.html.
16. Globus. Available at: http://www.globus.org.
17. MyProxy. Available at: http://Grid.ncsa.uiuc.edu/myproxy.
18. MyProxy and Grid Portals. Available at: http://Grid.ncsa.uiuc.edu/myproxy/portals.html.
19. MyProxy Upoad Tool. Available at: http://tiber.dl.ac.uk:8080/myproxy.
20. Resource Usage Service. Available at: http://www.ogf.org/Public_Comment_Docs/Documents/Oct-2005/draft-ggf-wsi-rus-15.pdf.
21. GESA. Available at: http://www.doc.ic.ac.uk/~sjn5/GGF/CompEconArch-GGF7.pdf.
22. Available at: DGAS, http://www.to.infn.it/Grid/accounting/main.html.
23. EGEE. Available at: http://www.eu-egee.org.
24. A. Barmouta and R. Buyya, "GridBank: A Grid Accounting Services Architecture (GASA) for distributed systems sharing and integration," in *Proceedings of the IEEE International Parallel and Distributed Processing Symposium*, Nice, France, 2003.
25. P. Gardfjäll, E. Elmroth, L. Johnsson, O. Mulmo, and T. Sandholm. "Scalable grid-wide capacity allocation with the SweGrid Accounting System (SGAS)," *Concurrency and Computation: Practice and Experience*, 20 (18), 2089–2122, 2008.
26. T. O'Reilly, "What is Web2.0." Available at: http://www.oreillynet.com/pub/a/oreilly/tim/news/2005/09/30/what-is-Web-20.html?page=1.
27. DWR. Available at: http://getahead.org/dwr.
28. Dojo. Available at: http://dojotoolkit.org.
29. Google Web Toolkit. Available at: http://code.google.com/Webtoolkit.
30. MicroSoft Atlas. Available at: http://ajax.asp.net/Default.aspx.
31. Open Rico. Available at: http://www.openrico.org.
32. Yahoo Ajax Library. Available at: http://developer.yahoo.com/yui.
33. Zimbra's Kabuki AJAX Toolkit. Available at: http://www.zimbra.com/community/kabuki_ajax_toolkit_download.html.
34. X. Yang, T.V. Mortimer-Jones, D.J. Wilson, M.T. Dove, and L. Blanshard, "Integration of AJAX into MaterialsGrid portal for material properties query," in *Proceedings of the UK e-Science All Hands On Conference*, pp. 599–606, Nottingham, UK, 2007.
35. GridLab. Available at: http://www.Gridlab.org.

36. A. Akram, D. Chohan, X. Wang, X. Yang, and R. Allan "A service oriented architecture for portals using portlets," in *Proceedings of the UK e-Science All Hands On Conference, All Hands On Meeting,* 2005.

37. The eXo platform. Available at: http://www.exoplatform.com.

38. B. Mestrallet, T. Nguyen, G. Azarenkov, F. Moron, and B. Revenant, "eXo platform v2, portal, JCR, ECM, groupware and business intelligence." Available at: http://www.theserverside.com/articles/article.tss?l=eXoPlatform.

39. Liferay. Available at: http://www.liferay.com.

40. StringBeans. Available at: http://www.nabh.com/projects/sbportal.

41. X. Yang, D. Chohan, X. Wang, and R. Allan, "A Web portal for National Grid Service," presented at the GridSphere and Portlets Workshop, March 3, eScience Institute, Edinburgh, 2006.

42. uPortal project. Available at: http://www.uportal.org.

43. Pluto. Available at: http://portals.apache.org/pluto.

44. Jetspeed-1. Available at: http://portals.apache.org/jetspeed-1.

45. Jetspeed-2. Available at: http://portals.apache.org/jetspeed-2.

46. OGCE. Available at: http://www.ogce.org.

47. GridPort. Available at: http://gridport.net/main.

48. IBM WebSphere Portal. Available at: http://www-306.ibm.com/software/genservers/portal.

49. IBM Academic Initiative. Available at: http://www-304.ibm.com/jct09002c/university/scholars.

50. Workflow Management Coalition (WfMC). Available at: http://wfmc.org.

51. Taverna. Available at: http://taverna.sourceforge.net.

52. T. Glatard, J, Montagnat, and X. Pennec, *An optimised workflow enactor for data-intensive Grid applications,* Project RAINBOW, Rapport de recherché, ISRN I3S/RR-2005-32-FR, 2005. Available: http://www.i3s.unice.fr/~mh/RR/2005/RR-05.32-T.GLATARD.pdf.

53. J. Yu and R. Buyya, "A taxonomy of workflow management system for grid computing," Technical Report, Grid and Distributed Systems Laboratory, University of Melbourne, 2005.

54. G. Singh, C. Kesselman, and E. Deelman, "Optimizing grid-based workflow execution," *Journal of Grid Computing,* 3 (3–4), 201–219, 2005.

55. Condor DAGMan. Available at: http://www.bo.infn.it/calcolo/condor/dagman.

56. A. Slomiski, "On using BPEL extensibility to implement OGSI and WSRF grid workflows," *Concurrency and Computation: Practice and Expereince,* 18 (10), 1229–1241, 2006.

57. B. Ludäscher, I. Altintas, C. Berkley, D. Higgins, E. Jaeger, M. Jones, E.A. Lee, J. Tao, and Y. Zhao, "Scientific workflow management and the Kepler system," *Concurrency and Computation: Practice and Experience,* 18 (10), 1039–1065, 2006.

58. D. Churches, G. Gombas, A. Harrison, J. Maassen, C. Robinson, M. Shields, I. Taylor, and I. Wang, "Programming scientific and distributed workflow with Triana services," *Concurrency and Computation: Practice and Experience,* 18 (10), 1021–1037, 2006.

59. T. Oinn et al., "Taverna: Lessons in creating a workflow environment for life sciences," *Concurrency and Computation: Practice and Experience,* 18 (10), 1067–1100, 2006.

60. G. Papakonstantinou, M.P. Bekakos, G.A. Gravvanis, and H.R. Arabnia, *Grid Technologies: Emerging from Distributed Architectures to Virtual Organizations.* WIT Press, 2006.
61 H. Zhu, K. Chan, L. Wang, W. Cai, and S. See, "A prototype of distributed molecular visualization on computational grids," *Future Generation Computing System*, 20 (5), 727–737, 2004.
62. S. Penugonda and P.K. Dash, "SCA application development, Part 1: An overview of service component architecture," IBM developerWorks: IBM Resource for Developers and IT professionals, August, 2006. Available at: http://www.ibm.com/developerworks/java/.
63. R. Peterson, "Get started with WebSphere integration developer," *IBM WebSphere Developer Technical Journal*, December, 2005. Available at: http://www.ibm.com/developerworks/websphere/techjournal/0602_gregory/0602_gregory.html.
64. G. Adams, R. Gregory, J. Fung, and R. Giffen, "A guided tour of WebSphere integration developer," *IBM WebSphere Developer Technical Journal*, March, 2006. Available at: http://www.ibm.com/developerworks/websphere/techjournal/0602_gregory/0602_gregory.html.
65. MaterialsGrid. Available at: http://www.materialsGrid.org.
66. eScience Program. Available at: http://www.rcuk.ac.uk/escience/default.htm.
67. Grids and eScience. Available at: http://www.parliament.uk/documents/upload/postpn286.pdf.
68. GridPP. Available at: http://www.Gridpp.ac.uk.
69. National Grid Service (NGS). Available at: http://www.Grid-support.ac.uk.
70. eMinerals. Available at: http://www.eminerals.org.
71. MaterialsGrid. Available at: http://www.materialsGrid.org.
72. Pipeline Pilot. Available at: http://www.scitegic.com/products/overview/index.html.
73. P. Murray-Rust, C. Leach, H.S. Rzepa, and S. Henry, "Chemical markup language," in *Book of Abstracts, 210th ACS National Meeting*, Chicago, August 20–24, 1995 (Pt. 1, COMP-040).
74. M.D. Segall, P.J.D. Lindan, M.J. Probert, C.J. Pickard, P.J. Hasnip, S.J. Clark, and M.C. Payne, "First-principles simulation: Ideas, illustrations and the CASTEP code," *Journal of Physical Condensed Matter* 14, 2717–2744, 2002.
75. NW-Grid. Available at: http://www.nw-Grid.ac.uk.

12

Grid Security

Richard O. Sinnott

CONTENTS

12.1 Introduction

Security is essential for interorganizational collaborative e-Research. Without robust, reliable, the easy to understand and manage e-Research security models and their implementations, many communities and wider industry will simply not engage. To support interorganizational, interdisciplinary research it is essential that e-Research security infrastructures support several key (defining) characteristics.

- They should be easy to use by end users, who should not have to learn complex new systems or adopt technological solutions that are not aligned with the fundamental reasons for engaging in e-Research—namely, to undertake research.

- They should have single sign-on to distributed resources; that is, once authenticated (and/or authorized) they are able to access and use a range of distributed resources without the need for further authentication.

- Sites should be able to allow or deny access to their resources for given collaborators at their own discretion; that is, they should be autonomous, and tools support should be available to help support this process.

- Systems should scale to potentially support establishment and management of very-large-scale virtual organizations (VO) involving many collaborators from numerous institutions with different privileges.

- Security infrastructures should incorporate or at least reflect existing (legacy) security infrastructures and policies of collaborators to ensure that e-Research-specific security infrastructures do not violate/weaken existing policies on access and usage of resources.

- Given the e-Research vision to support models where new collaborations can be rapidly supported, or where new institutions/users can be added, removed, or have their privileges changed "on the fly" to existing collaborations, security infrastructures should be able to support such dynamic scenarios.

It is clear that no single security model or policy will suffice for all e-Research collaborations. Different domains have their own requirements on access to and use of resources by collaborators. However, it is essential that best practice in supporting collaborations is achieved. It is the case that currently the predominant way in which security is currently addressed in the grid community is through public key infrastructures (PKIs) [1] to support authentication. While PKIs address some user identity issues, authentication in itself does not provide the fine-grained control over what users are allowed to do on remote resources (authorization). Furthermore, mainstream models of PKIs using centralized certificate authorities for authentication have their own associated problems. Instead, what are required are finer-grained security infrastructures that support e-Research and are aligned with the ways in which the e-Research community are comfortable with working. E-Research is—or should be!—about *research* and not learning about the nuances and/or complexities of different infrastructures, middleware or the associated mechanisms they require in place; for example, X.509 digital certificates [2].

In this chapter, we present an overview of PKIs and their limitations, and highlight how recent work in UK academia allows user-oriented security models that are aligned with access to Internet resources more generally through the UK Access Management Federation based upon the Internet2 Shibboleth technologies [3]. We present a variety of case studies showing how the world of e-Research and non-grid-based access to and usage of secure Internet resources can be aligned. Key to the success of this is shielding users from the underlying middleware, certificates, and features that are not directly supporting their primary reason for engaging in e-Research to begin with. Demonstrations of these solutions across a variety of application domains are given to offer both a snapshot of state-of-the-art security systems and a vision for an integrated and secure grid system of the future.

12.2 Authentication and Grid Systems

Fundamentally, grids are about sharing resources. With this in mind, it is essential that security is ensured, both of the underlying systems and of the grid infrastructures and applications running on top of them. This is especially the case as the grid community moves from the academic, research-oriented background to more commercial arenas, and especially when one moves toward more security-focused domains such as finance and e-Health. It is the case in computer security that the weakest link rule applies; this fact is magnified by grid infrastructures due to their collaborative openness. Highly secure multimillion pound compute facilities can be compromised by inadequately secured remote laptops. Rigorous security procedures at one site can be made redundant through inadequate procedures at another collaborating site.

This problem is due, in part, to the lack of granularity in how security is currently considered. Grid security is still primarily based on PKIs, which support the validation of the identity of a given user requesting access to a given resource—so-called *authentication*.

12.2.1 Public Key Infrastructure

Cryptography is one of the main tools available to support secure infrastructures. Using cryptographic technology, confidentiality can be established by encrypting and decrypting messages and their contents. Encryption and decryption are done using keys. When these keys are the same, this is called symmetric-key cryptography.

Public-key cryptography uses different keys: private and public keys. Messages encrypted with a public key can only be read by an individual who possesses the private key. Any user can direct a message to a known destination, knowing that it cannot be read by anyone else, simply by encrypting it using the public key of that destination. The owner of the private key can encrypt messages with that key, and the receiver of the message can be sure that it was sent by the owner of the private key. Both public key agreement and public key transport need to know who the remote public key belongs to; that is, who has associated private key. The public key certificate is the mechanism used for connecting the public key to the user with the corresponding private key. Public key certificates include a distinguished name (DN) that can be used for identifying a given user.

A PKI is responsible for deciding policy, managing, and enforcing certificate validity checks. The central component of a PKI is a certification authority (CA). A CA is a root of trust that holders of public and private keys agree upon. CAs have numerous responsibilities including issuing of certificates, often requiring delegation to a local registration authority (RA) used to prove the identity of users requesting certificates. CAs are also required, among other things, to revoke older or compromised certificates through issuing certificate revocation lists (CRL). A CA must have well-documented processes and practices that must be followed to ensure identity management. Various PKI architectures are possible, the selection of which depends upon numerous factors. Are numerous CAs to be trusted? How important is it to be able to add new CAs? What kind of trust relationships exist between CAs?

The simplest PKI involves a single CA that is trusted by all users and service/resource providers. This model has been chosen for UK e-Science and is based on a statically defined centralized CA with direct single hierarchy to users. Getting a certificate can be seen as the starting point in accessing and using grid-based resources such as the UK e-Science National Grid Service (NGS) [4]. The typical scenario for getting a certificate is as follows: researchers wishing to gain access to grid resources such as the NGS apply to the centralized CA at Rutherford Appleton Laboratory (RAL) [5]. The CA will then contact their local RA, who will in turn contact the user and request some form of photographic identification (such as a passport photo or a university card) to ensure that they are legitimate. Once the identity of the user has been ratified, the RA contacts the CA, who subsequently informs the user (via email) that their certificate is available for download. The user downloads the certificate and associated certificate revocation lists into their Internet browser. Once in their browser they are required to export it to forms appropriate to the grid middleware.

The main benefit and reason for the widespread acceptance of PKIs within the grid community is their support for single sign-on. Thus, since all grid sites in the UK trust the central CA at RAL, a user in possession of an X.509 certificate issued by RAL can send jobs or access resources more generally across all sites, or more precisely to all sites where a user has requested and been granted access to those sites. Typically with middleware solutions such as Globus [6], *gatekeepers* are used to ensure that signed grid requests are valid (i.e., from known collaborators). When this is so, that is, the DN of the requestor is in a locally stored and managed *grid-mapfile*, the user is typically given access to the locally set-up account as defined in the *grid-mapfile*.

12.2.2 Problems with PKIs

The above process is off-putting for many of the wider less-IT-focused research communities since it requires them to convert the certificate to appropriate formats understandable by grid middleware using complex, cryptic openSSL commands. Such requirements dissuade less IT-savvy researchers from engaging—especially as openSSL is not commonly available on platforms such as Windows. It is possible for Windows-based PC users to install openSSL-based solutions but this in turn requires them to install and configure additional software. In many cases, this is not possible; for example, if they do not have sufficient privileges on their PC (root access, etc.)—a not uncommon practice in departments and faculties at many universities in the UK. In this case the researchers will instead have to refer to a local system administrator to help with the installation and configuration.

Assuming researchers have managed to obtain a certificate that they have converted into the appropriate format, they are then expected to remember strong passwords for their private keys with the recommendation to use upper and lower case and non-alphanumeric characters. The temptation to write down such passwords is apparent and an immediate and obvious potential security weakness. Problems also arise with researchers from institutions that do not have RAs in place.

In short, this whole process does not lend itself to the wider research community that the e-Science and grid community needs to reach out to and engage with. It is a well-known adage that the customer is always right. Usability and addressing researcher requirements are crucial to the uptake and success of grid technology. End-user scientists require software that simplifies their daily research and does not make it more complex. Given the fact that the initial user experience of the grid currently begins with application for UK e-Science certificates, this needs to be made as simple as possible, or potentially removed completely.

There are other issues with PKIs and grid certificates as currently applied in the e-Research community. The fundamental issue with PKIs is trust. Sites trust their users, CAs, and other sites. If the trust between any of these is broken, then the impact can be severe, especially since users are in principle free to compile and run arbitrary code. Thus with PKIs there is no mention of what the user is allowed to do once they have gained access to the resource. For example, users can in principle run arbitrary applications, starting a variety of local processes. In reality, a set of existing applications and infrastructure are often pre-deployed across the grid nodes; hence the issue and risks of uploading executables are diminished. However, given the fact that compilers are commonly available on these resources, the possibility to compile arbitrary code and run executables spawning arbitrary processes exists. There is typically no security middleware enforcement on what processes can be started, by whom, and in what context, other than the local enforcement given by the privilege associated with the local account. As the grid community moves toward more security-focused domains such as e-Health, such a model will never be supported. Instead practices and solutions that help make grid infrastructures safer are required. Authorization-based systems offer one approach to improve this security model.

12.3 Authorization and Grid Systems

Once uses have had their identity validated at a remote resource, it is essential that their actions are restricted based on who they are, what they are trying to do, and in what context, and so on. There are various methods of enforcing this restriction, the simplest method being the use of an access control list (ACL), which lists what users have access to a privilege. Essentially, uses present their credentials at a gatekeeper to a resource, which consults a known list of users. This basic authorization structure extends the concept of authentication and no more. If the user cannot authenticate to the satisfaction of the gatekeeper then the resource request will be denied. The Globus GSI [7] software is an example of the classic ACL used to enforce authorization and provides a relatively coarse-grained approach to implementing security through the *grid-mapfile* mapping of DNs to local user accounts.

A problem that arises when trying to apply this method to a dynamic grid environment is that only one list exists, where there could be many privileges that require different ACLs. For example, a user might need access to a given resource for different purposes within a given VO. Having a single list with a predefined set of accounts and user DNs does not support this multirole approach. This is a solution that would not scale well in a large VO. A more sophisticated method of applying authorization controls is through the use of role-based access control (RBAC) mechanisms [8], which allow privilege management infrastructures (PMI). There are several RBAC middleware solutions available that support authorization that have been explored within the grid community. We provide an overview of some of the more prominent of those here.

12.3.1 Globus Security Infrastructure

The Globus toolkit [6] supports Globus Security Infrastructure (GSI)-based authentication and authorization. This includes the following

- WS-Authentication with support for both message-level and transport-level security. Message-level security is achieved through an implementation of the WS-Security standard that supports message protection at the simple object access protocol (SOAP) [9] message level. Transport-level security is achieved through the use of X.509 certificates to establish transport layer security (TLS) connections.

- WS-Authorization through an authorization framework based on the OASIS security assertion markup language (SAML) [10] authorization application programming interface (API). Through this SAML AuthZ API, a generic policy enforcement point (PEP) can be achieved, which can be associated with arbitrary services. Thus, rather than developers having to explicitly engineer a PEP on a per application basis, the deployment information associated with the service is used. Authorization checks on users attempting to invoke "methods" associated with Globus services are then automatically raised and forwarded to the policy decision point (PDP), which in the simplest case will respond with an allow/deny. However, we note that in recent versions of the Globus infrastructure it is now possible to configure a chain of authorization mechanisms together. We note that one issue that has been encountered with the SAML AuthZ profile is the lack of granularity in how users might invoke actions. For example, different actions may or may not be allowed, depending upon the data that they wish to access and potentially change. The SAML AuthZ profile does not currently allow actions to be distinguished based upon the parameters that might be associated with them. The grid standards community is working on addressing this deficiency.

- Credential management through MyProxy [11] is a credential storage and management system that has widespread acceptance as the way in which credentials should be managed within a grid environment. Instead of users managing their own private keys and credentials, they can delegate them to a MyProxy repository. Through username and password access to MyProxy repositories, short-lived proxy certificates can be created. MyProxy solutions are now being used in combination with portals for example, where users accessing a portal through a username and password will automatically have short-lived proxy certificates created, which can subsequently be used for grid-based job submission. Of all of the authorization infrastructures, GSI is arguably the most straightforward to establish and use. This is unsurprising, since GSI has been developed as an integral part of the Globus development. That said and as noted, the ACL-based approach offered by *grid-mapfiles* is a limited form of authorization. However, recent enhancements such as through the SAML authorization API now offer richer possibilities for finer-grained access control.

12.3.2 Community Authorization Service

Community Authorization Service (CAS) [12] implements access control using a centralized authorization model. The main idea behind CAS is that a resource owner delegates the allocation of authorization rights to a community administrator and lets the community administrator determine who can use this allocation. This is achieved by the administrator by having a CAS server, which acts as a trusted intermediary between VO users and resources. The CAS server decides whether a given user has sufficient privileges depending on the community policy and, if so, gives the user the right to perform the requested actions depending on the user's role in the community.

To achieve this, CAS keeps track of its community membership information. It also contains the access control policy statements that define policies along the lines of "who is allowed what type of access on which resources." To help manage this, CAS introduces the concept of rights in the form of a *capability*. Through possession of a particular capability (which is itself stored in a database associated with the CAS server), a user can show that they are allowed to access and use a particular resource.

To access a CAS-managed resource, a user has to first request a capability to use that resource. If this is the case, the CAS server responds with an appropriate capability. This capability corresponds to the intersection of the set of rights granted to the community by the resource provider and the set of rights defined by the capabilities granted to the user by the community. Following this, the user presents the capability to the resource provider responsible for that resource. The resource provider verifies the rights for both the community and the capability to grant access to the user to the resource.

For finer-grained access control and ensuring site autonomy, local resource providers can additionally apply their own local policies to determine the amount of access granted to users with presenting particular capabilities. This substantially reduces the work of resource administrators. The CAS architecture itself builds on the authentication and delegation mechanisms provided by the Globus GSI. In using CAS, a user will generate a proxy credential signed by his/her own user credential. The proxy credential is presented to the CAS sever, which returns a new credential, known as CAS proxy credential. This credential contains the CAS policy assertions to represent the user's capabilities and restrictions as an extension. SAML authorization decision statements are used to express the CAS policy assertions. The CAS proxy credential is presented to the resource provider. The resource provider then verifies the validity of the proxy credential and parses the CAS policy assertions to obtain the restrictions imposed by the CAS server. Thus, the CAS credential facilitates the mapping of the user to a local account, and the restrictions determine the operations the user is allowed to perform.

CAS provides scalability in terms of the number of users and VOs. Each user needs to be known and trusted by the CAS server (but not by each provider). Similarly, each resource provider needs to be known and trusted by the CAS server (but not by each user). However, the centralized model of a single CAS server, as with many other distributed system examples, leads to scalability and fault tolerance limitations. Many users requesting access to CAS will result in potential bottlenecks. Furthermore, the failure of the CAS server implies that no VO-wide resources enforcing access control based upon CAS capabilities will be available. This is further exacerbated since the VO administrator may need to maintain all VO-wide users' capabilities.

12.3.3 Virtual Organization Membership Service

Virtual Organization Membership Service (VOMS) [13] is a system for managing authorization data within VOs. VOMS has been developed as part of the European DataGrid (http://edg-wp2.web.cern.ch/edg-wp2). VOMS provides a centralized database of user roles and capabilities, and a set of tools for accessing and manipulating the database and using the database contents to generate grid credentials for users when needed.

The centralized VOMS model requires all sites to agree upon the roles and privileges that are to be used throughout a particular VO. In this model, all sites agree in advance on the definition and names of the roles that are applicable to their particular VO, and the privileges that will be assigned to them. A single VO administrator is then appointed who will typically assign these roles to individuals on a case-by-case basis when users ask to be granted particular roles or permissions in the VO. The VO administrator may appoint other administrators to help him in this task, but all administrators are conceptually equal, in that each can in principle, override the decisions made by the others. The VOMS model has gained widespread acceptance due to the simple model for defining the roles specific to a particular VO and how they can be used/enforced. Sites themselves are responsible for configuring their resources to use these roles. With VOMS, this is implemented with tools such as the Local Centre Authorization Service (LCAS) and the Local Credential Mapping Service (LCMAPS) [14], which map the user role information into group identities (*gid*), user identities (*uid*), and associated local pool accounts established on the local cluster for that particular VO. Refinements can be made to this model in order to allow more local control over the use of resources; for example, applying file store limits to a particular VO. We note that this local enforcement is not explicitly defined within the VO policy (given by the definition of the roles in the VOMS server). Rather, this is left up to local administrators to decide how the particular roles and privileges associated with that VO should be interpreted when accessing the resource.

VOMS offers tools that allow users to generate local proxy credentials based on the contents of the VOMS database and embed these within X.509 proxy credentials. This credential includes the basic authentication information that standard grid proxy credentials contain, as well as role and capability information from the VOMS server. One of the benefits of VOMS is that grid applications can use the credential without using the VOMS data. Alternatively, VOMS-aware applications can use the VOMS data to make both authentication and authorization decisions regarding user requests. Given the background and history to VOMS, the focus of authorization has primarily been at the level of mappings to local groups and accounts on clusters, but it is quite possible to use VOMS credentials to make finer-grained access control decisions as we shall see in the case studies. One way in which such finer-grained access control can be supported is through the Privilege and Role Management Infrastructure Standards Validation (PERMIS) technology [15,16].

12.3.4 Privilege and Role Management Infrastructure Standards Validation

The PERMIS project (www.openpermis.org) was an EC project that built an authorization infrastructure to realize a scalable X.509 attribute certificate (AC)-based privilege management infrastructure. Through PERMIS, an alternative and more scalable approach to centrally allocated X.509 public key certificates can be achieved through the issuance of locally allocated X.509 ACs.

The PERMIS software realizes an RBAC authorization infrastructure. It offers standards-based APIs that allows developers of resource gateways (gatekeepers) to enquire if a particular access to a resource should be allowed. The PERMIS RBAC system uses XML-based policies defining rules, specifying which access control decisions are to be made for given VO resources. These rules include definitions of subjects that can be assigned roles, source of authority (SoA) (e.g., local managers trusted to assign roles to subjects), roles and their hierarchical relationships, what roles can be assigned to which subjects by which SOAs, target resources and the actions that can be applied to them, which roles are allowed to perform which actions on which targets, and the conditions under which access can be granted to roles.

Roles are assigned to subjects by issuing them with X.509 ACs. Once roles are assigned and policies are developed, they are digitally signed by a manager and stored in one or more Lightweight Directory Access Protocol (LDAP) repositories. The process to set up and use PERMIS can be split into two parts: *Administration* and *Use*. To set up and administer PERMIS requires the use of an LDAP server to store the attribute certificates and reference the SoA root certificate. A local CA is required to be set up, which designates that the SoA and all user certificates created from this CA must have a DN that matches the structure of the LDAP server. The DN of the user certificate is used to identify the client making the call on the grid service.

From the user's perspective, once the administrator has set up the infrastructure, the PERMIS service is relatively easy to use. Unique identifiers are placed as parameters to services when they are deployed. These are the object identification (OID) number of the policy in the repository, the URI of the LDAP server where the policies are held, and the SoA associated with the policy being implemented. Once these parameters are input and the service is deployed, the user creates a proxy certificate with the user certificate created by the local CA to perform strong authentication. The client is run and the authorization process allows or disallows the intended action.

The PERMIS infrastructure offers very-fine-grained authorization capabilities in terms of both policy expression and enforcement. The policy editing tools allow for easy development of the XML-based policies. With support for the SAML authorization API, PERMIS allows direct linkage between grid services and authorization infrastructures.

PERMIS is perhaps the most advanced authorization infrastructure with software that meets the needs of the wider e-Research communities. It provides tools to support the definition and seamless enforcement of authorization policy. Recent enhancements to PERMIS and associated grid standards now allow PERMIS to work with a variety of grid middleware and other authorization technologies including VOMS and XACML [17].

Such authorization technologies are essential for site administrators in providing secure access to their resources. However, the purpose of such authorization technologies is not solely on protecting access to systems, but in allowing access to systems. That is, the end users must be able to access and use protected resources. Furthermore, the vast majority of researchers are unaware of X.509 attribute certificates and their use in supporting privilege management infrastructures. Rather they are more focused upon research. Thus technologies are required that hide these technological solutions as much as possible from the end user and ideally are aligned with the way in which researchers themselves wish to access distributed and heterogeneous resources. The Internet2 Shibboleth technology [3] is currently being rolled out across UK academia and provides the opportunity for such hiding of authorization technology from end users while providing a common way in which resources can be accessed and used more generally.

12.4 Shibboleth and Grid Security

With the Shibboleth model of resource access, sites are expected to trust remote security infrastructures; for example, in establishing the identity of users (authentication) and their associated privileges (authorization). To support this, the Shibboleth architecture and associated protocols identify several key components that should be supported including federations, identity providers, service providers, and optionally "where are you from?" (WAYF) services. Through these components, end users have single usernames and passwords from their home institutions (which they are more familiar with than PKIs!), which will provide for seamless access to a range of resources at collaborating institutions and service providers. Local security policies at service provider sites can then be used to restrict (authorize) what resources authenticated users are allowed access to. To support this, federations are established, which are used to agree on and enforce common policies and technical standards to provide a common infrastructure for managing access to resources and services in a uniform way. Numerous international Shibboleth-based federations exist, including InCommon (http://www.incommonfederation.org), the federation formed by the Internet2 community in the United States, InQueue (http://inqueue.internet2.edu/) for sites wishing to test and explore the Shibboleth federated trust model, the SWITCHaai federation of the higher education system in Switzerland (http://www.switch.ch/aai/), the HAKA federation developed by the Finnish universities and polytechnics (http://www.csc.fi/suomi/funet/middleware/english/), with more in the pipeline such as the Meta Access Management System (MAMS) in Australia (https://mams.melcoe.mq.edu.au/zope/mams/kb/shibboleth/). In 2006, the UK established the UK Access Management Federation for Education and Research (http://www.ukfederation.org.uk). Through the UK Federation common access to a wide range of resources is now possible covering a wide spectrum of the research community, from the arts, social sciences, and education, to the physical, engineering, and life sciences.

To understand the impact of Shibboleth technologies on grid security, it is first necessary to appreciate the interactions that typically arise with

Shibboleth. When a user attempts to access a Shibboleth-protected service or service provider (SP) more generally, they are typically redirected to a WAYF server that asks the user to pick their home identity provider (IdP) from a list of known and trusted sites. The service provider site already has a pre-established trust relationship with each home site, and trusts the home site to authenticate its users properly.

After the user has picked their home site, their browser is redirected to their site's authentication server, for example, an LDAP repository, and the user is invited to log in. After successful authentication, the home site redirects the user back to the SP and the message carries a digitally signed SAML authentication assertion message from the home site, asserting that the user has been successfully authenticated (or not!) by a particular means. The actual authentication mechanism used is specific to the IdP.

If the digital signature on the SAML authentication assertion is verified and the user has successfully authenticated themselves at their home site, then the SP has a trusted message providing it with a temporary pseudo-nym for the user (the handle), the location of the attribute authority at the IdP site and the service provider URL that the user was previously trying to access. The resource site then returns the handle to the IdP's attribute authority in a SAML attribute query message and is returned a signed SAML attribute assertion message. The Shibboleth trust model is that the target site trusts the IdP to manage each user's attributes correctly, in whatever way it wishes. So the returned SAML attribute assertion mes-sage, digitally signed by the origin, provides proof to the target that the authenticated user does have these attributes. We note that later versions of the Shibboleth specification have introduced a performance improve-ment over the earlier versions, by allowing the initial digitally signed SAML message to contain the user's attributes as well as the authentica-tion assertion. Thus the two stages of authentication and attribute retrieval can be combined.

We note that the connection from the IdP to the service provider can also be optionally protected by SSL in Shibboleth. Here SSL is used to provide confidentiality of the connection rather than message origin authentica-tion. In many cases a confidential SSL connection between the IdP and SP will not be required, since the handle can be opaque/obscure enough to stop an intruder from finding anything about the user, while the SAML signature makes the message exchange authentic. However, the message exchange should be protected by SSL if confidentiality/privacy of the returned attributes is required. The attributes in this assertion may then be used to authorize the user to access particular areas of the resource site, without the service provider ever being told the user's identity. Shibboleth has two mechanisms to ensure user privacy. First, it allows a different pseudonym for the user's identity (the handle) to be returned each time. Second, it requires that the attribute authorities provide some form of control over the release of user attributes to resource sites, which they term an attribute release policy. Both users and administrators should have some say over the contents of their attribute release policies.

Shibboleth offers numerous possibilities and potential advantages in the context of the grid. Single sign-on via authentication at a home site and subsequent acceptance and recognition of the authentication and associated attributes released to remote sites is the most obvious advantage. Thus users need not remember X.509 certificate passwords but require only their own institutional usernames and passwords. Institutions can establish their own trust federations and agree and define their own policies on attribute release, and importantly SPs can decide upon what attributes and attribute values are needed for authorization decisions.

The uptake and adoption of Shibboleth technologies within a grid context is not without potential concerns however. Sites need to be sure that collaborating sites have adopted appropriate security policies for authentication. Strength of user passwords and unified account management across sites is needed. Shibboleth is also by its very nature much more static than the true vision of the grid, where VOs can be dynamically established linking disparate computational and data resources at runtime. Instead Shibboleth requires agreed sets of attributes that have been negotiated between sites. The UK Federation, for example, is based around the exchange of a small agreed set of eduPerson attributes [18] between IdPs and SPs in the federation.

It is important to note that these attributes are typically statically defined and agreed upon between the institutions prior to joining the federation, and hence before any formulation of VOs or requests to access grid resources; that is, they are based upon statically defined PMIs. This is often sufficient to allow access to certain resources; that is, a given e-journal, for example, requires the SP only to know that the individual accessing the resource is from an institution that has paid their subscription for that journal. In the context of the grid, membership of an institution will not typically be sufficient information for a decision on access to a specific grid service hosted and managed by a given VO. Rather, VO-specific attributes are needed. This requires more dynamic models of attribute creation and assignment.

The JISC-funded DyVOSE project [19] developed solutions that allow for the dynamic creation and acceptance of attributes targeted to the specific needs of different VOs. This is more aligned with the dynamic creation of VOs across grid infrastructures where dynamic delegation of privilege is supported. As the complexity and number of security policies increase, the ability of a given SoA to delegate responsibility to others is necessary. Through extensions to the PERMIS software, the DyVOSE supported dynamic delegation of authority whereby grid sites were able to allow an attribute authority controlled by an external SoA to be delegated the ability to assign roles meaningful to a home SoA. Through this, a remote grid user could hold a role based in the home institution that will allow access to potentially remote service provider grid resources.

We note that in static delegation, the roles at the remote institution would need to be handwritten into the policy at the home institution. Dynamic delegation factors shift the role-assigning powers to subordinate authorities, which may delegate the ability to assign local roles to remote attribute authorities, and vice versa. Thus a Glasgow "student" role may be assigned to an Edinburgh computing science user, so they may access the Glasgow resource without the Glasgow SoA knowing about any Edinburgh roles. This trust relationship is agreed beforehand, where it is implicit that the role of "student" at Glasgow and "trainee," say at Edinburgh, are equivalent. Complex delegation allows new intermediate roles with less privilege than their superior role to be defined and assigned to remote attribute authorities. The DyVOSE Delegation Issuing Service supported such dynamic creation and recognition of attribute certificates and is described in detail in [20].

One of the key issues that has still to be resolved with attributes for the grid community is related to the attribute release policy. At present, an SP will request the attributes associated with the potentially opaque identifier (handle) that is returned from an IdP. If a user from the University of Glasgow is involved in numerous grid projects and VOs, and all of this information on what VOs this person is involved in, and what their role is in that VO, and so on are encoded in the core set of attributes, then it is difficult to restrict the information being released. Thus an SP may receive more information than they might actually need to make an authorization decision; for example, if this SP was just one of the many VOs that the user was involved in, then this SP would know more about all VOs the user was involved in. Of course, these attributes will be encoded; however, the SP will be able to decode the attributes due to the trust relationships and certificates previously put in place. It is possible to have a richer array of attributes other than the core set of eduPerson attributes, but for interoperability and simplicity having a core set is beneficial. Given that the focus of much of the grid community as being represented by the NGS does not focus upon privacy or confidentiality, such issues are not immediately important. Once more security-focused groups are involved; however, attribute release policies will become more important and only those attributes absolutely needed will be released.

To support the definition and enforcement of attribute release and acceptance policies, the Open Middleware Infrastructure Institute (OMII) Security Portlets simplifying Access to and Management of Grid Portals (SPAM-GP) project [21] is developing tools (portlets) that support attribute release and attribute acceptance policies, as well as portlets that support the configuration of content within portals appropriate with the attributes it receives from different IdPs. Initial results in applying these portlets across a range of VOs are described in [22].

Alternative models, for example, based upon agreeing upon a centralized attribute authority (e.g., VOMS) for a specific VO and using VOMS attributes for access decisions specific to VO-resources is another model. As noted previously, VOMS has traditionally been used for access to and usage of HPC resources using LCMAPS/LCAS and not specifically for fine-grained security access to services. The JISC-funded Integrating VOMS and PERMIS for Superior Grid Authorization (VPman) project [23] is exploring these issues. Through enhancements to the grid standards and implementation of technologies allowing for the pushing or pulling of attributes needed for access to a range of grid services (including services using the Globus/OMII middleware), a range of authorization scenarios have been demonstrated. These are described in detail in [24] along with the advantages and disadvantages of centralized versus decentralized security models. These scenarios show, for example, how VOMS attributes can be passed (pushed) and used to enforce access control (by PERMIS) on access to and usage grid services providing clinical data. Other scenarios focus on showing how VOMS attributes can be pulled and used to decide upon access control (by PERMIS) on access and usage of restricted electronics applications.

12.5 Case Studies in User-Oriented Grid Security

The vision of the grid in seamlessly accessing and using a range of resources is a compelling one, but one that depends on supporting technologies. Single sign-on to resources is one of the fundamental requirements to the realization of this vision. As noted, different domains will have their own requirements and needs on how this is achieved and in turn on the kinds of security infrastructures and associated policies that need to be enforced. Arguably, the domain that places greatest emphasis on security is the life sciences, especially when dealing with personal clinical and genetic datasets.

In the post-genomic era, data are growing exponentially. Numerous public genomic, proteomic, and metabolomic resources are arising for researchers interested in different organisms (humans rats, arabidopsis, etc.), different diseases (cancer, diabetes, etc.), different biochemical and cell signaling pathways, among numerous other areas. Linkage of individual genetic data with clinical data, including, for example, a given patient or family medical history, can be used to detect early onset of hereditary diseases, to suggest to patients ways of decreasing the likelihood of certain diseases arising, or even to treat patients in a personalized manner where different drugs can be targeted—not just to demographic/phenotypic descriptions of people where, for example, a young male might be given a different drug than an older grandmother for a similar disease, but actually on the genetic differences of the individuals themselves.

The life sciences in the post-genomic age offer huge potential benefits to mankind. At the same time, there is considerable concern regarding the use of these datasets for other, potentially nonclinical uses. Insurance agencies or employers would be keen to know the likelihood of those they are offering policies/jobs to are of developing a chronic disease. Public association with certain diseases or medical histories more generally leaves individuals rightly concerned that their data are not accessible to the masses.

Within this context, the National e-Science Centre (NeSC) [25] at the University of Glasgow has been exploring the development and management of a wide range of e-Research infrastructures. At the heart of these infrastructures are fine-grained security, usability for the end user, and support of interdisciplinary e-Research. Development of a new drug or exploration of a given disease requires a wide array of researchers and research domains to be involved. This can include biologists, bio-informaticians, clinicians, statisticians, chemists, physicists, pharmacologists, and epidemiologists, among numerous others. Only through their successful inputs can the whole research process be supported. To explore how we have developed infrastructures supporting interdisciplinary research, we describe several completed and ongoing projects, including the MRC-funded Virtual Organisations for Trials and Epidemiological Studies (VOTES) [26] project, the BBSRC-funded Grid Enabled Microarray Expressions Profile Search [27] project, and the DTI-funded Biomedical Research Informatics Delivered by Grid Enabled Services (BRIDGES) project [28].

For demonstration purposes we outline how this infrastructure can support research into cancer. However, the same infrastructure can (and is) be used across a wide range of other clinical areas. We also note that the following are based upon representative datasets only.

12.5.1 VOTES Project

The VOTES project was funded by the MRC to develop a grid framework through which a multitude of clinical trials and epidemiological studies could be supported. Thus, rather than engineering bespoke solutions for a given trial or study, VOTES focused on providing an infrastructure where a multitude of trials and studies could be developed and supported, each with their own particular nuances in terms of the data that are being accessed, by whom, and the security policies that apply, and so on.

At the heart of clinical trials are three key processes: patient recruitment, data collection, and study management. Recruitment is primarily concerned with identifying patients who are *potentially* suitable for a clinical trial, or even whether sufficient patients exist who meet the criteria for a given trial to take place, as might be the case when conducting a feasibility study. Once identified it is essential that patients are advised about all matters related to the clinical trial, including potential benefits, potential dangers, how the collected information will be used within the trial, and potentially any plans for future use of the data. One of the challenges here is that it is essential that patient consent is obtained before any access to identifying patient data is made.

Once a set of patients have been identified, invited to join a particular trial, and subsequently accepted, the next phase is typically focused on the actual undertaking of the trial itself. This includes collecting data on the patients throughout the course of the trial. If the trial is concerned with drug evaluation, say, then it is typically necessary to randomize patients with some patients being given the drugs to be evaluated and others a placebo: double-blinding is often used here, where neither the patients nor those running the trial are aware of who is receiving the drug/placebo to avoid potential bias. Instead, a trusted third party is used to keep this information. Tracking and monitoring the patients through-out the trial is typically essential to ensure that all necessary information is collected. In some circumstances, this monitoring and follow-up can run for many years—and can often lead to insights on the long-term effects of particular treatments, for example.

Throughout the whole trial process it is essential that the trial is con-ducted according to a strictly defined protocol. This will typically describe what information is being collected, for what purpose, as well as how it will subsequently be used within the trial and any potential follow-up on trials or studies. In all of these phases it is essential to ensure that the different people with different roles within the trial can only access and use the datasets and services associated with their particular role in the trial.

The grid infrastructure developed within VOTES has been described in [29,30]. In brief, the VOTES infrastructure is based upon a portal that provides access to a range of distributed services through which various clinical datasets can be accessed and used. The services themselves have been implemented using grid middleware such as Globus Toolkit version 4 and the OGSA-DAI software. The current implementation combines access to and usage of a range of software and datasets in widespread use across the NHS in Scotland. These include access to Scottish Morbidity Records (SMR)—one of the most comprehensive clinical data repositories in the UK. The SMR datasets are constructed in conjunction with the General Register Office (GRO) for Scotland. For the purposes of the VOTES project, a subset of over 4 million records of the data has been provided by the NHS covering over 30 years of patient data and care related to hospital discharges, psychiatric admissions and discharges, cancer registrations, and deaths. It is worth noting that the NeSC employees working on this project have been granted NHS Honorary accounts. Close collaboration with the NHS is essential in supporting the development of services pro-viding access to clinical data.

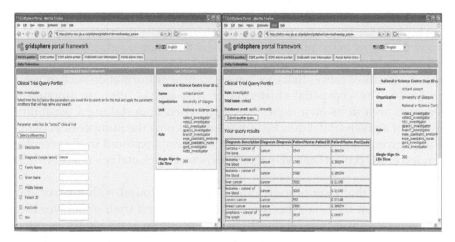

FIGURE 12.1   VOTES portal showing cancer clinical data query.

The access to the portal is through Shibboleth following the interactions described previously. Figure 12.1 (left) shows the interface to one study where information on cancer patients is being returned (the names are for demonstration purposes only and not real identities). The results of this query are returned on the right of Figure 12.1. Key to this is the various attributes that are returned (right-hand side of browser interfaces). These are delivered via Shibboleth interactions and used to personalize the access to different services through the portal; that is, the portlets that are accessible are based upon having the roles to see them. This model corresponds to security models along the lines of "what you can see is what you can do."

However, remote data providers are unlikely to simply allow access to their datasets for someone who has authenticated and provided the right roles to a remote portal. They will want to make their own authorization decisions. To support this, the remote services providing access to data are also protected with PERMIS and have their own local security policies on access and usage. When a user issues a query is federated to a remote service provider, their authorization infrastructure (PERMIS) is configured to pull the X.509 attribute certificate associated with that user request to make their own local authorization decision. In supporting this, the attributes delivered via Shibboleth are kept in a VO-specific attribute authority (LDAP server) associated with the portal. Thus all service providers know where to go to obtain the attributes that they need and have agreed upon when requests for secure access to their datasets are made. When pulled, these attributes are checked for authenticity and validity, and if acceptable, the query is run and resultant datasets are returned. We note that the infrastructure does not allow arbitrary querying. Rather, the queries are agreed *a priori* and configured in the portal so that a limited form of querying is possible. Thus a nurse or investigator can only query datasets based upon the forms in the portal, which can be parameterized.

The results of the given query shown here depict the cancer patients: the specific forms of cancer that they have and other information such as their postcode. The geospatial clustering of diseases is often an important research area in its own right. Similarly, when undertaking a feasibility study for a given trial it may be important to know the number of patients with a given condition. Such scenarios are being explored in the recently funded ESRC project Data Management through e-Social Science (DAMES) project [31].

Assuming that a set of individuals have been found with a particular condition, that they have been recruited to a given trial, understanding the genetic similarities and differences of these individuals is often required. Do they have the same gene mutations? Are there genetic differences that can help shed insight into why this individual developed this particular disease? To address these kinds of questions requires access to services that deal with individuals' genetic datasets. One way in which genetic data are established and analyzed is through microarray analysis. The BBSRC GEMEPS project developed various services that allowed secure access to both non-public and public microarray datasets.

12.5.2 GEMEPS Project

The GEMEPS project developed a grid infrastructure for discovery, access, integration, and analysis of microarray datasets. Through the GEMEPS infrastructure scientists were able to ask the following questions and obtain appropriate results based upon their privilege: Who has run a microarray experiment and generated results similar to mine? Who has undertaken experiments and produced data relevant to my own interests; for example, for a particular phenotype, for a particular cell type, for a particular pathogen, on a particular platform or microarray chip set? Show me the results from a particular collaborator, and show me the conditions and analysis associated with experimental results similar to mine.

In all of these scenarios, the premise was that sites should keep and maintain their own data and define their own security policies on access and usage. Since scientists are often reluctant to publish their data in public repositories until they have published results in recognized journals, which can, depending on the journal, be a long and protracted affair, many datasets remain inaccessible to the wider research community. These issues are discussed in detail in [32].

Scientists in the first instance would like to be able to query across a range of experiments based on any one or more of a variety of search terms. Thus scientists are unlikely to be interested comparing experimental results from *homosapiens* and *barley,* for example. We note that at the gene name level, however, it is often the case that common gene name clashes do exist across species. To support this basic metadata querying, the GEMEPS project implemented a simple user-oriented portlet that allows for a variety of these kinds of information to be used for querying over available (subject to authorization privileges) datasets as indicated on the left of Figure 12.2. This was compliant with the bioinformatics standards in this area. The portal itself was Shibboleth protected. Thus when the user from the VOTES portal wished to access and use the GEMEPS portal, they could simply redirect their browser to the GEMEPS portal. Access to and usage of this portal and the services that it makes available are done without the need for further reauthentication—one of the key characteristics identified for users: "single sign-on." Figure 12.2 (left) shows a user searching for experiments containing information on *homosapiens*-related microarray experiments conducted on the *GPL570* platform targeting the specific condition *cancer* with *standard Affymetrix procedures* used for hybridization. The resultant experiments that have been conducted and are available to that user are shown on the right of Figure 12.2. We note that this is just one example of a query that can be supported.

As well as querying over metadata of experiments, GEMEPS developed services that allowed quantification of the similarity of experiments themselves through comparison of the gene expression levels across experiments. Different statistical similarity models were used for this purpose and are described in [33].

Assuming that a researcher has identified a set of genes that they believe play a role in cancer for patients identified through the VOTES portal, they may often want to understand the protein and nucleotide sequences associated with those genes to compare how similar/dissimilar they are. The bioinformatics application Basic Local Alignment Search Tool (BLAST) [34]

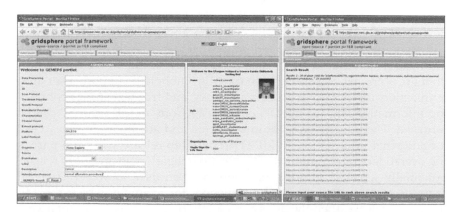

FIGURE 12.2   GEMEPS showing cancer microarray experiment query.

is often used for this purpose. This can be a computationally expensive task. The BRIDGES project developed services that support this directly.

12.5.3 BRIDGES Project

The BRIDGES project was successfully completed at the end of 2005. Its remit was to provide a grid infrastructure to support research into genetic causes of hypertension, one of the main causes of cardiovascular mortality. Before BRIDGES, many of the activities that the scientists undertook in performing their research were done in a time-consuming and largely nonautomated manner. This was typified through "internet hopping" between numerous life science data sources. To address this, BRIDGES developed a security-focused data and compute grid infrastructure. The data grid that was developed within BRIDGES is described in [35,36].

The BRIDGES compute grid used PERMIS to make/enforce distinctions between different privileged and nonprivileged users. In particular, the policies defined and enforced with PERMIS as follows:

- If they are *unknown* users the job would only be submitted to the local "free" Condor pool at NeSC Glasgow.
- If we recognized the users but they do not have a local account on HPC resources at Glasgow, the job would only be submitted to the Condor pool and the NGS.
- If we recognized the users and they have an account at Glasgow HPC resources, then the job would be sent to the Condor pool, the NGS, and ScotGrid.

These decisions on user identity used the DN of the individuals returned by Shibboleth from their IdP, although it is equally possible to use some other eduPerson attributes instead. We note that this raises issues in the application of Shibboleth itself in the grid domain. Shibboleth has been developed to support user anonymization and privacy when accessing and using resources across a federation. However, with the grid model, knowing which user is accessing a resource, especially in the biomedical domain, is crucial. We also note that while Shibboleth supports user anonymization and privacy it is not mandatory, and free text strings containing information such as the DN of the user from an IdP to an SP can be returned. The policies on what information and attributes an SP can ask for and what information an IdP is prepared to release will form part of the overall federation contract. There is no obligation on an IdP to release potentially sensitive information about a given user. However, if an SP requests certain attributes to be returned, for example, which the IdP refuses to release, then the SP is completely free to refuse to grant access to their own resource. SP autonomy is thus assured.

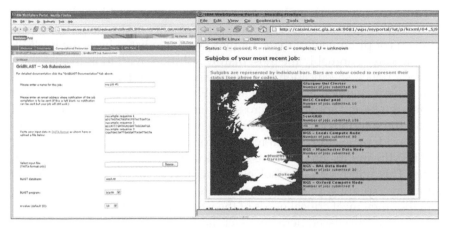

FIGURE 12.3   BRIDGES Grid BLAST front end and running/monitoring jobs.

The actual selection of where to submit a user job was based on the avail-ability of resources (which was established dynamically). In accessing and using the BRIDGES portal, users could simply redirect their browser from the GEMEPS or VOTES portal to the BRIDGES portal and automatically access and use services without the need for further reauthentication. The front end to the Grid BLAST service is accessible as shown on the left of Figure 12.3. This provided access to a range of genomic and microbial data resources predeployed on major HPC resources. To support large-scale BLAST usage, users were able to select options that allowed them to be emailed the results when the jobs completed, or they could interactively see the status of the jobs across the various grid resources (whether they are queued, completed, or running), as shown in the right of Figure 12.3.

12.6 Conclusions and Recommendations

In this chapter we have explored the current limitations of authentication-only-based solutions with PKIs as the basis for grid security. Lack of gran-ularity of authorization will dissuade large groups of researchers from engaging. Perhaps more of an issue are the initial steps through which researchers are asked to proceed before they are able to do anything "on the grid." X.509 certificates and the process of acquiring one and convert-ing it to appropriate formats is a hurdle that a large swathe of the non-grid research community will not overcome. It needs to be made simpler and ideally removed completely. Shibboleth offers one possibility through which the usability factor for end users can be addressed.

There are numerous authorization frameworks available today and we have tried to give an overview of the functionality and suitability of some of the most prominent of these. Of those we have listed here, the PERMIS middleware is arguably the most mature solution, with tools available for security policy specification and enforcement, for linkage to grid services in a generic manner, and for linkage to Shibboleth. It is still the case that wider uptake and application across a range of different scenarios is needed before the solutions can be hardened into real products however. For example, considerable effort is still required for deployment and configuration of PERMIS and its interworking with Globus, OMII, and Shibboleth solutions. This will no doubt resolve in time, but requires more community effort in exploring practical experiences of these solutions and seeing whether they meet critical needs of the research community.

Issues not discussed here but essential to consider include fabric management. Grids will always be seen as a threat if they ignore the issue of fabric management. A unified treatment and associated framework for analyzing the security of grid applications, grid middleware, and the underlying OS is needed. If VOs are to be truly secure, then blindly trusting partners to take necessary steps is naïve. Instead, tools are needed to assess the security infrastructures and software across sites before VOs are established. Will site X want to collaborate with site Y if they allow telnet access, or are they using an older version of software with known security holes? Configuration management needs to be brought to the fore in understanding the establishment, management, and monitoring of VO fabrics. This may well include patterns of usage, for example. Exploratory work in this area is presented in [37].

Shibboleth represents a clear opportunity to overcome the current issues with PKI-based security. Trust federations at an institutional level where users can authenticate at their home site and have appropriate attributes released to service providers (which will use them to make authorization decisions) change the dynamic of security. There has always been a large degree of trust in the grid community: trust of users, trust of sites, trust of CAs, and so on. Hence Shibboleth does not add a new trust requirement especially. Instead trust is moved to IdPs (and ensures that they have appropriately strong authentication and authorization schemes) and WAYFs (which ensure that the "correct" IdPs are identified and matched with SPs).

Understanding what attributes are needed in the grid community is essential. Many solutions may only require that the DN is passed over, for example, so that accounting and logging of the resource usage for that individual can be achieved; that is, no further attributes are needed to make an authorization decision. Other more prescriptive VOs may require more information such as VO membership, role of the user, and so on. The ability to map such attributes into a form that Shibboleth can make use of is needed. Once such scenarios can be supported, more understanding of the attribute release policies, attribute acceptance policies, and how they might be implemented can be achieved.

In addition to these efforts, the web service standards community OASIS, IETF, and W3C among numerous others are producing a plethora of specifications, which, in principle, could help simplify grid security; however, there is still considerable fluidity in these developments with partial/draft specifications, full specifications, and a variety of implementations existing. Single sign-on solutions to services at numerous sites, and complementary efforts within the Liberty Alliance consortia offer potential solutions of direct relevance to the grid community in its move toward web-based solutions and service-oriented architectures. Similarly, the Web 2.0 community through efforts such as OpenId [38] is proposing yet more security solutions. The security future thus remains in considerable flux.

References

1. R. Housley and T. Polk, *Planning for PKI: Best Practices Guide for Deploying Public Key Infrastructures*, Wiley Computer Publishing, New York, 2001.
2. ITU-T Recommendation X.509 (2001) Information Technology—Open Systems Interconnection—Public-Key and Attribute Certificate Frameworks. ISO/IEC 9594-8: 2001.
3. Internet2 Shibboleth Initiative. Available at: http://shibboleth.internet2.edu.
4. UK e-Science National Grid Service. Available at: http://www.ngs.ac.uk.
5. UK e-Science Certification Authority. Available at: http://www.grid-support. ac.uk/ca.
6. Globus project. Available at: http://www.globus.org.
7. Globus Grid Security Infrastructure (GSI). Available at: http://www.globus. org/toolkit/docs/4.0/security.
8. R. Sandhu, E.J. Coyne, H.L. Feinstein, H.L. and C.E. Youman, "Role-based access control models," *IEEE Computer*, 29 (2), 38–47, 1996.
9. OASIS, Assertions and Protocol for the OASIS Security Assertion Markup Language (SAML) v.1, September 2003. Available at: http://www.oasis-open.org/specs/#samlv1.0.
10. W3C Simple Object Access Protocol. Available at: http://www.w3.org/TR/ soap.
11. MyProxy Credential Management Service. Available at: http://grid.ncsa.uiuc. edu/myproxy.
12. L. Pearlman, V. Welch, I. Foster, C. Kesselman, and S. Tuecke, "A community authorization service for group collaboration," in *Proceedings of 3rd IEEE International Workshop on Policies for Distributed Systems and Networks*, pp. 50–59, Monterey, CA, 2002.
13. R. Alfieri et al., *Managing Dynamic User Communities in a Grid of Autonomous Resources*, Conference for Computing in High-Energy and Nuclear Physics (CHEP), La Jolla, San Diego, March, 2003.

14. Local Centre Authorization System. Available at: http://hep-project-grid-scg.web.cern.ch/hep-project-grid-scg/lcas-lcmaps.html.
15. D.W. Chadwick and A. Otenko, "The PERMIS X.509 role based privilege management infrastructure," *Future Generation Computer Systems*, 936, 1–13, 2002.
16. D.W. Chadwick, A. Otenko, and E. Ball, "Implementing role-based access control with x.509 attribute certificates," *IEEE Internet Computing*, 7 (2), 62–69, March–April 2003.
17. OASIS eXtensible Access Control Markup Language (XACML). Available at: http://www.oasis-open.org/committees/tc_home.php?wg_abbrev=xacml.
18. eduPerson Specification. Available at: http://www.educause.edu/eduperson.
19. Dynamic Virtual Organisations in e-Science Education (DyVOSE) project. Available at: http://www.nesc.ac.uk/hub/projects/dyvose.
20. R.O. Sinnott, J. Watt, D.W. Chadwick, J. Koetsier, O. Otenko, T.A. Nguyen, "Supporting decentralized, security focused dynamic virtual organizations across the grid," *Proceedings of the 2nd IEEE International Conference on e-Science and Grid Computing*, Amsterdam, 2006.
21. Open Middleware Infrastructure Institute (OMII) Security Portlets simplifying Access to and Management of Grid Portals (SPAM-GP) project. Available at: http://www.nesc.ac.uk/hub/projects/omii-sp.
22. J. Watt, R.O. Sinnott, J. Jiang, G. Stewart, A. Stell, D. Martin, and T. Doherty, "Federated Authentication and Authorisation for e-Science," in *Proceedings of Australian Partnership for Advanced Computing (APAC) 2007 Conference*, Perth, Australia, September 2007, http://www.apac.edu.au/apac07/.
23. Integrating VOMS and PERMIS for Superior Grid Authorization (VPman) project. Available at: http://sec.cs.kent.ac.uk/vpman.
24. R.O. Sinnott, D. Chadwick, T. Doherty, D. Martin, A. Stell, G. Stewart, L. Su, and J. Watt, "Advanced security for virtual organizations: Exploring the Pros and Cons of Centralized vs Decentralized security models," in *Proceedings of the 8th IEEE International Symposium on Cluster Computing and the Grid (CCGrid)*, pp. 106–113, Lyon, France, May 2008.
25. National e-Science Centre. Available at: http://www.nesc.ac.uk.
26. Virtual Organisations for Trials and Epidemiological Studies (VOTES) project. Available at: http://www.nesc.ac.uk/hub/projects/votes.
27. Grid-Enabled Microarray Expression Profile Search (GEMEPS) project. Available at: http://www.nesc.ac.uk/hub/projects/gemeps.
28. Biomedical Research Informatics Delivered by Grid Enabled Services (BRIDGES) project. Available at: http://www.nesc.ac.uk/hub/projects/bridges.
29. R.O. Sinnott, O. Ajayi, and A.J. Stell, "Supporting grid based clinical trials in Scotland," *Health Informatics Journal*, 14 (2), pp. 79–95, June 2008.
30. A.J. Stell, R.O. Sinnott, and O. Ajayi, "Supporting Nationwide e-Clinical Trials and Studies," in *Proceedings of 15th Mardi Gras Conference*, Baton Rouge, LA, February 2008, ACM Press, ISBN: 978-1-59593-835-0, http://www.mardigrasconference.org/conf_2008/.
31. Data Management through e-Social Science (DAMES) project. Available at: www.dames.org.uk.

32. R.O. Sinnott, M.M. Bayer, J. Koetsier, and A.J. Stell, "Grid infrastructures for secure access to and use of Bioinformatics Data: Experiences from the BRIDGES Project," in *Proceedings of the 1st International Conference on Availability, Reliability and Security (ARES'06)*, pp. 950–957, Vienna, Austria, April 2006.

33. R.O. Sinnott, C. Bayliss, and J. Jiang, "Security-oriented data grids for microarray expression profiles," in *Proceedings of HealthGrid 2007 Conference*, pp. 67–77, Geneva, Switzerland, April 2007.

34. S.F. Altschul, W. Gish, W. Miller, E.W. Myers and D.J. Lipman, "Basic local alignment search tool," *Journal of Molecular Biology*, 215, 403–410, 1990.

35. R.O. Sinnott and D. Houghton, "Comparison of data access and integration technologies in the life science domain," in *Proceedings of UK e-Science All Hands Meeting*, September 2005, Nottingham, England. Available at http://www.allhands.org.uk/2005/proceedings/index.html.

36. R.O. Sinnott, M.M. Bayer, J. Koetsier, and A.J. Stell, "Advanced security on grid-enabled biomedical services," in *Proceedings of UK e-Science All Hands Meeting*, September 2005, Nottingham, England. Avalilable at http://www.allhands.org.uk/2005/proceedings/index.html.

37. R.O. Sinnott, J. Muhammad, and Y. Wu, "Deployment of grids through integrated configuration management," in *Proceedings of 26th International Conference on Parallel and Distributed Computing and Networks (PDCN)*, Innsbruck, Austria, February 2008. Available at http://www.actapress.com/Content_of_Proceeding.aspx?proceedingID=469.

38. Open Id. Available at: http://openid.net.

13

Modeling P2P Grid Information Services with Colored Petri Nets

Vijay Sahota and Maozhen Li

CONTENTS

13.1 Introduction

The past few years have witnessed a rapid development of grid computing systems and their applications. Information services play a crucial role in grid computing environments to facilitate the discovery of resources and services. Existing work on information services, notably the Monitoring and Discovery System (MDS4) [1] of the Globus middleware [2], has advanced the research in this field. MDS4 adopts a hierarchical tree structure to distribute its monitoring load across a virtual organization (VO), in which every node is running an index service monitoring its resources and pushing this information up to a master index server. Thus a VO has a central point for registering the state information of computing resources involved in the VO.

However, due to the dynamic nature of the grid, keeping an accurate and up-to-date source of state information poses many challenges in terms of scalability, resilience, and performance. In the case of MDS4, its hierarchical structure and centralized management has an inherent delay associated with it, limiting its scalability. The work presented in [3] shows that MDS4 has a slow response even with a load of 100 nodes. MDS4 also lacks a mechanism to deal with index service node failures, which could break the information service network into isolated subnets.

Many factors exist that can influence the performance of an information service system. For example, slow network connections may produce traffic bottlenecks in communication between nodes, and frequently changing resources require more updates to be sent out to the information service network. The current workload being undertaken by key index nodes could also slow down the updates of information, while the type of index nodes in question could also be varied in terms of computing capabilities. These are just some of the obstacles that could hinder the accuracy of information and consequently limit the efficiency of large-scaled grid environments.

Peer-to-peer (P2P) systems have shown enhanced scalability and fault tolerance in file sharing under a highly dynamic computing environment. Distributed hash table (DHT)-based second-generation P2P systems such as Chord [4], Pastry [5], and CAN [6] use DHTs to speed up the process of message lookup. A message lookup can be completed within a guaranteed number of hops; for example, both Pastry and Chord have a maximum of $O \log(n)$ hops, with CAN differing with $O(dn1/d)$ hops. However, the common weakness exhibited by using a DHT is the lack of ability to deal with churn situations in which nodes join or leave at high rates [7]. One side-effect of the churn situation is that it could lead to a high fluctuation in both communication and computational cost, causing extra heavy overhead to a P2P network.

Building on grid and P2P technologies, P-Grid [8] bases its network on a binary tree with each leaf of the tree representing a computer node. Data are distributed using hash keys that dictate where the data will be stored on the network. These hash keys are then used to search for data, and at every junction of the tree a routing decision is made using the keys prefix compared with the junctions at the binary branches. However, P-Grid requires that extra processing for the data be moved to nodes. Another side-effect in using a binary tree is that a search must always transverse the tree's fixed paths producing slow and frequent traffic.

To emphasize, the large scale coupled with the dynamic nature of a grid environment poses many challenges for information services. In this chapter, we present PIndex, a grouped P2P model for scalable grid information services. PIndex aims to resolve the issues that arise within a highly dynamic grid environment with its salient features listed below:

- It introduces the concept of peer groups so that a large number of peer nodes in a grid environment can be aggregated into groups to speed up the lookup process in resource discovery.
- It builds on the dynamic aggregate framework of Globus MDS4 to handle dynamic node presence and data.
- It enhances fault resilience through the replication of state information among nodes within a peer group.
- It reduces the effects of churn, limiting them locally to a peer group level and isolating the effect to the rest of the network.

We build models using colored Petri nets (CPNs) to evaluate PIndex, and the simulation results show that PIndex is scalable with a large number of peer nodes.

The remainder of this chapter is organized as follows. The next section gives an overview of Globus MDS4. This is followed by an introduction to PIndex, and an explanation of its main concepts. We then describe the design and implementation of the CPN models for PIndex evaluations. Finally, we present the simulation results demonstrating the effectiveness of PIndex for scalable grid information services.

13.2 An Overview of Globus MDS4

Owing to its high dynamism, a grid environment can produce large amounts of state information that needs to be handled promptly. The Globus MDS4 is a suite of Web Services Resource Framework (WSRF) [9] compliant services that can be used to monitor and discover resources through notifications. MDS4 enables one to advertise large amounts of state information to users within a VO, and the ability to monitor and query these resources. This section gives an overview on MDS4, along with an extrapolation of its performance when placed in a highly dynamic grid environment.

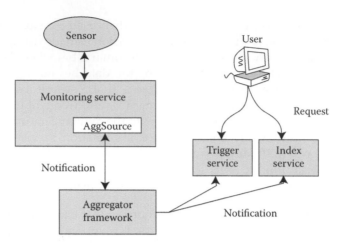

FIGURE 13.1   MDS4 structure.

As shown in Figure 13.1, MDS4 consists of two main services, the index service and the trigger service. The index service keeps an index of current state information relative to all the resources running under its container dynamically. The trigger service waits for predefined conditions before an event is triggered; for example, if CPU usage is more than 80%, then a trigger to stop accepting more jobs is executed. Both the index and trigger services are aggregator services in that they are created using the aggregator framework. It is the aggregator framework that provides the predefined methods to register, monitor, notify, and declare the data types of the resources. Aggregate sources (AggSource) collect and act on data from a WSRF supported resource property (WS-Property). Three types of aggregate sources exist. First, a query source updates its state information on a regular basis through polling operations. Second, a subscription source updates its data though the WS-Base notification mechanism. Finally an execution source executes an external program to update the values of resource properties. Both the index and trigger services monitor registered resources.

Upon registering, each source specifies the data type, source type, how, and where to get its data from, and its registration intervals. Registration intervals are used as heartbeats to make this aggregator service self-cleaning; that is, automatically removes expired resources. Figure 13.2 depicts the index services arranged in a hierarchical structure for a VO of a grid environment. The state information of resources can be pushed up or down to a designated index service node creating a tree structure. A single index service node may be designated to keep track of all resources available in a VO. Having a centralized server, users need only to query this node to discover and monitor all of the available resources for that specific VO.

Due to its rigid hierarchical structure, MDS4 typically works in a static grid environment. Although MDS4 can tolerate some node failures, if any

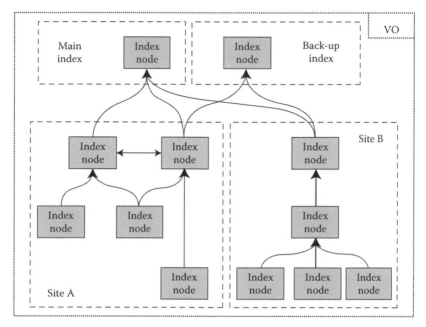

FIGURE 13.2   Hierarchical MDS4 structure in a VO.

of these failures occur at a main index service, it can cause the index service network to break into isolated subnets requiring manual maintenance. MDS4 also requires that its structure as shown in Figure 13.2 be predefined on container startup, imposing a fixed and nondynamic architecture that will unquestionably fail if the information service network experiences any form of *churn*. Messages flowing up or down along the hierarchical structure will also experience delays as traffic involved in keeping an up-to-date index increases.

An index service is generally configured to update its data at predefined intervals, which can be large (e.g., 5 minutes) in some cases. This can cause havoc with any form of time-sensitive services. On the other hand, reducing these intervals could create an expected increase in traffic and processing usage. To increase the performance of MDS4, the associated hierarchical structure must be replaced with a more scalable, dynamic, and robust structure.

13.3 PIndex for Scalable Grid Information Services

The main goal of PIndex is to enhance information services for grid environments through P2P techniques. However, it has been revealed by Foster and Iamnitchi [10] that a pure P2P approach cannot operate satisfactorily under the harsh conditions of high dynamism produced by grids. In order to build a P2P-based information service that can survive a highly dynamic environment posed by grids, a solution needs to be:

- Dynamic and robust, accommodating for high rates of *churn* and dynamic information updating.
- Responsive, being able to discover resources in the least time possible.
- Scalable, being able to offer information services for a large number of computing nodes.

In this section, we introduce PIndex as a scalable solution for information services. By structuring nodes into independent peer groups that communicate via SOAP [11] messages, PIndex is able to carry out distributed tasks more robustly. In the following sections, we explain in depth PIndex architecture, and its fast P2P lookup algorithm.

13.3.1 PIndex Architecture

One key feature of PIndex is its capability to group nodes into peer groups (PGs), while the entire network can be queried through a fast P2P structure between PGs. Each PG operates independently with its monitored data limiting the effect of *churn* to a PG. Grouping nodes into PGs brings in three major benefits. First, the cost of updating dynamic state information is greatly reduced since the update in a PG does not have to traverse the whole network. Second, when a node fails, its effects are kept local with only the nodes in the same PG needing to be informed to update their information. Finally, peer nodes might join or leave a PG freely, but a PG is a stable element in the PIndex network allowing searches to be performed relative to PGs. This removes the dependency on any particular node for routing.

Figure 13.3 depicts the architecture of PIndex. Each PG has a unique identifier (hash key) that is used as a common address in PIndex. PIndex is structured in such a way that each peer node in a PG has all the information about other peer nodes within the same PG. Upon receiving a request, a peer will be able to answer on behalf of the whole PG, while at the same time localizing monitoring data to a PG such that any changes to peers need to be updated only within the PG but not the rest of the whole P2P network to reduce network traffic.

As every node in PIndex has information about the resources available within a PG, fault tolerance can be enhanced through replication. In addition, the query process can also be sped up since only a single node need be queried per PG. When a peer node fails, all the references made to it will be removed dynamically. This is done by an index service using registration renewal messages as heartbeats. In addition, each node must

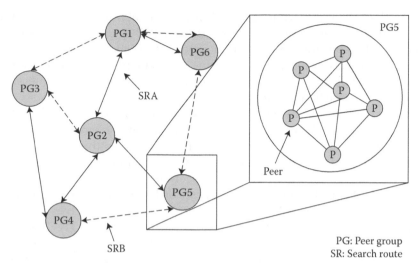

FIGURE 13.3   Peer groups in PIndex.

also have a minimal amount of random addresses to external peer groups called Table of External Contacts (TECs) for communication with the rest of the network. TECs can be self-cleaning and dynamic using the heartbeat mechanism provided by the index service.

Since each TEC is unique for each node, many routing paths exist which are dependent on where the search begins. As each search follows a different path, three main advantages are apparent. First, intrinsic to an unstructured P2P algorithm, the reliance on a central server or a hierarchical structure is eliminated along with their inherent delays and traffic costs. Second, search loads are distributed across the network, eliminating the case where a single node can be overloaded and act as a bottleneck. Finally, having many search paths also means that if a node fails only a small number of paths are affected, with the majority of network operating unhindered. This indirectly creates a virtual overlay for each node as depicted in Figure 13.3. From Figure 13.3, search route A differs from search route B because in route A, a node in PG1 has contacts existing in PG2 and PG6 which then extend to PG3, PG4, and PG5 through PG2 and PG4. However, in route B a different node in PG1 has contacts existing in PG6 and PG3, which respectively have contacts to PG5, PG4, and PG2. Depending on which node in PG1 a search starts, a route will be dynamically determined and taken.

13.3.2 Querying Resources in PIndex

Besides a robust, dynamic, and well-balanced architecture, PIndex needs a fast yet efficient search algorithm to query resources. PIndex localizes the bulk of the monitoring load within a PG, enabling one to query the whole network through a fast and efficient P2P query mechanism. By dynamically splitting up the search space into sections, PIndex enables parallel searches to be carried out using contacts in a node's TEC. At the same time, it appends upper and lower PG ID bounds for each message being sent. These bounds limit the destinations where these requests can be forwarded to, effectively splitting the search space dynamically at the time of the search. However, as the sizes of sections become large, delays within each section could also grow. PIndex resolves this by using the same splitting algorithm to split a section at every hop, as the one used to split the whole search space into many sections.

Figure 13.4 depicts how a single request can be dynamically reproduced to split a search space. As shown in Figure 13.4, PG1 receives a search request with an upper bound of 14 and a lower bound of 1 (14–1). It then forwards three copies to PG2, PG7, and PG11 respectively with new bounds (sections) 6–2, 10–7, and 14–11 respectively. PG2 then sends another two more search requests within its range of 6–2 to PG3 and PG5, who in turn forward copies to the remaining PG4 and PG6. Thus all the PGs 1–14 can be queried in 13 hops, which is the same cost as if a simple forwarding to the nearest neighbor method was used. However, Figure 13.4 also groups these hops that occur at the same time (hops 1, 2, and 3). As these hops work in parallel, the time taken to query all the PGs will be equivalent to three hops. Producing more parallel hops can reduce search times, resembling a controlled form of flooding of messages (avalanching).

As shown in Figure 13.4, messages must be restricted in movement to one direction to ensure that searches begin at the start of the section. By

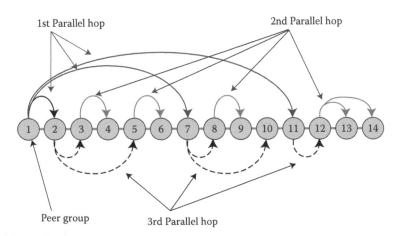

FIGURE 13.4 Dynamic sectioning.

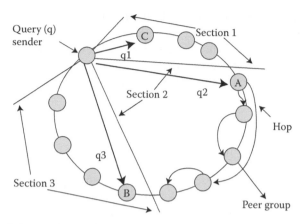

FIGURE 13.5   Splitting the search space.

restricting the direction, the receiving node will never have to check if it has dealt with this request before to avoid redundant searches. Figure 13.5 depicts a search space being split with node q, sending three parallel search queries (q1, q2, and q3) to PGs A, B, and C, with each query having upper and lower bounds defining searching sections 1, 2, and 3 respectively. Avalanching is also depicted in Figure 13.5, where q2, the query received by PG_A, is then allowed to further split the search space within searching section 2. Using its TEC, a node in PG_A forwards the search request simultaneously to two more peers, which in turn forward it to the remaining PGs. For completeness, in cases where a peer receives a request that it cannot deal with (no TEC entry within the search space), the request will then be forwarded to another peer in the same PG to deal with it, so as not to violate the sectioning boarders or send a request backwards.

As shown in Figure 13.5, the actual size (range) of a section must be known to calculate the average delays incurred and the total messages sent. Using Figure 13.4 as an example, this can be calculated with formula (1) subtracting the ID numbers of PG_A and PG_B:

$$\text{Range} = [\text{ID}(PG_B) - 1] - \text{ID}(PG_A) \tag{13.1}$$

In cases where there is no further sectioning (PIndex simple), the total time it takes to execute a request can be computed using formula (13.2). We assume the splitting is done evenly over the network. We also assume that there are no processing delays incurred to send the initial request; their delays are considered to be zero. Knowing that a processing delay (D) will occur at every PG, multiplying D with the range of each section would give the total delay per section. Since each section is queried in parallel, this becomes the total delay. By multiplying the range with the number of initial requests, we obtain the total cost of messages being sent using formula (13.3).

Let D be the average delay incurred at each PG, X the total number of PGs, R the range per section, N the initial number of requests sent, S the number of splitting requests sent within a PG, and Z the number of hops per section.

$$\text{Total delay} = (R - 1) \times D \tag{13.2}$$

$$\text{Total messages sent} = N \times R \tag{13.3}$$

In a case where avalanching (dynamic sectioning) is permitted, we assume that all sections are equal in range and each PG sends S splitting requests. The total time it takes to execute a request can be computed using formula (13.4). Similar to formula (13.2), the range is now replaced with Z hops per section, which can be calculated using formula (13.6). Since every parallel hop produces multiple requests, the total messages sent at every hop can be computed using formula (13.5).

$$\text{Total delay} = Z \times D \tag{13.4}$$

$$\text{Total messages sent} = N \times \sum_{n=1}^{z} S^n \tag{13.5}$$

$$Z = \ln\left(\left(1 + \left(\frac{1}{S}\right)\right) \div (X/N)\right) \div \ln\left(\frac{1}{S}\right) \tag{13.6}$$

13.4 Modeling PIndex with Colored Petri Nets

Since PIndex is designed for large-scale grid information networks, simulations are required to observe its features as proposed. A simulator or a model that can be used to simulate a PIndex network should have the following features:

- Performing the specific algorithms used by PIndex
- Performing parallel PG operations
- Prioritizing tasks within PIndex
- Modeling many thousands of nodes
- Modeling time intervals of around 100 m
- Supporting different types of probability distributions of events.

13.4.1 Existing Grid Simulators

This section gives a brief review of some existing simulators that might be suitable for simulations of PIndex.

13.4.1.1 *SimGrid*

SimGrid [12] offers the freedom to design a network topology. However, its definition of service times and use exists in only two static classes, which may become problematic when modeling PIndex. PIndex has multiple PGs and vast simulated population data that will become a bottleneck using SimGrid. In addition, the nature of PIndex is dynamic which cannot be modeled through the rigid definition of only tasks and resources; an accommodation of P2P aspects such as dynamic updates, search algorithms, and dynamic node presences must be held.

13.4.1.2 *GridSim*

GridSim [13] builds on top of SimJava [14], which is primarily an API to Java threads with a built-in discrete clock. However, the constraints placed in accessing these threads pose some flexibility issues when trying to implement prioritizations for different job types and to mimic the parallel nature of the PGs of PIndex. In addition, similar to SimGrid, having a central controller class may reduce performance when simulating many thousands of nodes as required by PIndex.

13.4.1.3 *GangSim*

Although created specifically for Ganglia [15], GangSim [16] does highlight the need for using real data to improve the plausibility of simulated results. However, it is unsuitable for simulating PIndex as its precision is derived on the assumption that jobs run for a minimum of 100 seconds, which is too long a period for an information service.

13.4.1.4 *CPN Tools Package*

Based on the theory of colored Petri nets (CPNs), the CPN Tools package [17] enables one to create a mathematical model from a bipartite graph. However, PIndex's multi-PG nature cannot be simulated with this package, as the package was designed for standard flat Petri nets, where a single model represented the whole network dictating that PIndex should have a static architecture, which is not the case.

Summarizing, existing simulators usually work under the assumption that a grid structure is flat and lacks dynamism (i.e., fixed and predefined).

PIndex requires a highly dynamic structure, making most grid simulators unsuitable. A more specific simulator is needed, which can be tailored to model a dynamic environment with many thousands of nodes. Although P2P network simulators [18,19] are available, they only work for specific P2P network structures, which cannot be easily modified for PIndex.

13.4.2 Choosing Colored Petri Nets for PIndex Modeling

A Petri net is a modeling language that can be used to graphically depict the structure of a distributed system. Consisting of two main elements, places and transitions, many complex processes can be modeled. More importantly, it enables one to easily depict parallel processes. A Petri net has place nodes, transition nodes, and directed arcs connecting places with transitions. Petri nets have tokens that are represented as dots indicating the occupation of a place caused by a previous transition. These tokens either occupy a place or not (akin to binary 0, 1), leading to the representation of Petri nets as matrices.

Much work has been proposed to extend the original Petri net concepts. For example, stochastic Petri nets (SPNs) [20] allow timings to be implemented to Petri nets giving more realistic complex models, and include the ability to invoke models of probability distributions of events. Another concept is colored Petri nets (CPNs) [21], which allow tokens with distinct characteristics to be represented by colors simplifying the depiction of a complex Petri net for various functions. The following main points have been made to further demonstrate the rationale in choosing Petri nets to model PIndex.

- The nodes in PIndex may have many different states. These can be depicted with colored tokens.

- The primary concern of PIndex is to perform through the use of messages; this provides a level of coarse granularity, which can be represented by transitions in Petri nets.

- As many PGs operating in parallel and interact with each other in PIndex, the ability to model concurrent events is needed. This requirement is intrinsic to Petri nets as they were designed with concurrency in mind.

- Since PIndex is a robust and fault-tolerant network, these features must be able to be simulated at any time. Petri nets have the ability to change states during a transition, allowing for failures to occur randomly.

- Each node in PIndex must operate independently, which can be mimicked by the use of tokens as objects in CPNs.

13.4.3 Modeling PIndex with Colored Petri Nets

13.4.3.1 Model Design

As discussed in Section 3, PIndex is scalable through PGs operating independently and concurrently. However, although Petri nets have been designed with concurrency in mind, running multiple instances that interact with each other is not a part of Petri nets, although one may include a combination of many Petri nets into a single Petri net to mimic multiple instances and derive a mathematical model. This would not suit the needs to model PIndex since (1) the number of PGs would be fixed, fixing the size of the network to be simulated, disabling the model in investigating PIndex's scalability; and (2) creating a fixed model will limit its flexibility in investigating different network and routing conditions, such as forcing a certain type of PG grouping using a node probability of failure as a metric, which would have to rewrite the CPN model and implementation.

Due to the multiple PGs nature of PIndex, representing a PG as a thread would take advantage of their parallel execution to produce an object-orientated CPN simulation model, which enables multiple instances of a PG to run concurrently, truly modeling PIndex in a multinetwork distribution (PIndex over many independent sites). Not only could one set each thread with a priority, but this further gives an advantage in that PGs may have a higher priority than the simulated clock, ensuring that all required tasks are executed before the clock turns over. In addition the use of threads allows the model to follow and act in a more parallel manner, reflecting real-life situations while conforming to the real-life paradigm, with only ever one instance of a PG created. Note that each issued state has a time stamp (start and stop) relative to the simulated clock. Having instantiated these threads, the colored tokens (nodes) are processed as objects. This means that in order to vary the network population, we only need to change the number of token objects created.

13.4.3.2 Model Implementation

PIndex has many states when placed in a real working environment. Before a CPN model can be established, firstly each state must be recognized. In the case of PIndex, a token represents a computing node that has colors for different states as shown in Table 13.1.

From the description given previously, a generic flow chart of PIndex is produced as shown in Figure 13.6, from which our CPN model is derived as shown in Figure 13.7. When a token is fired it leaves the "start" place and transits to the "submit a task (color)" place from which an inhibitor is placed to the next stage if the "buffer" is occupied. If so, the token will transit to the "buffer" place, otherwise the transition is made to the

TABLE 13.1

Colored Tokens for Modeling PIndex

States (Colors)	Description
Free	The node is free and able to process a message.
Dead	The node is dead; its position in the PG is vacant for joining nodes.
Busy	The node is busy processing a message, but will place a message in its buffer.
Leaving	The node will leave soon; it will not accept any more messages but will complete the ones in its buffer. It will also tell its PG and referencing nodes to update their TECs.
Search request	The node is sending a search request; the receiving node responds and forwards it according to the PIndex search algorithm.
Query response	The node is sending a response back, to any form of query.
Join requests	The node is requesting to join a PG (receiving); if space is vacant it will join, otherwise the request is forwarded to a neighboring PG.
Contact updates	The node is updating its TEC replacing its existing contacts with new ones; done when a node has newly joined a PG.
Resource updates	The node is updating its resource information, and informs the nodes in its PG to do the same along with nodes in its TEC.

"execute stage" consuming the required node and message usage from their respective places. Finally after the delayed transition (relative to the job) and the condition that the buffer is empty, the token can be transited to the "end" place.

13.5 PIndex Evaluation Using Colored Petri Nets

In this section, we simulate PIndex under a large network population to give an insight into the performance and behavior of PIndex. A description of the simulated conditions is given followed by their results.

13.5.1 Simulation Conditions

Having implemented the PIndex CPN model, simulations were performed under varying network conditions, while keeping the node firing rates (search, update, leaving, failure, and joining) and their probabilities (firing and failure) constant. The rates used in the PIndex simulator were spread over a Poisson distribution, while the probabilities were spread over a normal distribution. The rates used in the simulations for search, update, leave, failure, and join rates were respectively 4, 20, 0.33, 0.16, and 0.5 per second. It is noted that the search and update rates were used for each PG, while the remaining rates were for the whole PIndex network.

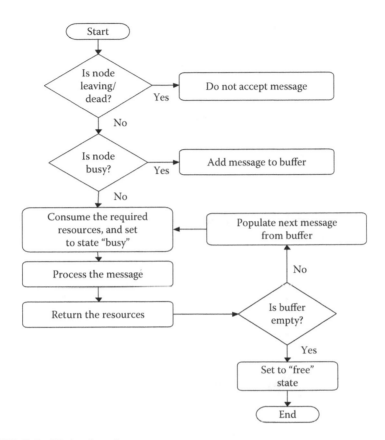

FIGURE 13.6 PIndex flow chart.

In addition, the dynamic sectioning of the PIndex algorithm was limited to two per hop (the allowed minimum for dynamic sectioning).

First, the scalability of PIndex was evaluated. The network population was varied with different numbers of PGs, of which each PG consists of 100 computing nodes. The behavior of PIndex under increasing loads could be observed. Increments of 10, from 10 to 100 PGs, were simulated. However, for clarity only simulations of 10, 20, 60, and 80 are presented.

Second, an investigation into improving the behavior of PIndex was carried out. By grouping nodes in accordance to the activity (failure probability), the behavior of PIndex under varying network conditions could be observed. Not only would this investigation find a recommended PIndex configuration, but by grouping nodes with high failure rates it tests the ability of PIndex to handle churn. In this simulation the number of PGs was fixed at 100, each containing 100 nodes. Three grouping methodologies were investigated: (1) alike—grouping nodes with the same failure rates; (2) even—grouping nodes with a normal distribution;

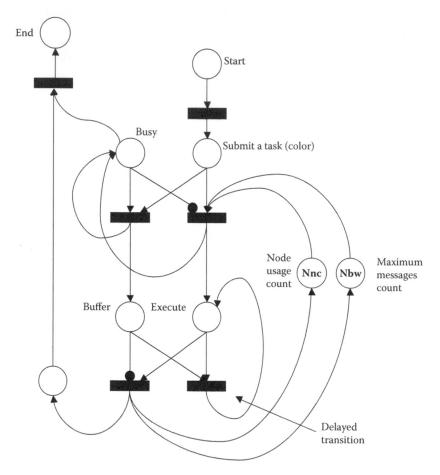

FIGURE 13.7   A CPN representation of PIndex.

and (3) mixed—this involved the grouping of nodes that gave a flat distribution over two PGs through complementary distributions. As the network population was already distributed over a normal distribution, the complementary distributions used took the form of a small bell shape and its invert.

13.5.2 Simulation Results

Figure 13.8 shows the number of messages being sent as the network population is increased. It can be seen that the number of messages sent is linear with network size; we see a steady increase in message consumption. Taking for example the peak message usage for 10 PGs to be 17, and that for 80 PGs to be 125, this gives a gradient of 7.3/8 ~ 1. This proves that there is a predictable cost when network population increases in PIndex.

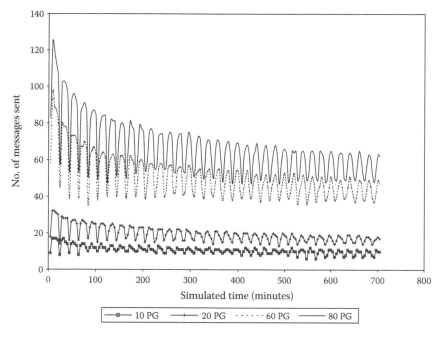

FIGURE 13.8 Message usage in PIndex.

Furthermore, on observing the results no sudden "bursts" in increased message consumption occur that can be associated with a network compensating for a failed node (churn) or the creation of a network bottleneck. Note that the leveling off of message usage after its initial startup reflects a state of stability.

Figure 13.9 shows the number of nodes being used as the network population is increased. Again a linear rise of nodes being used is seen as network population increase, proving an expected cost when implementing PIndex and that the creation of "key routing node" is not formed. Although a general increase is seen with node population, it is worth noting that its gradient is not 1. Considering the case where peak node usage for 10 PGs is 52, and for 80 PGs is 326, this gives a gradient of 6.24/8 ~ 0.78. This means we get six times the node usage for eight times the population, and in correlation with messages being sent no sudden "bursts" in increased node usage can be observed. This result is most likely due to how the PIndex search algorithm works, in that the more population that exists, the more parallel message can be sent. This increases the overall messages being sent but lessens the time spent on these searches, and hence the time computing nodes spend on processing a search.

Figure 13.10 shows the plot of number of messages sent given the different grouping methods implemented, as having an even distribution

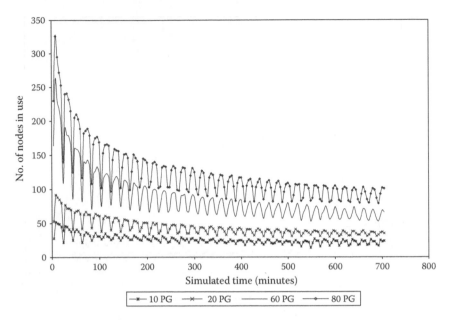

FIGURE 13.9   Node usage in PIndex.

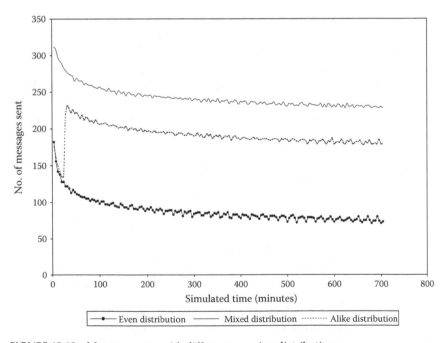

FIGURE 13.10   Message usage with different grouping distributions.

of nodes per PG gives results with the lowest message usage. This is mostly due to the fact that in both the remaining cases the most likely nodes to fail were grouped together creating highly unstable PGs. Not only will this increase the updating activity within the PG but all of their TEC contacts have to update their tables too, creating a constant surge of traffic in these locations. Despite the creation of highly active/unreliable PGs, (churn) observations show that the network still proves to be stable and robust, which is further verified by the absence of bursts in usage.

13.6 Conclusion

This chapter has presented PIndex as a viable peer-to-peer-based solution for grid information services, which builds on top of MDS4. Colored Petri nets were used to create a simulation model that could simulate many thousands of nodes in the PIndex network. Having simulated network populations of up to 10,000 computing nodes and investigated a variety of peer group grouping methods, the simulated results show PIndex to be highly salable and stable in addition to handling node failures.

Having discovered that PIndex remains stable despite the grouping of computing nodes with high failure rates, and that an expected rise of message usage and node usage has respective peak gradients of 1 and 0.78, this proves that PIndex is a highly scalable network structure and a viable solution to the grid information service problem. However, having modeled PIndex using CPN that included Poisson distribution rates and normal distribution probabilities, these can only give an indication of the performance PIndex will have under these conditions. The CPN model did not cover the underlying network infrastructure that PIndex was running on and at best related to a local network configuration. However, given the results the following two improvements to PIndex could be made.

Having already stated that the underlying network was not included in the CPN model, an additional improvement that could be made to PIndex is to include proximity in the PIndex algorithm such that the geographically local computing nodes will belong to one PG.

Although in the simulated results we observe a startup flux, this levels off to a stable state. On closer inspection oscillations are seen that can only be explained by PGs updating at regular intervals, which creates variations in resource usage. By implementing a time slotting mechanism for PG updates, resources can be effectively utilized.

References

1. Globus MDS4. Available at: http://www.globus.org/toolkit/mds/ #mds_gt4.
2. Globus Middleware. Available at: http://www.globus.org.
3. J.M. Schopf, M. D'Arcy, N. Miller, L. Pearlman, I. Foster, and C. Kesselman, "Monitoring and discovery in a Web services framework: Functionality and performance of the globus toolkit's MDS4," Technical Paper, Argonne National Laboratory, 2006. Available at: http://www-unix.mcs.anl.gov/~schopf/Pubs/mds4.hpdc06.pdf.
4. I. Stoica, R. Morris, D. Liben-Nowell, D. Karger, M. Kaashoek, F. Dabek, and H. Balakrishnan, "Chord: A scalable peer-to-peer lookup protocol for internet applications," *IEEE/ACM Transactions on Networking*, 11, 2003, 17–32.
5. A. Rowstron and P. Druschel, "Pastry: Scalable, decentralized object location and routing for large-scale peer-to-peer systems," in *Proceedings of the IFIP/ACM International Conference on Distributed Systems Platforms*, 2001, Heidelberg, Germany, pp. 329–350.
6. S. Ratnasamy, P. Francis, M. Handley, R. Karp, and S. Shenker, "A scalable content-addressable network," in *Proceedings of ACM Special Interest Group on Data Communications*, 2001, San Diego, CA, pp. 161–172.
7. S. Rhea, D. Geels, T. Roscoe, and J. Kubiatowicz, "Handling churn in a DHT," in *Proceedings of the USENIX Annual Technical Conference*, 2004, pp. 127–140.
8. V. Gehlot, and A. Hayrapetyan, "Systems modeling and analysis using colored petri nets: A tutorial introduction and practical applications," in *Proceedings of ACM Southeast Regional Conference*, 2007, 514, Winston-Salem, NC, p. 514.
9. K. Aberer, P. Cudré-Mauroux, A. Datta, Z. Despotovic, M. Hauswirth, M. Punceva, and R. Schmidt, "P-Grid: A self-organizing structured P2P system," presented at the *Sixth International Conference on Cooperative Information Systems (CoopIS)*, Trento, Italy, 2001.
10. I. Foster and A. Iamnitchi, "On death, taxes, and the convergence of peer-to-peer and grid computing," in *Proceedings of the 2nd International Workshop on Peer-to-Peer Systems (IPTPS)*, 2003, Berkeley, CA, pp. 118–128.
11. SOAP Version 1.2 Part 1: Messaging Framework, W3C Recommendation, 24 June 2003. Available at: http://www.w3.org/TR/soap12-part1.
12. H. Casanova, A. Legrand, and L. Marchal, "Scheduling distributed applications: The SimGrid simulation framework," *Proceedings of Cluster Computing and the Grid*, 2003, Tokyo, Japan.
13. R. Buyya and M. Murshed, "GridSim: A toolkit for the modeling and simulation of distributed resource management and scheduling for grid computing," CoRR cs.DC/0203019: 2002.
14. W. Kreutzer, J. Hopkins, M. van Mierlo, "SimJAVA—A framework for modeling queuing networks in java," in *Proceedings of Winter Simulation Conference*, 1997, Atlanta, GA, pp. 483–488.
15. Ganglia. Available at: http://ganglia.sourceforge.net.
16. C. Dumitrescu and I. Foster, "Gangsim, a simulator for grid scheduling studies," in *Proceedings of CCGrid*, 2005.

17. A.V. Ratzer, L. Wells, H. Lassen, M. Laursen, J. Qvortrup, M. Stissing, M. Westergaard, S. Christensen, and K. Jensen, "CPN tools for editing, simulating, and analysing coloured petri nets," in *Proceedings of ICATPN*, 2003, Eindhoven, The Netherlands, pp. 450–462.
18. M. Schlosser and S.D. Kamvar, "Modeling interactions in a P2P network," Technical Report, Stanford University, 2003.
19. N. Ting, and R. Deters, "Peer-to-peer network simulation," *International Conference on Enterprise Information Systems*, 4, 2004, pp. 84–91.
20. M. Marsan, "Stochastic petri nets: An elementary introduction," In Rozenberg, G. (Ed.), *Lecture Notes in Computer Science, Vol. 424; Advances in Petri Nets 1989*, Berlin: Springer-Verlag, 1990, pp. 1–29.
21. Web Services Resource Framework (WSRF). Available at: http://www.globus.org/wsrf.

Part III

Grid Applications

14

WISDOM: A Grid-Enabled Drug
Discovery Initiative against Malaria

Vincent Breton, Doman Kim, and Giulio Rastelli

CONTENTS

14.1 Introduction

The goal of this chapter is to present the WISDOM initiative, which is one of the main accomplishments in the use of grids for biomedical sciences achieved on grid infrastructures in Europe. Researchers in life sciences are among the most active scientific communities on the EGEE infrastructure. As a consequence, the biomedical virtual organization stands fourth in terms of resources consumed in 2007, with an average of 7000 jobs submitted every day to the grid and more than 4 million hours of CPU consumed in the last 12 months. Only three experiments on the CERN Large Hadron Collider have used more resources. Compared to particle physics, the use of resources is much less centralized as about 40 different scientific applications are now currently deployed on EGEE. Each of them requires an amount of CPU that ranges from a few to a few hundred CPU years. Thanks to the 20,000 processors available to the users of the biomedical virtual organization, crunching factors in the hundreds are witnessed routinely. Such performances were already achieved on supercomputers but at the cost of reservation and long delays in the access to resources. On the contrary, grid infrastructures are constantly open to the user communities.

Such changes in the scale of the computing resources made continuously available to the researchers in biomedical sciences open opportunities for exploring new fields or changing the approach to existing challenges. In this chapter, we would like to show the potential impact of grids in the field of drug discovery through the example of the WISDOM initiative.

14.2 Grid-Enabled Drug Discovery

14.2.1 *In Silico* Drug Discovery: Requirements and Grid Added Value

The pharmaceutical R&D enterprise presents unique challenges for information technologists and computer scientists. The diversity and complexity of the information required to arrive at well-founded decisions based on both scientific and business criteria are remarkable and well recognized in the industry.

Drug discovery is the process by which drugs are discovered and/or designed. Drug candidates are inputs to the drug development process. Recent progress in genomics, transcriptomics, proteomics, high-throughput screening, combinatorial chemistry, molecular biology, and pharmacogenomics has radically changed the traditional physiology-based approach to drug discovery where the organism is seen as a black box. *In silico* drug discovery contributes to increasing biological system knowledge. The efficiency gains of such an integrated knowledge system could correspond to save 35% costs, or about US$300 million, and 15% time, or two years of development time per drug.

In silico drug discovery is one of the most promising strategies to speed up the drug discovery process. It is important to know and control the *in silico* process, which is described below. Figure 14.1 shows the different phases of a drug discovery process with their approximate duration, success rate, and corresponding *in silico* contributions.

A target is a cellular or genetic molecule that is believed to be associated with a desired change in the behavior of diseased cells and on which drugs usually act. The target identification and validation aims to isolate and select it. *In silico* drug discovery contributes to the target discovery by gene expression analysis, target function prediction, and target three-dimensional (3D) structure prediction for postprocessing.

To identify a lead compound, a substance affecting the target selected in a drug-like way, two different *in silico* pipelines can be used that speed up the process and reduce costs avoiding useless *in vitro* tests: *de novo* design and virtual screening. *De novo* design builds iteratively a compound from the structure of a protein active site. Virtual screening selects *in silico* the best compound from a molecule database.

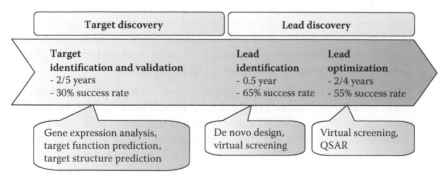

FIGURE 14.1 Representation of the different phases of the drug discovery process with their duration, success rate, and corresponding *in silico* contributions.

Lead optimization addresses the development from the most promising lead compounds to a safe and effective drug. Instead of expensive and longer *in vitro* and *in vivo* tests, evaluation of basic chemical properties can be achieved by virtual screening and quantitative structure activity relationship (QSAR). QSAR is a quantitative correlation process of chemical structure with well-defined methods, such as optimization for pharmaceutical properties [absorption, distribution, metabolism, excretion and toxicity (ADMET)] or efficacy against the target organism.

In silico drug discovery contributes to increasing biological system knowledge, to managing data in a collaboration space, to speeding up analysis, and consequently to increasing the low success rate of the traditional "wet" approach. The efficiency gains of such an integrated knowledge system could correspond to save 35% costs, or about US$300 million, and 15% time, or two years of development time per drug.

Nevertheless, in spite of increasing levels of investment in *in silico* techniques, there is a steady decline in the number of new molecules that enter clinical development and reach the market. Many factors have changed over the past ten years, particularly the domination of the target-based drug discovery paradigm, favoring screening and rational drug discovery programs. A new approach aims to integrate rational drug discovery with a strong physiology and disease focus.

14.2.1.1 Requirements

Reducing the research time and cost in the discovery stage and enhancing information about the leads are key priorities for pharmaceutical companies worldwide [1]. To achieve this goal, *in silico* drug discovery must meet the following requirements:

- The *in silico* drug discovery process includes the management of a large variety and quantity of scientific data; for example, images, sequences, models, and databases. Data integration is thus a challenge to increase knowledge discovery but also to ease the complex workflow. This implies data format standardization, dataflow definition in a distributed system, infrastructure and software providers for data storage, services for data and metadata registration, data manipulation, and database updates.

- The *in silico* drug discovery process also includes the management of a large variety and quantity of software. Software integration is another challenge to build efficient and complex workflows and to ease data management and data mining. Software can be provided in a distributed environment such as a Web server on the Internet. Different experts are absolutely necessary to maintain and update software and workflows to propose new methods or pipelines, to use remote services, exploit outputs, and finally to propose compounds for assay. A software workflow will assist the scientist and the decision-maker in organizing their work in a flexible manner, and in delivering the information and knowledge to the organization.

- Deploying intensive computing is a challenge for *in silico* drug discovery. For instance, computing 1 million docking probabilities or modeling 1000 compounds on one target protein requires in the order of a few TFlops per day. Very large computing resources are also needed to describe accurately protein structure models by computational methods based on all-atom physics-based force fields including implicit solution. Computing power is also required for bioinformatics resource centers where server access is saturated by the large number of short tasks requested by users.

- Joining new information technologies with life sciences to enable *in silico* drug discovery requires strong remote collaboration between different public and private experts when addressing neglected and emerging infectious diseases. It also involves strong sharing of resources: data and knowledge, software and workflow, and infrastructures such as computing, storage, and networks. The collaboration space needs experts to maintain the resources. Having tools and data accessible to everyone in collaboration requires intuitive interfaces that need to be maintained. These interfaces reduce the development time of new methods. They also help the integration of data and software from *in silico* drug discovery but also from experimental processes.

- Security is a key challenge for pharmaceutical industries but also for academic institutes in most cases. Effective protection of intellectual properties and sensitive information requires, for instance, authentication of users from different institutions, mechanisms for management of user accounts, and privileges and support for resource owners to implement and enforce access control policies.

In summary, the main requirements to develop *in silico* drug discovery are data and software integration, intensive computing deployment, remote collaboration and resources sharing, and, of course, security. Thus, there is a need for powerful and secured environment sharing and integrating remote resources such as tools, data, computing, and storage.

14.2.1.2 Grid Added Value

The grid added value in the development of *in silico* drug discovery for neglected and emerging infectious diseases has multiple dimensions:

- Grids offer unprecedented opportunities for resource sharing and collaboration.

- Grids open exciting perspectives for handling information flows.
- Grids provide the resources to speed up the execution of time-consuming software.

Grids offer unprecedented opportunities for resource sharing and collaboration:

- The sharing of resources in a cross-organizational collaboration space between the pharmaceutical industry and academic research institutions, and between developed and least developed countries
- The creation of a virtual laboratory for the different actors, increasing cooperation and communication between partners
- The mobilization of resources routinely or in an emergency
- The sharing of diverse, complex, large, and distributed information for collaborative exploration and mutual benefit
- The use of new information technology such as large databases or time-consuming software
- The optimal exploitation of resources by taking advantage of spare computing cycles or by maximizing the use of high-performance computing platforms usage
- The reduction of hardware costs

Grids also open exciting perspectives for handling information flows:

- The deployment of services for healthcare and research centers in endemic regions
- The deployment of infrastructures to collect data and improve disease surveillance and monitoring
- The building of knowledge space with genomics and medical information (epidemiology, status of clinical tests, drug resistances, etc.)
- Access to relevant data, periodically updated databases, and publications
- The federation of regional or international databases for disease study and monitoring of vector control, clinical trials, and drug delivery
- The provision of transparent and secure access to storage and the archiving of large amounts of data in an automated and self-organized fashion
- Connection, analysis, and structuring of data and information in a transparent mode according to predefined rules (science or business process based)

Finally, grids provide the resources to speed up the execution of time-consuming software:

- Access to large computing resources for *in silico* drug discovery, data analysis and mathematical modeling
- The application of high-performance computing to new areas
- The production of additional or more accurate analyses
- The facilitation of the exchange of tools and workflows between scientists
- The performance of computing intense tasks in a transparent way by means of an automated job submission and distribution facility
- Access to services and resources 24 hours a day
- The running of the same job on many platforms across different sites
- Access to computing resources by a single efficient path

Grids are unique tools for collecting and sharing information, networking experts, and mobilizing resources routinely or in an emergency. A grid is thus an appropriate environment to develop *in silico* drug discovery.

14.2.2 Grid-Enabled Virtual Screening

Virtual screening is about selecting *in silico* the best candidate drugs acting on a given target protein [2]. Screening can be done *in vitro* but it is very expensive as there are now millions of chemicals that can be synthesized [3]. A reliable way of *in silico* screening could reduce the number of molecules required for *in vitro* and then *in vivo* testing from a few millions to a few hundreds. Docking is only the first step of virtual screening since the docking output data have to be processed further [4].

However, *in silico* virtual screening requires intensive computing, in the order of a few TFlops per day, to compute 1 million docking probabilities or for the molecular modeling of 1000 compounds on one target protein. Access to very large computing resources is therefore needed for successful high-throughput virtual screening [5].

14.2.2.1 The Virtual Screening Pipeline

Screening of chemical compounds against a target is an important step in the drug discovery process. Virtual screening is the process that screens the chemical compounds *in silico* against a target. The prerequisite to set up a virtual screening experiment is knowledge on the target, against which the screening has to be performed, and on the chemical compound libraries. Most of the information related to the targets is available in the literature, whether it is digital or paper based.

Docking is the method of first choice for rapid *in silico* screening of large ligand databases for drug research, since it is based on a rational physical model. Basically, protein-compound docking is about computing the binding energy of a protein target to a library of potential drugs using a scoring algorithm. The target is typically a protein that plays a pivotal role in a pathological process; for example, the biological cycles of a given pathogen (parasite, virus, bacteria, etc.). The goal is to identify which molecules could dock on the protein active sites in order to inhibit its action and therefore interfere with the molecular processes essential for the pathogen. Libraries of compound 3D structures are made openly available by chemistry companies that can produce them. Many docking software are available either open-source or licensed.

However, there is very often a compromise between speed and accuracy of results (in terms of the actual binding mode as well as the calculated affinity values) concerning the best scoring docking solutions. Docking methods usually generate a number of possible orientations of ligands in the binding site of the receptor, and the "correct" one (e.g., the orientation observed in the crystal structure) may not necessarily be ranked among the first docking solutions. This is due to deficiencies in orientation sampling and to the approximate nature of the scoring functions. Thus, it seems reasonable to subject screening results to postdocking refinement; for example, using molecular dynamics or similar methods that can describe biomolecular structure and energy in more details.

An example of workflow to deploy virtual screening that additionally includes postdocking refinement goes through the following steps:

Step 1 Selection of the target, the chemical compound database, and the docking software.

Step 2 *Preparation.* If the selected target is an X-ray crystal structure with a bound ligand, then it requires preparing the binding site of the protein by taking 6-8 Angstroms from the cocrystallized ligand, ensuring that significant amino acids for the activity are included in the binding site. Information on the significant amino acids can be obtained either from the literature or from the Brookhaven protein database. Regarding the preparation of compounds to dock, open-source databases providing ready to dock molecules are made available on the Internet by companies selling the compounds, but both target and compound have to be prepared according to the needs of the software.

Step 3 *Docking.* Access to data analysis and visualization software is required at this point.

Step 4 *Postdocking analysis.* Results are analyzed based on the docking energy score and binding mode of the compound inside the binding site.

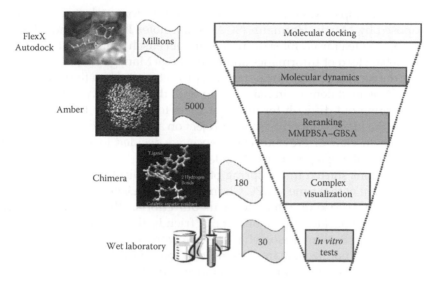

FIGURE 14.2   Example of a virtual screening workflow (credit: A. Da Costa).

Step 5   Refinement of docking orientations using molecular dynamics and reranking of the best compounds with more accurate scoring functions.

Step 6   *Postmolecular dynamics* analysis. Access to data analysis and visualization software is required at this point.

Step 7   Selection of a few hundreds of compounds for *in vitro* testing.

Figure 14.2 illustrates this workflow with some popular software packages.

14.3   Virtual Screening on the Grid: The WISDOM Initiative

14.3.1   Historical Perspective

The WISDOM initiative was born from discussions that took place in July 2003 at the PharmaGrid conference in Welwyn, UK. Organized yearly since 2001 by the PRISM forum [6], the PharmaGrid cycle of conferences was aimed at steering exchanges between the leaders of the IT departments of the pharmaceutical laboratories and the developers of the grid technology. During the conference in Welwyn, the idea was discussed to identify a few topics where private and public partners could work together on the development and deployment of pilot projects on the grid.

One of the topics was *in silico* drug discovery for neglected diseases. Proposed by Martin Hofmann (SCAI, Fraunhofer Institute) and Vincent Breton, the idea raised interest from Manuel Peitsch who was at that time the Global Head of Informatics and Knowledge Management at Novartis. After the conference, Novartis started a project focused on finding *in silico* new hits against dengue in partnership with the University of Basel. As it was turning out difficult for other partners to contribute to this dengue project, the idea to launch a second project on an open-grid infrastructure emerged. In view of previous experience, we decided to start the project without involving a pharmaceutical laboratory.

Discussions started in the spring of 2004 with Nicolas Jacq, who was a PhD student at that time at LPC Clermont-Ferrand. The first decision was to choose a disease on which to focus our efforts. We did not have any *a priori*, so we decided to ask one of our friends, the African pastor Joany Bazemo from Burkina-Faso, what was the worst plague in Africa. When asked, he did not hesitate in replying that malaria was the worst plague because it was killing children in hundreds of thousands. Once the disease was chosen, the next decision was to choose a biological target. This was entrusted to Vinod Kasam [27], a masters student at SCAI Fraunhofer, who studied the literature and identified the family of plasmepsins as a new potentially promising target. Plasmepsins are involved in the degradation of hemoglobin by the malaria vector, *Plasmodium falciparum*. Structures for plasmepsin II and IV were available in the Protein Data Base so that all the ingredients were there for launching our first large virtual screening deployment on the grid. Large-scale deployments are called data challenges in EGEE jargon.

14.3.2 First Data Challenge on Malaria

14.3.2.1 Preparation

Having identified the disease and targets, there was still a lot of work to do in order to prepare the data challenge. On the biochemical side, three ingredients were required:

- 3D structures of targets
- A docking software
- A database of drug-like molecules to dock against the targets

We had already identified the 3D structures of interest in the PDB [7] database.

The docking software initially selected was Autodock [8], an open-source algorithm developed by the Scripps Research Institute. SCAI Fraunhofer is known worldwide for developing one of the best docking algorithms, FlexX [9], but FlexX could not be deployed on multiple grid sites because it is licensed software. We had planned, however, that the best compounds selected with Autodock would be reprocessed on the Fraunhofer cluster using the FlexX algorithm to compare Autodock and FlexX scores.

Libraries of compound 3D structures are made available open source by chemistry companies that can produce them. ZINC is one of these open-source databases of compounds [10], which is constantly growing and was already providing the 3D structures of more than 3.4 million compounds in 2005.

Tests performed in December 2004 showed that docking the whole of ZINC against one target using Autodock represented about 35,000 jobs of approximately 20 hours, corresponding to about 80 years CPU. It soon appeared relevant from a biochemical perspective to perform docking not only on one but rather on three plasmepsin II structures (1lee, 1lf2, 1lf3) and one plasmepsin IV structure (1ls5) obtained from the Brookhaven Protein Data Base. In view of this increase of the number of targets, it was decided to focus on a subset of ZINC, the ChemBridge database collecting "only" 500,000 compounds. Finally, thanks to our collaborators from SCAI Fraunhofer, the BioSolveIT company distributing the FlexX software made graciously available up to 3000 free licenses of the software for deployment on the grid. This was extremely important as we had only a limited experience with Autodock.

Preparation of the data challenge involved, on the biochemical side, a number of steps in relation to the preparation of targets and the validation of the docking procedure. On the grid side, this was the first time a biomedical application was going to require about 100 years of CPU time. At that time, only the LHC experiments had been deploying data challenges and the know-how for such large-scale deployments was shared by only a few experts. Preparation of the deployment included the development of an environment for job submission and output data collection. This environment had to be able to handle the submission of about 70,000 15-hour long jobs and the collection of the output data. A major issue was to handle job resubmission whenever a job failed for any reason, as the grid success rate was typically of the order of 80–85%. Large-scale tests were made on the French regional grid AuverGrid to validate the environment and to identify potential issues and bottlenecks. Other issues were raised by the data challenge, like the usage of licensed software on the grid or the need for a high-throughput job submission scheme.

Based on the experience acquired during the testing phase on AuverGrid, the WISDOM production system (see Figure 14.3) [11] was developed in Perl, except for the multithreaded job submission tool in Java. Two packages, wisdom_install and wisdom_test, were developed for installing the application components on the resources and for testing these components, together with the resources and grid services.

For the user submitting jobs, the entry point was a simple command line tool. Its users during the data challenge were members of the

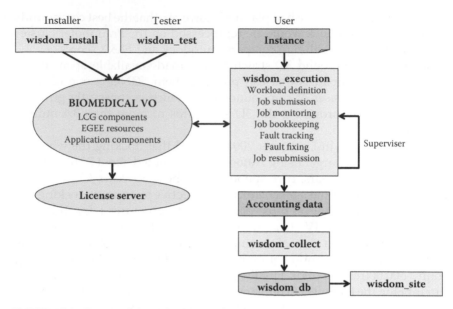

FIGURE 14.3   Design of the WISDOM production system.

Biomedical Task Force, which gathers a team of engineers with recognized expertise in application development and deployment. This software environment was developed to allow the submission and monitoring of job sets, which were called *instances*. The different jobs of a given instance have the same target input and docking software. They only differ by the molecules of the compound library that are docked. Tasks needed to submit an instance were automatically executed by the WISDOM execution system. The user, authenticated by a proxy certificate, had to start his or her instance execution following a precise submission schedule to avoid too much competition between the computation participants, which would lead to a grid overload. Once the computation had started, the WISDOM environment took care of monitoring jobs and registering results. The user only had to check regularly if the process ran correctly, up to the end of all the jobs belonging to the instance. The overall process progression could be monitored through an output file for follow-up messages and an error file in case of any problems.

For each instance, a configuration file contained the instance information (software, target, database, parameter settings) and the grid parameters (number of jobs for the instance, resource brokers, computing elements, storage elements). A shell script and a job description language file were created for each job and used by the submission tool.

On the worker node, after the environment was configured, the shell script downloaded the database file from a storage element chosen by the information system using the large hadron collider computing grid application programming interface (LCG API). Binaries were then called with the target and parameter settings transferred with the job. The compressed result was stored on a storage element and registered in the grid file catalog. A backup copy was also generated on another storage element. For the sake of simplification, the most relevant metadata relative to the output (software, parameter settings, compounds database, etc.) were stored in the name of the output itself. Output, errors, and accounting messages were transferred on the user interface.

Execution of FlexX jobs required handling floating licenses. Each job using FlexX software was contacting the Flexlm server at the beginning of the job and asked for a license, namely an ASCII file with specific keys generated for this server. Then the job was able to run without connection to the license server. Accessing floating licenses on the grid behind firewalls required known IP and the opening of two specific ports for institutes hosting worker nodes.

14.3.2.2 Deployment

The deployment took place in July and August 2005. During this period, ten users launched jobs from five user interfaces, monitoring the process with the help of the WISDOM environment and interacting with the user support of the EGEE project and the nodes administrators. For a total of 80 CPU years, 72,751 jobs were launched, producing 1 TB of data (500 GB, doubled for the backup). Detailed analysis of the deployment can be found in [11], from which Table 14.1 is extracted.

TABLE 14.1

Main Metrics from WISDOM First Data Challenge

Metrics	Total	FlexX Phase	Autodock Phase
Cumulated number of docked compounds (in millions)	4127	3141	987
Effective duration	37 days	22 days	15 days
Number of docked compounds/hour	46,475	59,488	27,417
Crunching factor	662	411	1031
Number of jobs submitted	72,751	41,520	31,231
Number of grid computing elements used	58	56	57
Number of resource brokers used	12	12	11
Maximum number of jobs running in parallel on the grid	1643	1008	1643
Volume of output data	946 GB	506 GB	440 GB
Total CPU time	80 years	29.5 years	50.5 years
Effective CPU time used by successful jobs	67.2 years	24.8 years	42.4 years

The data challenge was very useful to identify the limitations and bottlenecks of the EGEE infrastructure. The WISDOM production system developed to submit the jobs on the grid accounted for a small fraction of the failures as well as the grid management system. On the other hand, the resource brokers were observed to significantly limit the rate at which the jobs could be submitted. Another significant source of inefficiency came from the difficulty for the grid information system to provide all the relevant information to the resource brokers when they distributed the jobs on the grid. As a consequence, job scheduling was a time-consuming task for the WISDOM users during all the data challenge due to the encountered limitations of the information system, the computing elements, and the resource brokers.

14.3.2.3 Results

Postprocessing of the huge amount of data generated was a very demanding task as millions of docking scores had to be compared. At the end of the large-scale docking deployment, the best 1000 compounds based on scoring were selected thanks to postprocessing ranking jobs deployed on the grid. They were inspected individually. Several strategies were employed to reduce the number of false positives. A further 100 compounds were selected for postprocessing. These compounds had been selected based on the docking score, the binding mode of the compound inside the binding pocket, and the interactions of the compounds to key residues of the protein.

Several scaffolds were identified in the 100 compounds selected for postprocessing. The urea, thiourea, and guanidino scaffolds were the most frequently observed in the top 1000 compounds. Some of the compounds identified were similar to already known plasmepsin inhibitors, like the urea analogs that were already established as micromolar inhibitors for plasmepsins (Walter Reed compounds) [12]. These results were already an indication that the overall approach was sensible and that large-scale docking on computational grids had the potential to identify new inhibitors, such as the guanidino analogs. To confirm these results, a new step had to be implemented: the refinement of the compound selection using molecular dynamics computations.

14.3.3 Molecular Dynamics on the Grid

14.3.3.1 Introduction

While docking methods have been significantly improved in the last few years by including more thorough compound orientation searches, additional energy contributions, and/or refined parameters in the force field, it is generally agreed that docking results need to be postprocessed with more accurate modeling tools before biological tests are undertaken. Molecular dynamics (MD) [13] has great potential at this stage: first, it enables a flexible treatment of the compound/target complexes at room temperature for a given simulation time, and therefore is able to refine compound orientations by finding more stable complexes; second, it partially solves conformation and orientation search deficiencies that might arise from docking; third, it allows the reranking of molecules based on more accurate scoring functions.

Just at the time when the results of the first WISDOM data challenge were being analyzed at SCAI Fraunhofer by Vinod Kasam, the BioinfoGRID project was launched in January 2006. LPC Clermont-Ferrand contribution to the project focused on the deployment of MD computations on the grid.

14.3.3.2 Deployment

Work started on the identification of the relevant molecular dynamics software and on the choice of the grid infrastructure on which to deploy the computations. Contacts were established with two groups at SCAI Fraunhofer and University of Modena, which expressed interest in reranking the best hits coming out of the first WISDOM data challenge. Both groups were using the Amber software for molecular modeling which raised again the problem of deploying a licensed software on a grid. Contacts were established with the institution distributing Amber regarding the license policy on the grid.

The outcome of the negotiation was that we were allowed to deploy Amber on the grid under the following conditions:

- Each cluster deploying Amber had to have at least one license.
- Grid users allowed to use Amber had to come from one of the laboratories owning an Amber license.
- Grid users allowed to use Amber under the conditions described above could deploy their computations on all the grid clusters.

Regarding the choice of infrastructure, contacts were established with the EGEE biomedical virtual organization and several groups involved in DEISA. A clear preference for deployment on EGEE was expressed by the groups collaborating with us at the University of Modena and SCAI Fraunhofer. Both had developed MD procedures that were well fitted for running on clusters. As a consequence, we focused our efforts on deploying these procedures on EGEE.

From September 2006, we started to investigate in detail the deployment of the MD procedure designed by Giulio Rastelli (University of Modena and Reggio Emilia) on EGEE and based on the Amber software package. Amber [14] is a suite of different tools that carry out molecular dynamics simulations. The simulations in Amber can be divided into three phases and different programs of the Amber distribution are responsible for performing these steps:

- Preparatory phase
- Simulatory phase
- Analysis phase

Encoding these steps in separate programs has some important advantages. First, it allows individual pieces to be upgraded or replaced with minimal impact on other parts of the program suite. Second, it allows different programs to be written with different coding practices: LEAP is written in C using X-window libraries, ptraj and antechamber are text-based C codes, mm-pbsa is implemented in Perl, and the main simulation programs are coded in Fortran 90. Third, this separation often eases porting to new computing platforms: only the principal simulation codes (*sander* and *pmemd*) need to be coded for parallel operations or need to know about optimized (perhaps vendor-supplied) libraries. The preparation and analysis programs are carried out on local machines on a user's desktop, whereas time-consuming simulation tasks are sent to a batch system on a remote machine; having stable and well-defined file formats for these interfaces facilitates this mode of operation.

14.3.3.3 Molecular Dynamics Refinement and Rescoring Procedure

As shown in Figure 14.4, an automated multistep procedure for the refinement and rescoring of docking screening results was designed and

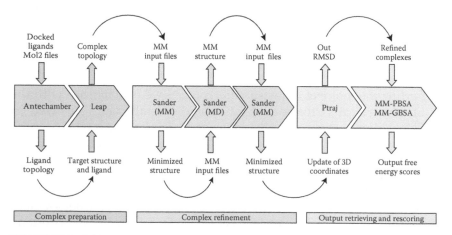

FIGURE 14.4 Schematic representation of the automated multistep refinement and rescoring procedure of the ligand–target complexes based on Amber. (From Rastelli, G. et al., *Bioorganic and Medicinal Chemistry*, 15, 7865–7877, 2007.)

validated at the University of Modena in Giulio Rastelli team [15]. The workflow, based on the Amber package, is able to automatically and efficiently refine docking poses, which sometimes may not be accurate, and rank the compounds based on more accurate scoring functions. The procedure called BEAR (Binding Estimation After Refinement) requires as input a pdb file containing the structure of the protein and a mol2 file containing the coordinates of the docked ligands. The coordinates of the docked ligand and target structure are merged to create the complex.

The topology files are created using antechamber. The ligand atoms are described with GAFF (general Amber force field) atom types and AM1-BCC charges. In order to avoid the time-consuming procedure of charge calculation, atomic charges of the ligand are read from the original mol2 file and not computed during the procedure; this choice adds the advantage that charge calculation of ligands can be performed only once, and obtained mol2 files can be used for many other target proteins. The ligand charges are calculated using antechamber by means of a separate script. Interestingly, the same set of AM1-BCC charges can also be exploited for automated ligand docking and MD. Missing GAFF parameters for the ligand are automatically assigned by parmcheck. The ligand, receptor, and complex topologies are written using leap utility (Amber 9). Minimization, MD, and final reminimization of the complexes are performed using *sander* with a distance-dependent dielectric constant $\varepsilon = 4r$. For each of these steps, the procedure enables the user to set ad hoc refinement options depending on the application. After refinement of the complex, a pdb file is generated as output and the final coordinates of the ligand, receptor, and complex are updated and used to compute the binding free energy evaluation using Amber MM-PBSA and MM-GBSA. The free energy results (ΔG_{MM}, ΔG_{solv}, and $\Delta G'_{bind}$) are written to an output file and compounds are ranked on the basis of their binding free energy [15].

This workflow was implemented in the EGEE grid using the WISDOM production environment designed for the large-scale docking experiments to deploy the MD procedure on the grid. The wide CPU availability of the EGEE grid allowed the submission of a large subset of best docking complexes to the MD refinement and rescoring procedure despite its high computational demand.

The first successful deployment of the MD procedure was achieved in December 2006 on fake data. At the preparation stage, all the required input files and Amber executables were stored on the storage element (SE) of the grid. At the execution stage, jobs lasting for approximately 20 hours were submitted corresponding to 50 compound subsets. Each subset with 50 compounds was submitted on one worker node along with the Amber executables, target structure, and the main script controlling the MD refinement and rescoring operations. After the jobs were finished the result files (structures and energy scores) were copied and stored on the storage element.

Following the successful test, the 5000 best compounds coming out of the first data challenge on plasmepsin II were refined and reranked. Computing resources from the EGEE Biomed virtual organization were used exclusively; 100 CPUs were used in parallel, all of them belonged to our local cluster at Clermont Ferrand, due to licensing issues. One single simulation was consuming ~20 CPU minutes on an Intel Xeon 3.05 machine. The estimated CPU time if the simulations were to be performed on one machine was therefore expected to be 124 days. By using EGEE infrastructure, the simulation time was significantly brought down to 7 days.

14.3.3.4 Results

After rescoring the 5000 best docking results by Molecular Dynamics with Amber and MM-PBSA and MM-GBSA, the next step was to select the best compounds in the perspective of *in vitro* tests. The starting points were the two ranked lists of compounds, one according to MM-PBSA and the other according to MM-GBSA free energies. One hundred complexes of each list were analyzed manually. Each complex was visualized in 3D with UCSF Chimera software in order to determine the molecular interactions between protein and ligand in the complex. The major criteria for selection were the ligand-making interactions to the two catalytic residues of plasmesin II, Asp 34 and Asp214. Second, the interaction with other key amino acids was checked: Gly36, Val78, Gly216, Ser79. The complexes with no interaction with at least one of the two amino acids of the catalytic dyad were rejected. The complexes that were kept had at least one main interaction to amino acids of the catalytic dyad. In total, 30 out of 200 compounds were selected for *in vitro* tests.

14.3.4 *In Vitro* Tests

A subset of the 30 compounds selected for testing showed submicromolar or nanomolar IC_{50} values against recombinant *P. falciparum* plasmepsin II, using the inhibition assay based on FRET substrate degradation, which is well documented in the literature [16,17]. In the parasite, plasmepsin II is translated as an inactive zymogen containing a 124 amino acid-long N-terminal prosequence that has a membrane-spanning domain. Within the food vacuole the prosequence is removed by a calpain-like maturase, and active plasmepsin II is released [18]. Plasmepsin II was expressed well from a pET3d construct that contained Glu124 after initiator Met, and the inclusion body of the protein was refolded and purified to near homogeneity as judged by SDS-PAGE. Based on molecular mass standards, recombinant plasmepsin II migrates at about 37 kDa.

Pepstatin A, a general inhibitor of aspartic proteases of microbial origin [19], was also reported to inhibit hemoglobin degradation by extracts of digestive vacuoles of *P. falciparum* [20]. In the *in vitro* inhibition test, the recombinant plasmepsin II activity without an inhibitor was used as a negative control and pepstatin A was used as a positive control. Pepstatin A showed inhibition (subnanomolar) against the recombinant plasmepsin II activity. Among 30 compounds selected for biological tests, 26 compounds revealed subnanomolar IC_{50} values, but four compounds had no inhibitory activity. And three compounds showed similar or better inhibitory activity compared to Pepstatin A. These results are extremely encouraging and suggest that the overall approach used to select the candidates is sensible for discovery of new plasmepsin inhibitors. The seven compounds are currently being evaluated for their antimalarial activities *in vitro*.

14.4 Second Data Challenge on Malaria

14.4.1 Introduction

With the success achieved by the first data challenge on both the computation and biological sides, several scientific groups around the world proposed targets implicated in malaria that led to the second assault on malaria. The WISDOM-II project dealt with several targets, which were both X-ray crystal structures and homology models. Targets from different classes of proteins were also being tested: reductases such as malarial dihydrofolate reductase (DHFR) and transferases such as glutathione S-transferase (GST) as can be seen from Table 14.2. The same procedure described previously for the first data challenge was applied for target preparation. We again had the privilege of being able to use the FlexX software, thanks to the generous support of BioSolveIT.

DHFR and dihydropteroate synthase (DHPS) are two enzymes that belong to the folate biosynthetic pathway. The antifolates are the most exploited class of antimalarials, to which belong well-known molecules like pyrimethamine and cycloguanil. To date, the most widely used antifolate is a combination of pyrimethamine, a DHFR inhibitor, and sulfadoxin, a DHPS inhibitor. Nevertheless, their synergic action that results in enhanced activity is seriously compromised by drug resistance and hypersensibilization. For example, drug resistance is due to point mutations of various amino acids in the DHFR and DHPS (*P. falciparum* and *P. vivax*) active sites, and severely decreases drug efficacy. Although most antimalarial research has been conducted on *P. falciparum* DHFR, there is growing interest in *P. vivax* DHFR, a less-studied target that is becoming increasingly important because mixed falciparum and vivax infections are increasing, and parasites have developed resistance to both. Therefore, there remains a pressing need of new molecules apt to selectively bind these targets. To date, six different malarial DHFR crystal structures are

TABLE 14.2

Structural Features of all the Targets Used in the Second WISDOM Data Challenge

Target	Activity	Structure	PDB id	Resolution (Å)	Ligand	Cofactor
GST	Detoxification	Dimer	1Q4J	2.2	GTX	NO
P. falciparum DHFR (wild-type)	DNA synthesis	Bifunctional (with TS)	1J3I	2.33	WR99210	NADPH
P. falciparum DHFR (quadruple mutant)	DNA synthesis	Bifunctional (with TS)	1J3K	2.10	WR99210	NADPH
P. vivax DHFR (wild-type)	DNA synthesis	Bifunctional (with TS)	2BL9	1.90	Pyrimethamine	NADPH
P. falciparum DHFR (double mutant)	DNA synthesis	Bifunctional (with TS)	2BLC	2.25	Des-chloropyrimethamine	NADPH
Tubulin (*Plasmodium*)	Cell division	Monomer	Homology model	—	—	GTP

available in the Protein Data Bank. These are the structures of wild-type *P. falciparum* DHFR (PDB code 1J3I) and of its C59R + S108N (1J3J) and N51I + C59R + S108N + I164L (1J3K) highly resistant mutants. In addition, four crystal structures of *P. vivax* DHFR are available, the wild-type structure (2BL9, 2BLB) and the S58R + S117N resistant mutant structures (2BLA, 2BLC). Hence, we thought it was a great opportunity to make use of these available resources to develop technology-based *in silico* screening on these targets. We chose *P. falciparum* DHFR (wt and the highly resistant quadruple mutant) and *P. vivax* DHFR (wt and the double mutant) structures (Table 14.2).

14.4.2 Evolution of the Production Environment

Following the experience acquired during the first data challenge on malaria and the first data challenge on avian flu [21], the WISDOM environment was reorganized in two different and independent tasks:

- The submission of the jobs
- The follow-up of the jobs, and eventually their resubmission, as well as the collection of the job status and their publication on a Web site

These two processes can be started and run simultaneously, the second one being fed from the information provided by the first one.

As seen during the deployment against avian flu [21], we removed the automatic resubmission of jobs in case of a job failure, because we thought that the overload generated by this automatic resubmission was a major cause for the poor reliability we observed during the first WISDOM deployment. Removing the automatic resubmission of jobs helped indeed a lot to improve overall reliability, but this induced a large amount of work for the users, who had to handle manual resubmissions of the instances after all the jobs had finished. The environment was further modified and enhanced to become more a "launch and forget" system, to carry all the tedious tasks and relieve the users whenever possible, such as automatic resubmission of jobs, automatic storage of job results in a relational database, automatic and real-time update of the experiment statistics and status, viewable through a Web site, and so on. As a consequence, the WISDOM environment was easily customized, tested, and successfully used by the EELA [22] and EUChinaGRID [23] virtual organization.

The main objective was also to improve the fault tolerance of the system, in implementing, for instance, a persistent environment that can be stopped and restarted at any time without the risk of losing important information. This also proved to be very useful as it enabled the whole maintenance of the scripts and code and improved interactivity with the user, as the user could also manage jobs in fine detail; for instance, force the cancellation and resubmission of a scheduled job. Along with this, we tried to minimize the cost of the environment in terms of disk space and CPU consumption for the user interface. While a 500 MB quota account was just enough to handle a single instance during the deployment on avian flu [21], it was now enough to handle no less than 20 instances simultaneously. Most of the job files were now generated dynamically, which also allowed the user to modify on the fly the configuration of the resource brokers and the job requirements and this way, the user was sure that the next submissions would take these modifications into account. Figure 14.5 describes the overall architecture of the environment and the deployment process:

- The user is interacting with the system through the two main scripts (wisdom_submit and wisdom_status) deployed on the user interface. These scripts will take care automatically of job files generation, submission, status follow-up, and eventually resubmission.

- The jobs are submitted directly to the grid workload management system, and are executed on the grid computing elements. As soon as it is running, a job transfers all the files stored on the storage elements via the data management system of the grid, and the FlexX software, which asks a floating license for the FlexLm license server, can start to process the dockings.

- During the job lifetime, the status is retrieved from the user interface and some statistics are generated and collected to a remote server that hosts a relational database and outputs these statistics through a Web site.

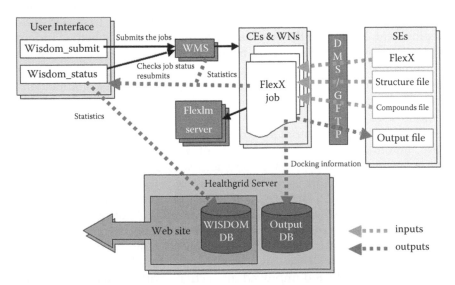

FIGURE 14.5 Evolution of the production environment.

- Once the job is finished, the outputs are stored back on the grid storage elements via the data management system and the useful docking results are inserted directly from the grid to a relational database where they can later be more easily queried and analyzed.

14.4.3 Data Challenge Deployment

The deployment was performed on several grid infrastructures (Auvergrid [24], EELA, EGEE, EUChinaGrid) and involved at least one manager to oversee the process on each of them. The three groups of targets (GST, *P. vivax*, and *P. falciparum* DHFR) were docked against the whole ZINC database (4.3 million ligands). The database was actually cut into 2422 chunks of 1800 ligands each. This splitting was chosen because we wanted to have an approximated processing time ranging from 20 to 30 hours for each job (one docking process takes from 40 seconds to 1 minute, depending on CPU power). The subsets had to be stored on the grid infrastructures. They were basically copied on a storage element and registered on the grid file catalog (LFC) and were also replicated on several locations whenever possible to improve fault tolerance. We defined a WISDOM instance as being a target structure docked against the whole ZINC database, with a given parameter set. Table 14.3 shows the instances deployed on the different infrastructures.

A total number of 32 instances were deployed, corresponding to an overall workload of 77,504 jobs, and up to 140 million docking operations. As shown in Figure 14.5, the environment included a FlexLm server that provided the floating licenses for the FlexX commercial software. During the first WISDOM deployment in 2005, the license server was identified as a potential bottleneck and point of failure because we had just one server

TABLE 14.3

Instances Deployed on the Different Infrastructures during the WISDOM-II Data Challenge

Target Structures	Number of Instances Deployed
GST (A chain)	4 on EGEE
GST (B chain)	4 on EGEE
2BL9 (*P. vivax* wild-type DHFR)	3 on EGEE, 1 on EELA
2BLC (*P. vivax* double mutant DHFR)	3 on EGEE, 1 on AuverGrid
Dm_vivax (*P. vivax* DHFR 2BLC minimized)	4 on EGEE
Wt_vivax (*P. vivax* DHFR 2BL9 minimized)	4 on EGEE
1J3K (*P. falciparum* quadruple mutant DHFR)	4 on EGEE
1J3I (*P. falciparum* wild-type DHFR)	3 on EGEE, 1 on EuChinaGRID

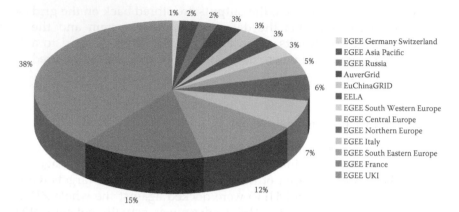

FIGURE 14.6   Distribution of jobs on the different grid federations.

available at this time. For WISDOM-II, we started the deployment with only one server as well. Very soon, up to three servers were made available at the SCAI Fraunhofer institute (http://www.scai.fraunhofer.de), with 3000 licenses available on each server. The FlexX software binaries were stored like all the inputs on the grid storage elements and were installed on the fly on each worker node at the beginning of the job.

As the average duration of a job was around 30 hours, we submitted one instance per day, with a delay of 30 seconds between each submission. As one instance was submitted in about 20 hours, the submission process was quite continuous during the first month of deployment. The jobs were submitted to 15 resource brokers in a round-robin order. At the end of a job, the results were stored on the grid storage elements, and directly into a relational database.

Figure 14.6 shows the repartition of jobs on the different grid federations. It also shows the AuverGrid, EuChinaGRID, and EELA infrastructures contribution. Each of these three infrastructures ran one single instance that corresponds to 3% of the total 32 instances.

14.4.4 Postdocking Analysis and MD Refinement

Analysis of the docking results was carried out on the four targets docked. The University of Modena took charge and analyzed the results obtained on DHFR targets from *P. vivax* and *P. falciparum*. CSIR, in collaboration with LPC Clermont-Ferrand, analyzed results on GST. Two criteria were employed for result analysis: statistical analysis based on scoring values and analysis of compounds making interactions to key residues of the receptor. Based on these two criteria, the best compounds coming out of the docking step were selected and refined using the MD procedure described previously [15].

The results refined with MD at the University of Modena for wild-type *P. falciparum* DHFR showed that the identification of novel families of anti-malarials whose structures are not related to classical antifolates is indeed possible. For wild-type *P. falciparum* DHFR, the best 15,000 compounds resulting from the docking screening (FlexX) have been refined with the MD procedure, and the molecules have been rescored using MM-PBSA and MM-GBSA. Hydrogen bond analyses of the 15,000 ligands interacting at the active site revealed a significant enrichment of hydrogen bonds with DHFR key residues Asp54, Ile14, and Ile164 (as well as their combinations) in the first ranking positions. This finding may validate the approach, because interaction with these residues is known to be required for biological activity.

In perspective, MD refinement will also be performed on the results obtained with the quadruple mutant DHFR structure and the results generated with *P. vivax* DHFRs. Once the analyses are complete, a final selection of potential inhibitors will be performed and the best candidates will be tested *in vitro* at Mahidol University (Thailand) in Worachart Sirawaraporn's laboratory. MD refinement of the results obtained on GST has been performed in collaboration with the University of Modena, and the results of this rescoring step are presently under analysis.

14.5 Conclusion and Perspectives

The WISDOM initiative has demonstrated how the grid can significantly change the approach to screening, which is a mandatory step on the road to drug discovery. In the last couple of years, a number of targets have been screened *in silico* using grid infrastructure resources. The molecules that have been selected through this process have shown significant inhibition activity *in vitro* and some of them are under a patenting process. The grid added value has been clearly demonstrated by the remarkable increase of the virtual screening throughput up to 100,000 docked compounds per hour using 5000 computers on the EGEE grid.

Following this success, the WISDOM initiative, started from discussions between Martin Hofmann and Vincent Breton at the PharmaGrid conference in Welwin City in July 2003, has turned into a collaboration that has seven partners today:

- CNRS in Clermont-Ferrand (http://clrpcsv.in2p3.fr), with expertise in grid technology and Web services for life sciences
- SCAI Fraunhofer institute in Bonn, with expertise in biochemoinformatics, and text-mining technology, and author of the FlexX docking software

- University of Modena, with expertise in molecular dynamics for virtual screening and drug design on malaria
- CNR-ITB Institute for Biomedical Technologies in Milano, with expertise in high-performance and GRID computing applications for bioinformatics in life sciences
- HealthGrid (http://healthgrid.org), with expertise in grid technology for life sciences and in infrastructure support
- Chonnam National University, with expertise in biochemistry and *in vitro* validation of virtual screening
- KISTI in Korea, with expertise in grid technology

Besides this core group, a number of biological laboratories have expressed interest to propose targets or to perform *in vitro* tests of the best molecules selected *in silico*:

- Mahidol University in Thailand
- University of Los Andes in Venezuela
- CSIR and University of Pretoria in South Africa

These very satisfactory results obtained both on the biochemical side and on the grid side of the project should not hide the fact that very significant progress can be achieved on both sides.

The virtual screening pipeline described in Figure 14.2 can be refined at several levels.

- In terms of target preparation, it is well known that the PDB is full of bugs, especially for the structures that have been downloaded many years ago.
- In terms of ligand selection, our approach has been very naïve as we have been docking all the compounds available in ZINC and Chembridge. Software packages exist to perform the first selection of the ligands on physical and chemical criteria to reduce the number of dockings.

The availability of three-dimensional macromolecular coordinates is a prerequisite for many types of studies. PDB contains many anomalies ranging from proteins with small deviations from normal geometry, to structures that fit their submitted experimental data very poorly.

Led by CMBI, a massive rerefinement project was launched to obtain a better match between the experimental data and the atomic parameters (coordinates, B-factors) in the structure models while not compromising the geometric quality of the structure models [25]. The rerefinement of such a vast number of structure models requires enormous amounts of computer power. The rerefinement of 18,738 PDB files with 18,738 independent jobs requiring from 1 to 24 hours of CPU each is a very good example of a parallel project ideally suitable for deployment on a computing grid.

The rerefinements of the structure models were performed on a hybrid computer environment consisting of the Biomed and EMBRACE [26] virtual organizations on EGEE grid infrastructure as well as several clusters of EMBRACE partners in Europe. Each grid job consisted of 20 proteins that would run for 20 hours approximately on queues of 72 hours and were managed using the WISDOM production environment. More than 90% of the total of 17 years of CPU time was finished in two months. The 18,738 rerefinements were completed in four months. The rerefined structure models are available from http://www.cmbi.ru.nl/pdb_redo/.

Selection of ligands can significantly reduce the number of compounds docked. It can be based on the toxicity or on the properties of the target binding site. We are presently investigating existing open-source software packages to filter the ligands before their docking on the grid.

Acknowledgments

The authors would like to acknowledge the contribution from numerous collaborators to the research activities described in this chapter. We would like to particularly thank Vincent Bloch, Ana Da Costa, Gianluca Degliesposti, Matteo Diarena, Géraldine Fettahi, Nicolas Jacq, Yannick Legrè, Simon Nowak, and Jean Salzemann.

The work described in this chapter was partly supported by grants from the European Commission (BioinfoGRID, EGEE, Embrace), the French Miinistry of Research (AGIR, GWENDIA), and the regional authorities (Conseil Régional d'Auvergne, Conseil Général du Puy-de-Dôme, Conseil Général de l'Allier).

The Enabling Grids for E-sciencE (EGEE) project is cofunded by the European Commission under contract INFSO-RI-031688. The BioinfoGRID project is cofunded by the European Commission under contract INFSO-RI-026808. The EMBRACE project is cofunded by the European Commission under the thematic area "Life sciences, genomics and biotechnology for health," contract number LHSG-CT-2004-512092. The SHARE project is cofunded by the European Commission under contract number FP6-2005-IST-027694.

AuverGrid is a project funded by the Conseil Regional d'Auvergne. The AGIR and GWENDIA projects are supported by the French Ministry of Research.

References

1. The Innovative Medicines Initiative (IMI) Strategic Research Agenda. Creating Biomedical R&D Leadership for Europe to Benefit Patients and Society, February 2008. Available at http://www.imi.europa.eu/docs/imi-gb-006v2-15022008-research-agenda_en.pdf.
2. Lyne, P.D. Structure-based virtual screening: An overview. *Drug Discovery Today* 7, 2002, 1047–1055.
3. Congreve, M., Murray, C.W., and Blundell, T.L. Structural biology and drug discovery. *Drug Discovery Today* 10, 2005, 895–907.
4. Ghosh, S., Nie, A., An, J., and Huang, Z. Structure-based virtual screening of chemical libraries for drug discovery. *Current Opinion in Chemical Biology* 10, 2006, 194–202.
5. Chien, A., Foster, I., and Goddette, D. Grid technologies empowering drug discovery. *Drug Discovery Today* 7, 2002, 176–180.
6. PRISM forum. Available at: http://www.prismforum.org.
7. Berman, H.M., Westbrook, J., Feng, Z., Gilliland, G. Bhat, T.N., Weissig, H., Shindyalov, I.N., and Bourne, P.E. The Protein Data Bank. *Nucleic Acids Research* 28, 2000, 235–242.
8. Morris, G.M., Goodsell, D.S., Halliday, R.S., Huey, R., Hart, W.E., Belew, R. K. and Olson, A.J. Automated docking using a Lamarckian genetic algorithm and empirical binding free energy function. *Journal of Computational Chemistry* 19, 1998, 1639–1662.
9. Rarey, M., Kramer, B., Lengauer, T., and Klebe, G. Predicting receptor-ligand interactions by an incremental construction algorithm. *Journal of Molecular Biology* 261, 1996, 470–489.
10. Irwin, J.J. and Shoichet, B.K. ZINC—a free database of commercially available compounds for virtual screening. *Journal of Chemical Information and Modeling* 45 (1), 2005, 177–182.
11. Jacq, N. et al. Grid-enabled virtual screening against malaria, *Journal of Grid Computing*, 6 (1), 29–43, 2008.
12. Silva, A.M., et al. Structure and inhibition of plasmepsin II, A haemoglobin degrading enzyme from Plasmodium falciparum. *Proceedings of the National Academy of Sciences USA* 93, 1996, 10034–10039.
13. Lamb, M.L. and Jorgensen, W.L. Computational approaches to molecular recognition. *Current Opinion in Chemical Biology* 1, 1997, 449.
14. Case, D.A. et al. The Amber biomolecular simulation programs. *Journal of Computational Chemistry* 26, 2005, 1668–1688.
15. Ferrari, A.M., Degiesposti, G., Sgobba, M., and Rastelli, G. Validation of an automated procedure for the prediction of relative free energies of binding on a set of aldose reductase inhibitors. *Bioorganic and Medicinal Chemistry* 15, 2007, 7865–7877.
16. Luker, K.E., Francis, S.E., Gluzman, I.Y., and Goldberg, D.E. Kinetic analysis of plasmepsin I and II aspartic protease of the *Plasmodium falciparum* digestive vacuole, *Molecular and Biochemical Parasitology* 79, 1996, 71–78.
17. Matayoshi, E.D., Wang, G.T., Krafft, G.A., and Erickson, J. Novel fluorogenic substrates for assaying retroviral proteases by resonance energy transfer. *Science* 247, 1990, 954–958.

18. Banerjee, R., Francis, S.E., and Goldberg, D.E. Food vacuole plasmepsins are processed at a conserved site by an acidic convertase activity in *Plasmodium falciparum*. *Molecular and Biochemical Parasitology* 129, 2003, 157–165.
19. Morishima, H., Takita, T., Aoyagi, T., Takeuchi, T., and Umezawa, H. The structure of pepstatin. *Journal of Antibiotics (Tokyo)* 23, 1970, 259–262.
20. Gluzman, I.Y., Francis, S.E., Oksman, A., Smith, C.E., Duffin, K.L., and Goldberg, D.E. Order and specificity of the *Plasmodium falciparum* hemoglobin degradation pathway. *Journal of Clinical Investigation* 93, 1994, 1602–1608.
21. Lee, H.-C., Salzemann, J., Jacq, N., Chen, H.-Y., Ho, L.-Y., Merelli, I., Milanesi, L., Breton, V., Lin, S.C. and Wu, Y-T. Grid enabled high throughput *in silico* screening against influenza A neuraminidase. *IEEE Transactions on Nanobioscience*, 5 (4), 2006, 288–295.
22. EELA. Available at: http://www.eu-eela.org.
23. EUChinaGRID. Available at: http://www.euchinagrid.org.
24. AuverGrid. Available at: http://www.auvergrid.org.
25. Joosten, R.P., Salzemann, J., Blanchet, C., Bloch, V., Da Costa, A.L., Diarena, M., Fabbretti, R., et al. Re-refinement of all X-ray structures in the PDB. *Proteins* (submitted).
26. EMBRACE. Available at: http://www.embracegrid.info.
27. Kasam, V., Maaß, A., Schwichtenberg, H., Zimmermann, M., Wolf, A., Jacq, N., Breton, V., and Hofmann, M. Design of plasmepsine inhibitors: A virtual high throughput screening approach on the EGEE Grid. *Journal of Chemical Information Modeling* 47 (5), 2007, 1818–1828.

15

Flow Networking in Grid Simulations

James Broberg and Rajkumar Buyya

CONTENTS

15.1 Introduction

Simulation tools play an essential role in the evaluation of emerging peer-to-peer, computing, service, and content delivery networks. Given the scale, complexity, and operational costs of such networks, it is often impossible to analyze the low-level performance, or the effect of new scheduling, replication, and organizational algorithms on actual test-beds. As such, practitioners turn to simulation tools to allow them to rapidly evaluate the efficiency, performance, and reliability of new algorithms on large topologies before considering their implementation on test-beds and production systems.

In particular, the study of grids is significantly aided by robust and rapid prototyping via simulation, due to the sheer scale and complexities that arise when operating over many administrative domains, which precludes easy prototyping on real test-beds. Grid computing [1] has been integral in enabling knowledge breakthroughs in fields as diverse as climate modeling, drug design, and protein analysis, through the harnessing of computing, network, sensor, and storage resources owned and administered by many different organizations. These fields (and other so-called "grand challenges") have benefited from the economies of scale brought about by grid computing, tackling difficult problems that would be impossible to feasibly solve using the computing resources of a single organization. However, when prototyping such applications and services that harness the power of the grid, it is beneficial to test their operation via simulation in order to optimize their behavior, and avoid placing strain on grid resources during the development phase.

Despite the obvious advantages of simulation when prototyping applications and services that run on grids, realistically simulating large topologies and complicated scenarios can take significant amounts of memory and computational power. For statistical significance, large numbers of simulation runs are needed to increase our confidence in the results we obtain from simulation platforms. This is particularly the case when studying applications and services that store and move significant volumes of data over the grid, such as data-grids or content and service delivery networks. Simulators that attempt to model the full complexity of TCP/IP networking in such environments scale poorly and often run significantly slower than real time, practically defeating the purpose of simulating such environments in the first place.

In this chapter we look at incorporating flow-level (or "fluid") networking models into grid simulators, in order to improve the scalability and speed of grid simulations by reducing the overhead of data- and network-intensive experiments, and improving their accuracy. Network flow models are used that closely approximate actual steady-state TCP/IP networking. We utilize the GridSim toolkit as a candidate implementation, and fully replace the existing packet-level networking model in GridSim with a flow-level networking stack. However, the principles outlined in this chapter could be applied to other simulation platforms.

The remainder chapter is organized as follows. The next section describes the GridSim Toolkit and gives a brief overview of its feature set. The existing packet-level networking implementation for the GridSim toolkit is then described, and some inefficiencies are identified that arise when doing large-scale network and data centric simulations. We then outline the basic principles behind modeling network traffic and transfers as flows or "fluid," rather than discrete packets. The bandwidth-sharing model utilized in our flow-level networking model is then described. The new flow-level networking implementation for the GridSim toolkit is then introduced, highlighting the additions made to GridSim in order to support the flow-based networking paradigm. We then describe the flow tracking and management algorithms required to compute the durations of network flows and to update them when conditions change during a simulation run. The performance improvements gained from the flow-networking model over existing packet-based implementation are highlighted. Finally, we conclude the chapter by taking a macroscopic view of the potential applications of flow-level networking in large-scale grid simulations.

15.2 The GridSim Toolkit

GridSim is a grid simulation toolkit for resource modeling and application scheduling for parallel and distributed computing [2]. The GridSim toolkit has been used extensively by researchers across the globe [3] to model and simulate data grids [4], failure detection [5], differentiated service [6], auction protocols [7], advanced reservation of resources [8], and computational economies in grid marketplaces.

GridSim has been designed as an extensible framework by following a multi-layer architecture as shown in Figure 15.1. This allows new components or layers to be added and integrated into GridSim easily. GridSim implementations use SimJava [9], a general purpose discrete-event simulation package for handling the interaction or events among GridSim components.

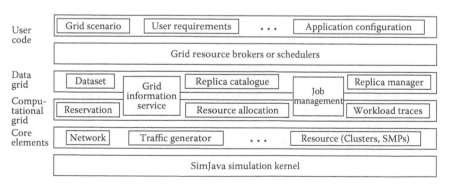

FIGURE 15.1 The GridSim architecture.

At a basic level, all components in GridSim communicate with each other through message passing operations defined by SimJava. The second layer models the core elements of the distributed infrastructure, namely grid resources such as clusters, storage repositories, and network links. These core components are absolutely essential to create simulations in GridSim. The third and fourth layers are concerned with modeling and simulation of services specific to computational and data grids [4], respectively. Some of the services provide functions common to both types of grids such as information about available resources and managing job submission.

From a networking perspective, the current version supports packet-based routing including background network traffic modeling based on a probabilistic distribution [6]. This is useful for simulating data-intensive jobs over a public network where the network is congested. The limitations of this network model are highlighted in the next section.

15.3 The GridSim Packet Networking Architecture

A typical dog-bone topology is shown in Figure 15.2 for a GridSim experiment using the existing packet-level network framework. Consider a user at user node 1 that wishes to send a 10 Mb file to resource node 6. In the current GridSim network model [6], the file would be packetized into MTU-sized packets by the *Output* class of the *NetUser* GridSim entity and sent over the links. Every packet but the last is an empty

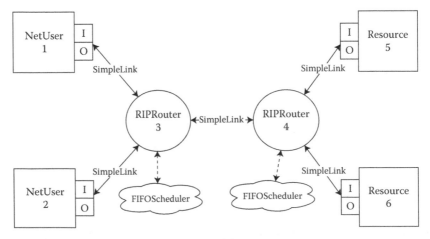

FIGURE 15.2   GridSim packet networking architecture.

packet (*GridSimTags.EMPTY_PKT*), with the last packet containing the actual data (*IO_data*). If the maximum transmission unit (MTU) was 1500 on all elements between the source and the destination, sending a 10 MB file would result in approximately 34,952 packets being generated. In GridSim, each packet is represented by a *NetPacket* Java object, thus creating a considerable amount of overhead for large data transfers. This can lead to lengthy simulation execution times for data or network-dependent simulations. The magnitude of this overhead will be quantified later in this chapter. In the next section, we describe the new flow networking implementation that seeks to minimize the overhead of network-dependent simulations.

15.4 Flow Networking Concepts

Rather than modeling each network transfer using packets, we wish to consider a network flow model that captures the steady-state behavior of network transfers. For convenience we will denote our Grid topology (such as that depicted in Figure 15.2) as a graph, $G = (V, E)$ where V is the set of vertices and E is the set of edges, consisting of two-element subsets of V. For instance, if vertices x and y are connected, then $\{x,y\} \in E$. In the system there exists flows $f = 1, 2, \ldots, F$, with each flow f having a source and destination. Each flow f describes a simple path of length k represented by a set of edges $\{(v_1, v_2), \ldots, (v_k, v_{k+1})\}$. The number of bytes in each flow f is denoted as $SIZE^f$.

Let us consider a simple topology where the two entities, node u and node v, are *directly* connected by an edge (u, v), with available bandwidth $BW_{u,v}$ (in bytes per second) and latency $BW_{u,v}$ (in seconds). Calculating the duration of a single network flow f with size $SIZE^f$ from u to v can be trivially computed as follows:

$$T_f = LAT_{u,v} + \frac{SIZE^f}{BW_{u,v}} \tag{15.1}$$

As an interesting aside, the above equation can be tested in a rudimentary fashion by utilizing the first networking example in the GridSim distribution (NetEx01).* An extremely coarse approximation of basic flow networking can be achieved with the current packet-level network framework in GridSim by setting the MTU to equal the size of the network flow to be transferred, causing only a single *NetPacket* to be generated, which is held at the *Output* of the *NetUser* GridSim entity for the appropriate duration. However, this does not model bandwidth sharing on the links in any way.

---

* Available at http://www.gridbus.org/gridsim/example/net_index.html.

More generally, a flow f with a source u and destination v that is not directly connected has an expected duration of

$$T_f = \sum_{(u',v') \in f} LAT_{u,v} + \frac{SIZE^f}{\min BW^f} \tag{15.2}$$

where $\min BW^f$ is the smallest bandwidth available on any edge on the path f between u and v (i.e., the bottleneck link), and latency $LAT^f = \sum_{(u',v') \in f} LAT_{u,v}$ is the sum of the latency of all edges (u', v') that connect the source u to the destination v.

We note that the above equations and discussions are only valid for a single active flow at a time, as it does not account for any bandwidth sharing between multiple flows on common (overlapping) links. Where multiple flows are active over links, then $\min BW^f$ is the smallest bandwidth allocated by edge (based on some bandwidth sharing model) on the path f between u and v. The implications of this will be discussed in the following section.

15.5 Bandwidth Sharing Models

Earlier, we examined a simple theoretical model to compute the duration of each flow in a system based on the bottleneck bandwidth. This approach significantly improves the speed of grid simulations by avoiding the need to packetize large network transfers, instead of taking a macro or fluid view of network traffic in a given topology.

In order for this approach to be effective we need to calculate the appropriate bandwidth given to flows on each segment of their respective route. More importantly, we must model how the bandwidth is shared when many flows are active over one or many links. As a proof of concept for the GridSim flow networking implementation, we have implemented simple MIN–MAX bandwidth fair sharing, where each flow that shares a link is allocated an equal portion of the bandwidth. That is, an edge (u, v) with available bandwidth $BW_{u,v}$ that has n active flows will allocate each flow $(BW_{u,v}/n)$ bandwidth. While it has been found that other bandwidth sharing models are closer to actual TCP/IP behavior [10], MIN–MAX bandwidth sharing is a useful candidate model with minimal state to track in the implementation.

We intend to include other bandwidth sharing models that more closely approximate TCP/IP in the near future, such as proportional bandwidth sharing that considers latency, round-trip times, and class-based priorities [11,12].

15.6 The New GridSim Flow Networking Architecture

In order to implement the flow-level networking model described previously, we need to make some fundamental changes to the existing packet-level network implementation in GridSim. More specifically, we need to replace the entire networking stack with flow-aware components due to the significant differences between the two approaches.

Figure 15.3 depicts a high-level class diagram showing the flow-aware networking stack that is to be added to GridSim to enable flow-level network functionality. The new support components are shown as dotted boxes to differentiate them from the existing packet networking stack. A summary of these additions (and changes) is listed in Table 15.1. Figure 15.4 shows an example GridSim topology that utilizes the new flow model.

To keep the flow-level network functionality logically separated, a new package was added, namely gridsim.net.flow. This will encapsulate all of the flow-level networking functionality to be added. A new interface, NetIO, was created to provide a common set of functions for the existing input and output classes, as well as the new flow-aware FlowInput and FlowOutput classes. These flow-aware input and output classes are automatically generated for GridSim entities by calling GridSim.initNetwork-Type (GridSimTags.NET_FLOW_LEVEL), before initializing a GridSim simulation.

The FlowOutput class performs a similar function to the existing output class, but instead of packetizing data that are sent by GridSim entities into MTU-sized chunks (as described previously), it creates a single FlowPacket that will represent an active flow for its lifetime. The FlowOutput class also supports background traffic, creating junk flows to simulate load on links and the Grid Information Service (GIS), which is an entity that provides grid resource registration, indexing and discovery services. These two features were available in the NetPacket implementation and are dependent on by GridSim users worldwide for simulating complex scenarios and topologies, and thus were supported in the flow implementation.

We still require an entity to represent the network flow. As such, for convenience we will leverage a subset of the existing packet implementation, extending it to create a FlowPacket class. This allows us to utilize the existing features of the packet class, while adding logic that will support an accurate flow-networking model for GridSim. A flow is then simply represented by a single FlowPacket, which exists as long as the flow is active. As it traverses along a GridSim topology from its source to destination, it maintains a list of the FlowLink entities it passes over, and more specifically the latency and bandwidth available on each of these links.

As a FlowPacket traverses a FlowLink, it is registered as an active flow on that link for the purpose of computing the bandwidth a FlowLink

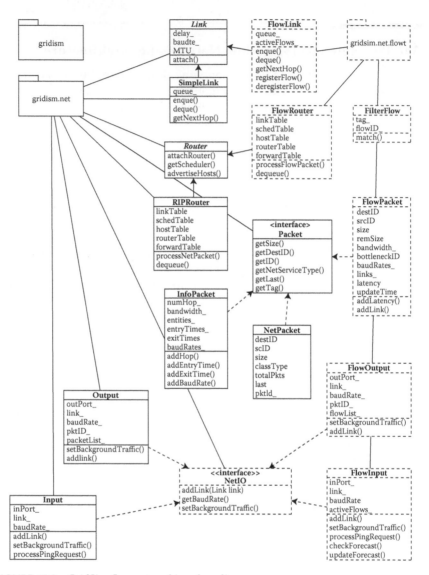

FIGURE 15.3   GridSim flow networking class diagram.

allocates a given FlowPacket, when multiple flows are active on a given link. If the bottleneck bandwidth of an existing flow is affected by a new flow becoming active or an existing flow becoming deactive, then FlowLink notifies the remaining active flows (which are held at the FlowInput of their destination) of their new bottleneck bandwidth.

Routers for flow-level networking are significantly less complicated than those supporting the packet networking model, as they have minimal

TABLE 15.1

Summary of Changes between GridSim Packet and Flow Implementations

Component	Packet Model	Flow Model
GridSim network type	GridSimTags. NET_PACKET_LEVEL	GridSimTags. NET_FLOW_LEVEL
Input/output	Input/Output extends Sim_entity	FlowInput/FlowOutput extends Sim_entity implements NetIO
Packet	NetPacket extends Packet	FlowPacket extends Packet
Link	SimpleLink extends Link	FlowLink extends Link
Router	RIPRouter/FloodingRouter/ RateControlledRouter extends Router	FlowRouter extends Router
Scheduler	SCFQScheduler/ FIFOScheduler/ RateControlledScheduler implements PacketScheduler	N/A
Event filter	N/A	FilterFlow
Package	gridsim.net	gridsim.net.flow

responsibility in the flow model. A new class, FlowRouter, has been added, which enables many-to-many (*m:m*) connections from GridSim entities but performs no actual scheduling itself. As such, there is no equivalent to PacketScheduler needed for the flow model.

Finally, the FlowInput holds the FlowPacket for the appropriate duration, based on Equation 15.2 and an appropriate bandwidth sharing

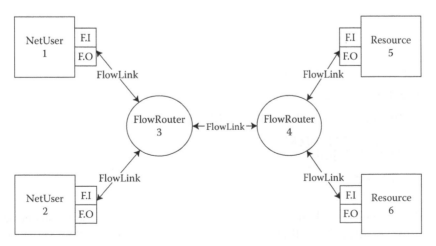

FIGURE 15.4 GridSim flow networking architecture.

model that determines the bandwidth assigned to each flow (as described previously). As stated previously, the bottleneck bandwidth of a flow can change during its lifetime due to the arrival of a new overlapping flow or the termination of an existing flow. When this occurs, affected flows are notified and the duration is updated, potentially being brought forward as available bandwidth increases or pushed back as available bandwidth decreases. This process is explained in more detail in the next section.

15.7 High-Level Flow Management Algorithms

When running any nontrivial GridSim scenario, it is obvious that more than one flow will be active at a given time, and that flows will overlap, begin, and end at different times. As such, the bandwidths assigned to each flow can change frequently. Therefore, flow management algorithms are employed to make an initial forecast based on the bottleneck bandwidth when the flow begins, and to update the forecast when the bottleneck bandwidth increases or decreases, decreasing or increasing the expected duration accordingly.

Let us consider a GridSim system connected in a dog-bone topology similar to that depicted in Figure 15.2. We wish to compute the end time T_{end}^f of a new flow f that arrives into a system, created at time T_{NOW}:

Algorithm 1: Initial Flow Forecast

$BW_{min}^f \leftarrow \infty$

$T_{start}^f \leftarrow T_{NOW}$

$T_{end}^f \leftarrow \infty$

$REM\_SIZE^f \leftarrow SIZE^f$

for each $BW_{u,v} \in BW^f$

   do $\{$if $BW_{u,v} < BW_{min}^f$ then $BW_{min}^f = BW_{u,v}\}$

$T_{dur}^f = LAT^f + \dfrac{REM\_SIZE^f}{BW_{min}^f}$

$T_{end}^f \leftarrow T_{start}^f + T_{dur}^f$

If the bottleneck link of an active flow f changes (i.e., it becomes larger or smaller) at time T_{NOW}, then the expected duration of that flow must be updated:

Algorithm 2: Update Flow Forecast

$BW_{old\,min}^{f} \leftarrow BW_{min}^{f}$

$BW_{min}^{f} \leftarrow \infty$

$T_{elap}^{f} \leftarrow T_{NOW} - T_{start}^{f}$

$T_{start}^{f} \leftarrow T_{NOW}$

$T_{end}^{f} \leftarrow \infty$

$REM\_SIZE_{old}^{f} \leftarrow REM\_SIZE^{f}$

$REM\_SIZE^{f} = REM\_SIZE_{old}^{f} - (T_{elap}^{f} - BW_{old\,min}^{f})$

for each $BW_{u,v} \in BW^{f}$

  do {**if** $BW_{u,v} < BW_{min}^{f}$ **then** $BW_{min}^{f} = BW_{u,v}$}

$T_{dur}^{f} = LAT^{f} + \dfrac{REM\_SIZE^{f}}{BW_{min}^{f}}$

$T_{end}^{f} \leftarrow T_{start}^{f} + T_{dur}^{f}$

15.8 Performance Comparison

In this section we quantify the performance improvements gained from modeling networking traffic as *flows* instead of packets. Using our candidate implementation of a flow model for the GridSim toolkit (described previously) we perform some numerical comparisons of specific scenarios. In each scenario we compare the existing NetPacket implementation and the new flow-level FlowPacket implementation described in this chapter. All tests are run using a Macbook with a 2 GHZ Intel Core 2 Duo and 2 GB of RAM. Each data point represents the average of 30 runs.

The first scenario is a classic dog-bone topology similar to that depicted in Figures 15.2 and 15.3. NetUser1/FlowUser1 and NetUser2/FlowUser2 each send three identically sized files to Resource6 and Resource5, respectively. The size of the files is varied from 0.5 to 500 MB. The links between the users and the first router are rated at 10 MB/s. The link between the two routers has a capacity of 1.5 MB/s, and is clearly the bottleneck link. The link between the second router and the resources is 10 MB/s. The files transfers are initiated in 10-second intervals. The latencies from the users to the router, between the routers, and from the router to the resources are 45, 25, and 30 milliseconds, respectively.

In Figure 15.5 we see a comparison of the CPU time taken for the simulation to execute when utilizing either the NetPacket or FlowPacket networking stack. The size of the files sent by each user is varied from 0.5 to

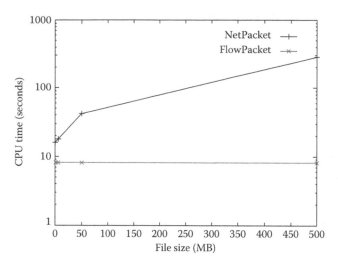

FIGURE 15.5   Comparison of CPU time needed for "dog-bone" simulation run.

500 MB. We can clearly see that as the size of the files increase, the overhead of the NetPacket implementation is demonstrated as the CPU time explodes. The plot is presented on a log-scale on the *y*-axis to highlight the huge difference in simulation running time. The FlowPacket implementation is totally impervious to the size of the files being transferred, as it has no effect whatsoever on the amount of state it maintains. When transmitting several 500 MB files, the FlowPacket implementation takes 0.432 seconds to execute, while the NetPacket implementation takes a staggering 6699 seconds, or approximately 111 minutes.

We can see a linear relationship between the amount of memory consumed and the size of the files being transmitted by the GridSim simulator when using the NetPacket implementation in Figure 15.6. When utilizing the FlowPacket implementation, the size of the files being transferred has no effect on the peak memory consumption, as it stores the same amount of state regardless of whether it is sending a 0.5 MB file or a 500 MB file.

In the second scenario we examine a similar dog-bone topology where users submit Gridlets for processing, instead of sending files. A Gridlet is a construct that contains all the information related to a grid job and its execution management details such as job length expressed in MI (Millions Instruction), and the size of input and output files. Individual users can model their application by creating Gridlets for processing on grid resources. These basic parameters are utilized to determine the execution time, the time required to transport input and output files between users and remote resources, and returning the processed Gridlets back to the originating user along with the results. We have two users submitting

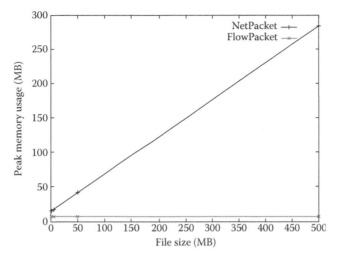

FIGURE 15.6 Comparison of peak memory usage for "dog-bone" simulation run.

Gridlets with 5000 byte input file, an service requirement of 5000 MI and return a 5000 byte output file to 10 available resources. The number of Gridlets sent by each user is varied from 5 to 40.

From Figure 15.7 we can see that as the number of Gridlets being sent by each user increases, the CPU time taken to execute the simulation increases exponentially (highlighted by the near-linear line on the plot, where the y-axis is a log scale). On the other hand, the FlowPacket implementation

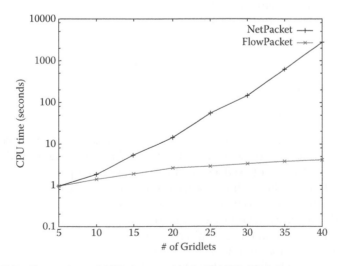

FIGURE 15.7 Comparison of CPU time needed for Gridlet simulation run.

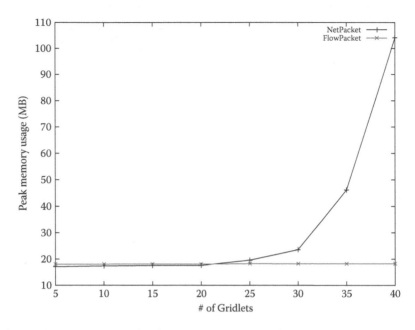

FIGURE 15.8   Comparison of peak memory usage for Gridlet simulation run.

only sees a nominal increase in simulation time as the number of Gridlets sent by each user increases.

In Figure 15.8 we see an examination of the peak memory usage of the NetPacket and FlowPacket implementations for GridSim, running the Gridlet scenario described previously. As the number of Gridlets sent by each user increases, we see a significant increase in memory utilization by the NetPacket implementation after an initial flat response. Conversely, the FlowPacket sees only negligible increases in memory utilization as the number of Gridlets increases.

15.9 Conclusion

We have explored the improvements in accuracy and reduction in complexity that can be achieved by utilizing flow-based networking in grid simulation, instead of packet-based networking, when attempting to model real TCP/IP networks. From the results obtained it is clear that significant improvements, in the order of many magnitudes, can be made in terms of speed and scalability when executing complex grid simulations involving large data transfers and large numbers of grid job submissions. It is clear that practitioners will not be limited in the size or complexity of the scenarios they wish to model, allowing them to simulate complex computational grid topologies, content and service delivery networks, and data grids, to name a few examples. This will hopefully lead to greater advances in solving the so-called "grand challenges" in areas such as climate modeling, drug design, and protein analysis, by allowing practitioners to prototype their solutions trivially and rapidly before committing time and resources in building complex software and network systems that operate on global grids.

Acknowledgments

This work was supported in part by the Department of Industry, Innovation, Science, and Research (DIISR) and the Australian Research Council (ARC) through their respective International Science Linkages (ISL) and Discovery Projects. We thank Anthony Sulistio for adding support classes to GridSim to enable this work, and to Srikumar Venugopal for valuable discussions regarding the implementation.

References

1. Foster and C. Kesselman, *The Grid: Blueprint for a New Computing Infrastructure.* Morgan Kaufmann, San Francisco, 1998.
2. R. Buyya and M. Murshed, "GridSim: A toolkit for the modeling and simulation of distributed resource management and scheduling for Grid computing." *Concurrency and Computation: Practice and Experience* 2002, 14 (13–15), 1175–1220.
3. R. Buyya and A. Sulistio, "Service and utility oriented distributed computing systems: Challenges and opportunities for modeling and simulation communities," in *Proceedings of the 41st Annual Simulation Symposium.* IEEE CS Press, Ottawa, pp. 68–81, 2008.
4. A. Sulistio, U. Cibej, S. Venugopal, B. Robic, and R. Buyya, "A toolkit for modelling and simulating data grids: An extension to GridSim," *Concurrency and Computation: Practice and Experience* 20 (13), 1591–1609, 2008.
5. A. Caminero, A. Sulistio, B. Caminero, C. Carrion, and R. Buyya, "Extending GridSim with an architecture for failure detection," in *Proceedings of the International Conference on Parallel and Distributed Systems,* Hsinchu, Taiwan, pp. 1–8, 2007.
6. A. Sulistio, G. Poduval, R. Buyya, and C.K. Tham, "On incorporating differentiated levels of network service into GridSim," *Future Generation Computer Systems* 2007, 23 (4), 606–615.
7. M. Dias De Assuncao and R. Buyya, "An evaluation of communication demand of auction protocols in grid environments," in *Proceedings of the 3rd International Workshop on Grid Economics & Business.* World Scientific Press, Singapore, pp. 24–33, 2006.

8. A. Sulistio and R. Buyya, "A grid simulation infrastructure supporting advance reservation," in *Parallel and Distributed Computing and Systems*. MIT, Cambridge, MD, pp. 1–7, 2004.
9. F. Howell and R. McNab, "SimJava: A discrete event simulation package for Java with applications in computer systems modelling," in *Proceedings of the First International Conference on Web-based Modelling and Simulation*. San Diego, CA, pp. 51–56, 1998.
10. D.M. Chiu, "Some observations on fairness of bandwidth sharing," in *Proceedings of the 5th IEEE Symposium on Computers and Communications (ISCC 2000)*. IEEE Computer Society: Los Alamitos, CA, p. 125, 2000.
11. S. Floyd and K. Fall, "Promoting the use of end-to-end congestion control in the Internet." *IEEE/ACM Transactions on Networking* 1999, 7 (4), 458–472.
12. M. Matthew, S. Jeffrey, and M. Jamshid, "The macroscopic behavior of the TCP congestion avoidance algorithm," *Proceedings of the SIGCOMM Computer Communication Review* 1997, 27 (3), 67–82.

16

Virtual Machines in Grid Environments: Dynamic Virtual Machines

Cong Du, Prerak Shukla, and Xian-He Sun

CONTENTS

16.1 Introduction

Grids [1] take advantage of widely available network resources and provide remote information services. They allow the network resources to be shared by users and institutions, and can provide information services remotely upon request. Current grid systems, however, provide limited separation of the service and the computing platform. They allow users to access shared network resources, but do not provide an infrastructure so that the users can customize their own private computing environments or enforce personalized policies. Recent studies show that system-level virtualization can help to solve these issues by emulating the resources and presenting a different view. In system-level virtualization the emulated resources are at the system level. A system-level virtual environment has many advantages. It allows users to work in a virtualized but familiar environment, provides an extra layer of hardware isolation and security to defend computing environments against possible attacks from malicious applications, and protects the privacy of virtual environments. Virtualization can be provided at different levels. It could be at the service level, such as grid service, at process/language level, such as Java virtual machine, or at the system level. We focus on system-level virtualization. Unless pointing out explicitly, virtualization means system-level virtualization in this chapter. The following section explains the term "system" or "system-level" virtualization.

16.1.1 Understanding System Virtualization

In a single statement, virtual machine technology allows the sharing of a physical hardware between various virtual machines, each running the same or different type/flavors of operating systems. The software that provides the features of virtualization is called a hypervisor or virtual machine monitor (VMM). The VMM can run directly on hardware or in conjugation with the operating system installed on that hardware. Usually virtual machines can emulate different Instruction Set Architectures (ISA) at the software level from what the original hardware gives them capability to host multiple operating systems on the same hardware. There are many virtualization techniques, but full virtualization and paravirtualization are the two most widely used ones. In a fully virtualized system the guest system (virtual machine) can have and run an unmodified operating system on it, whereas in paravirtualization technique the guest operating system is modified to efficiently work with and integrate into the VMM software. VMware provides virtualization solutions [2] generally based on the full virtualization technique and Xen [3] uses paravirtualization techniques.

If we gaze into history virtual machine (VM) technology was first developed in the 1960s by IBM to support virtualization for multiple users on mainframes. But due to low speed of mainframes at that time (by slow, we are talking about IBM System 370) and the later introduction of powerful operating systems like UNIX, which had a strong capability of resource isolation between users, the research and development of virtual machines were slowed down. Virtualization was effectively abandoned during the 1980s and 1990s when client-server applications and inexpensive x86 servers and desktops established the model of distributed computing [2]. But the development of grid computing, in addition to other forms of Internet-based computing, raises some important operational challenges like low infrastructure utilization, increasing physical infrastructure cost, increasing IT management cost, security, and insufficient failover and disaster protection [2]. VM technology [4] has recently been studied extensively to show its ability in supporting system-level virtualization of computing environments to solve the above-mentioned problems of resource abstraction and better management. VM and virtual network (VN), generates a virtual computing environment in which virtual resources are mapped to physical resources. A virtual environment (VE), also known as a private virtual grid or virtual private working space [5], has many advantages. It allows users to work in a virtualized but familiar environment, provides an extra layer of hardware isolation and security, defends computing environments against possible attacks from malicious applications, and protects the privacy of virtual environments. Owing to these advantages, virtualization has regained its popularity in recent years. Virtualization enables better resource management by marshalling users' configuration requests for physical resources. Although these works have shown the potential in supporting system-level virtual environments, they have not adequately addressed how to enhance the current VM technology so that it can adapt to a grid or a general shared cyberspace. The major problem is that in a general shared and distributed environment, resources are heterogeneous and their availability varies with time. In addition, users' requirements also change with time. These heterogeneity and dynamics demand virtual environments supporting not only swift instantiation but also the ability of reconfiguration and migration from time to time to adapt to the resource availability making conventional virtualization technologies inadequate to provide an efficient virtual environment (VE) on grid systems. Configuring an appropriate VE in a general parallel and distributed environment is at least as complex as that of a physical system. Furthermore, current virtualization platforms have mainly focused on functionality and have limited support for configuration management of virtual environments.

System-level virtualization is not only critical for grid computing, but also vital for the next generation network computing in general. Let us consider today's distributed environment and the IT industry in general, where computing paradigms such as service-oriented architecture, utility computing, business-on-demand, adaptive enterprise, and data center automation are becoming more and more popular. All these paradigms are loosely related to each other and we can say there are lots of similarities among them. To better understand the impact of system-level virtualization on these new computing paradigms we explain utility computing and on-demand computing briefly herein.

According to Foster and Tuecke [6], the term "utility computing" is often used to denote both a separation between the service provider and consumer and the ability to negotiate a desired quality of service from the provider. Here the provider may be an organization's IT department or an external utility provider and the coverage of the service may include storage, computing, or an application. Similar services are provided by Amazon's Elastic Cloud [7] that charges its user as per their use of CPU power, storage, and so on. Similarly, Salesforce.com [8] provides a Customer Relationship Management (CRM) application as an on-demand service. On-demand is a term used to denote technologies and systems that allow users or applications to acquire additional resources to meet changing requirements in a broader sense than utility computing.

Investigating more in grid infrastructure, system virtualization is a level below it and is designed to hide the idiosyncrasy of physical resources and myriad of different software platforms. As we know, the responsibility of virtualization is to provide isolation and provisioning. The task of grid infrastructure is to manage resources above the virtualization layer. This task may include and is not limited to provisioning (on higher level), monitoring, and QoS. The layer above is responsible to provide application-level services based on the grid infrastructure. These layers may include the workload managers to efficiently optimize the use of grids. This situation can be visualized in Figure 16.1, which shows the architecture of current distributed environment. Here, the top-level service is considered to be utility or on-demand as now these terms/concepts are becoming more and more important.

The above model seems to be correct in the first glance. But in practice, it has some unfinished issues. In an actual enterprise computing environment, different users may have different needs and may use different computing paradigms such as service-oriented architecture, utility computing, on-demand computing, cloud computing, and so on, each of which has its own characteristics and structure while all use the same underlying grid computing infrastructure to utilize distributed computing resources. That is, the users should have the ability to manage their virtualization environments, not a fixed mapping as given in Figure 16.1. This requires system-level virtualization. Traditionally, grid computing is based on service-level virtualization which is not well suited and not ready for this shift. The gap between the current virtualization layer and the grid infrastructure is big. This leaves application services to have an extra burden of configuring and managing low-level details when users require different system configurations. To solve this problem, we should have another layer of service that can well integrate in the current grid infrastructure and

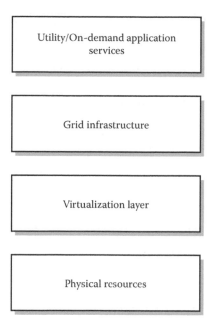

FIGURE 16.1 Architecture of current distributed environment.

provide automated and standard configuration and management of the virtualization layer below. Another important and desirable property of such a service layer is to provide simple and standard handles to effectively control and realize the entire architecture.

As this middleware service is provided at a high level, it should be a standard that can be well suited for any application type. The following figure shows the position and role of such a middleware service named as dynamic virtual machine (DVM) services. We call a VM with the support of DVM services a DVM.

As shown in Figure 16.2, a DVM should be well integrated with the existing grid infrastructure providing services. Another important aspect is that it should be tightly coupled with the virtualization layer and loosely coupled with the top application layer.

In this chapter, we discuss the development of the DVM middleware for the configuration, automatic instantiation, and deployment of VEs. A DVM introduces a new layer of services, which models and specifies the configuration of a VE. We devise a two-level language system and its associated precompiler tool chains to translate the high-level description of VEs' configuration requirements into a machine-readable format. A Web portal is designed to help users to specify a VM configuration requirement using a Web browser. To facilitate automatic instantiation and deployment, a run-time system is developed to conduct the VM incarnation automatically. Under a DVM, users can instantiate their virtual environments using existing

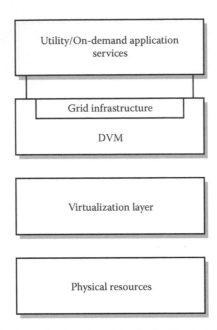

FIGURE 16.2　Role of a dynamic virtual machine in the distributed architecture stack.

templates. A VM only needs to be configured once with the help of the Web portal, and can be deployed swiftly to different physical environments. DVM's runtime system in turn facilitates automatic adaptation, reconfiguration, and migration at runtime. In essence, a DVM separates the user domain from the system management domain, and hence eases the task of managing various VE configurations. The new layer of abstraction does not only apply to the VE as a whole but also applies to each individual component, such as to a single virtual machine. This enables a VM or a group of VMs to be swiftly redeployed or migrated to different physical locations. This redeployment could also be accompanied with a different configuration specification. In this way, a DVM provides a better support of configuration management of both physical and virtual computing environments.

Our argument here is that the key to solving system manageability is to solve VM configuration management at and above the virtualization layer. We also show a threefold solution for a VM configuration through VM modeling, service provisioning, and mobility explained in detail in later sections. The next section explains technical challenges to deploy a DVM.

16.1.2 Technical Challenges

VM is a software emulation of a physical computing environment and is implemented by adding software to an execution platform to give it the appearance of a different platform, or of several different platforms [4]. It allows a virtual system to be temporarily associated with physical resources, and this association can change, decoupling software from the underlying resources. The current virtualization platform [9] provides a few predefined VM images. These VM images are statically configured with basic functionalities. As such, users can either use individual basic VMs and then configure them manually, or customize and port their applications to the virtual environment provided. In some cases, they need to do both. Configurations need to be versatile. This is because some applications are designed for specific system architectures, some demand customized configuration and administration policies, some run on a VE composed of different VMs, and some need to be deployed on different virtual platforms. Additionally, VMs and their connecting VN channels may have their own configuration requirements. One or several predefined VMs are insufficient to fulfill all applications' requirements in a parallel and distributed environment. In a large-scale distributed computing environment, such as Grid environment, resources are heterogeneous and their availability varies with time. To adapt to the dynamics, a VM needs to be dynamically initiated and swiftly deployed to various heterogonous platforms. To do so, a VM needs to be reconfigured and migrated from time to time to adapt the resource availability. A runtime system is needed to support the reconfiguration, redeployment, migration, and adaptation automatically. This is the same for massively parallel modern high-end computers, especially for the demands of swift deployment and fault tolerance. An effective method to model virtual environments, manage virtual configuration, and facilitate fast instantiation and migration of VMs is necessary. The major technical challenge of providing a configurable and manageable virtual computing environment can be summarized into the search of answers to the following four questions: first, how can we model a virtual environment? Each virtual environment needs to be configured for its target applications. However, common users usually may not have such knowledge, expertise, and/or time to create and configure their computing environments. Virtual environment modeling is thus a necessity to achieve automatic configuration management. Secondly, how can we instantiate and deploy a virtual environment efficiently? Since a VM image could be huge, moving a naive copy of the image to the targeted location will deteriorate the performance and overburden the network and storage. Thirdly, how can we support mobility and dynamic management of virtual environments? Mobility requires efficient migration and adaptation to the destination environment, which may be subjected to further security policy as well as physical constraints. In short, a virtual environment needs to have the ability to adapt the dynamics of the underlying physical environment, where the dynamics may be due to the dynamics of the cyberspace or the accumulated failure rate of massive parallel computers. Finally, how can we migrate an active VM in a concurrent and coordinated environment? Migrating and redeploying a VM in a parallel and distributed environment require the direction and redirection of VM communications during and after migration. The redirection of communications must also take into account of optimized group communications, in addition to point-to-point communications. Existing VM technologies are inadequate to address these challenges. Current virtualization technologies embed the configuration and management information within the VM image. The configuration information, therefore, cannot be easily changed to accommodate resource fluctuations, or be managed automatically. Users need to perform configuration and to set up the execution environment manually. VM creation, deployment, and migration are currently achieved via copying the VM image. This approach is very slow, problematic, and does not support adaptation. None of the existing systems supports communication direction and redirection [10]. These four challenges are largely unsolved with current virtualization technologies. We introduce cutting edge research in the rest of the chapter to address these challenges.

16.2 Recent Work in Virtual Machines for Grids

Some efforts in virtualization techniques focus on virtualization on a set of given computing [2,3,11–15] or communication [16,17] resources. These techniques, including VM and VN, emulate computing resources on top of a given physical computing infrastructure and form the basis of the virtualization technology. Since in these studies the virtual resources are bound to their underlying physical resources, these techniques do not address the issues of configuration description and dynamic configuration of virtual computing environments.

Several projects endeavor to enable virtualization in distributed computing environments. Figueiredo, Dinda, and Fortes explored the feasibility of virtual machines on grid computing and proposed a VM-based architecture for grid computing [18,19]. They aim to isolate the user and administrator's views, provide system security, and ease the administration. A virtual private workspace [5] focuses on the authentication, authorization, and resource management of a virtual environment in the grid. However, these projects are designed to use current virtualization technologies in a distributed environment, not to extend current virtualization technologies for system configuration management. They do not support the modeling, incarnation, and migration of a virtual environment, and cannot address the configuration and maintenance challenges.

Researchers from Stanford have performed some experiments on VM portability [20]. Other works include Internet Suspension and Resume [21] and Xen Checkpointing and Resume [3]. As their names suggest, they suspend or checkpoint a VM on one computer and resume it on another. Because a system-level VM has a huge image, it cannot be migrated directly in a timely manner. Even though migration performances are optimized in these projects, the migration overheads are still very high. A more efficient approach is the VM live migration work also from the Xen group [10]. This method optimizes the migration of the memory state, but its feasibility is limited to clusters with a global accessible file system. As all the checkpoints and resume methods, it does not address the issues of heterogeneity and communication. The image-based migration does not support reconfiguration and does not reduce the cost of configuration and deployment.

A virtual cluster (V-Cluster) is a small-scale cluster emulated by the virtualization technology. Projects on the V-Cluster include the Virtual Cluster by Purdue University and the Cluster-on-Demand (COD) [22] by Duke University. The former provides dynamic fine-grained virtual clusters by running unmodified commodity operating systems on a scalable shared-memory multiprocessor. The latter partitions a physical cluster into multiple independent V-Clusters, where the V-Clusters are created within the constraints of the physical cluster. These methods are designed to support a limited configuration management: configuring V-Clusters on a shared-memory multiprocessor or a physical cluster. They do not support the modeling, incarnation, and reconfiguration of a virtual environment.

In summary, most recent advances in virtualization technologies are in extending virtual technologies to distributed and parallel environments. There is little or no attention to system configuration management. The work that gives most attention to the topic has been carried out by researchers of Oak Ridge National Laboratory, who have developed a grid solution named OSCAR-V [23] that provides VM instantiation, management, and monitoring tools. But their solution does not provide support for VM migration capabilities and flexibility to work with other grid tools (like job schedulers).

The DVM middleware is a complement of existing research and they mutually support each other. As virtualization becomes increasingly popular, the configuration of virtual environments becomes increasingly timely and important. Meanwhile, a DVM solves the system configuration management issue via solving the configuration of virtual environments. It makes virtualization a better and more practical choice for the current distributed environment. Its impact is beyond system configuration management and extends to lift the competitiveness and applicability of the virtualization technologies as a whole. The DVM approach is explained in detail in the following section.

16.3 Dynamic Virtual Machine: The System Approach

A DVM middleware is proposed for configuring virtual computing environments and managing various system configurations. A DVM is designed to specifically address the four technical challenges as discussed previously. A DVM has three key functionalities: (1) VE modeling, (2) service provisioning, and (3) mobility. In DVM, a VM and its model are integrated to form a (integrated) VE. A VM model defined in an integrated VE is named a DVM. The DVM is a primary building block of an integrated VE and a fundamental unit of management. A VE can be a standalone VM or a networked distributed virtual environment with several VMs connected by overlay virtual networks. The integrated VE contains all virtual hardware and software idiosyncrasies and their configurations. A VE can be configured and reconfigured based on its model, which provides the means of VE configuration management. Within a VE, a VM or a set of VMs can be remodeled and redeployed, which provides the means of system configuration management. Through VM modeling and provisioning, a DVM provides a thorough solution to system configuration management. The VM modeling and service provisioning functions support basic VE lifetime managements. VM modeling defines the virtual computing environment for an application, including the communication platforms, DVMs, and their configurations in a platform-independent format. With VM modeling, the associated configurations and management policies become an integrated part of a VE, which makes swift redeployment and mobility possible. We address the instantiation and deployment challenge by incarnating a VM from its model, which is significantly more efficient than the existing method of copying the predefined virtual machine image. As shown in Figure 16.3, users compose a VM with the assistance of the VM Composition Service on the DVM Web portal when preparing a job submission. To facilitate the user configuration, predefined configuration templates are collected and stored in the Knowledge Base repository. The VE composition service may reference these templates when composing VMs. After configuration, a user submits the configuration together with its application and data to the VM Modeling Service. The Web portal then generates a platform-independent VM model accordingly. When a VM is scheduled to run on a physical platform, it is incarnated from its model into a VM with the user application by the VM Incarnation and Deployment Service. The realized virtual computing environment, as shown in Figure 16.3, contains one or several VM hosting server(s) with interconnecting networks mapped to physical resources. In this way, a DVM separates the user domain and management domain and eases the virtual configuration.

DVM mobility is designed to remove the dependency between the virtualization layer and the physical layer, and hence to allow a DVM to be scheduled and migrated dynamically. Existing solutions migrate a VM via moving its image over the network [3,21]. This method is straightforward but it is very slow, due to the huge size of a system-level VM image, and cannot support environment adaptation, which limits its applicability in a grid environment.

The DVM approach is innovative. In its design, a VM image is decomposed into several functional units, which are saved on a high-speed virtual

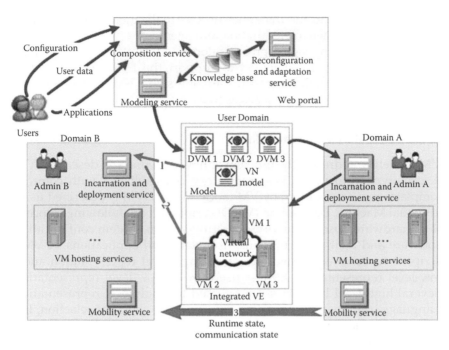

FIGURE 16.3 VM modeling and migration.

image server. As shown in Figure 16.3, VM migration is achieved by a VM mobility service. During a migration, a VM is reconstructed at the destination site according to its model, and linked with the functional units prestored on the virtual image server local to the destination. During the reconstruction some system parameters can be changed to adapt the new local environment. The reconstructed VM image is called a VM skeleton. The VM-skeleton reconstruction is achieved by an appropriate VM incarnation and deployment service. The current runtime state of user tasks or system services is then migrated to and restored on the VM skeleton. The runtime information transforms the skeleton into an active VM. In a coordinated concurrent environment, a DVM directs and redirects VM communications with a set of communication protocols during and after the migration to ensure correct live communications between VMs. The VM model is platform-independent, so it can be incarnated on various virtualization platforms and under different management policies. By modeling, incarnation, and redeployment, a DVM can naturally address the reconfiguration and adaptation challenge. If a reconfiguration is necessary, the VM model is adjusted to reflect the new configuration through the VM reconfiguration and adaptation service and then the new model is incarnated and deployed to the new virtualization platform. The reconfiguration could be at the VM level, such as remapping VMs to different resources and adapting VMs to the dynamics of a virtualization platform, which leads to a better system configuration management.

Configuration languages, precompilers, and runtime systems have been designed and implemented to support the DVM design and its functionalities. The following section explains how VM modeling is supported in a DVM to provide services like configuration and migration.

16.3.1 Virtual Machine Modeling

VM modeling addresses the problem of how to define and describe a VM. A prototype Knowledge Base repository is established with common templates that are required by VM modeling. A DVM is described using a Virtual Machine Description (VMD) script with the following information: hardware, system identity, operating system, system configuration, software and their configurations, user data, and applications. A VMD script is represented in two formats: a high-level representation and a low-level representation. XML is used for the low-level representation. Several high-level languages can be used for high-level representation. Language used to describe/define data grid is a choice of selection. It is descriptive and thus users can easily understand as well as manually create a script. The templates describe the resources and their configurations. They also describe and enforce related software and hardware dependencies. High-level description templates are used to represent a DVM when communicating with a user or with a virtual environment service and low-level representation of XML to represent a DVM during DVM instantiation and migration. A compiler is used to translate a VMD script in a high-level language into a script in XML. In this way, a VMD script is sufficient for DVM composition, instantiation, and migration. Altogether, the templates (e.g., VMD scripts) form an extensible knowledge base of hardware and software resources that may be referenced when composing a DVM.

As shown in Figure 16.4, a DVM script is separated into hierarchical functional units. Each unit, such as hardware, software, network, and so on, is defined with a template to describe its characteristics and configuration. Figure 16.4 only shows the basic template structure. The Knowledge Base repository is highly extensible and flexible. Third parties can contribute to the Knowledge Base. For instance, virtual computing environment providers can define their hardware templates specifying their own policies. VM image service providers can provide more choices in operating systems and their configuration by providing their own OS templates. Software developers may support their software by providing a software template that includes all the possible configuration customization and rules to configure them. The service providers also provide some frequently used software templates to provide a starting point to a user; the Knowledge Base is open to accept other templates. The user configuration

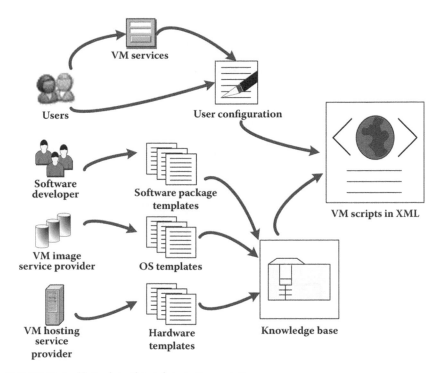

FIGURE 16.4 Virtual machine description scripts.

script can be automatically generated by a DVM composing service through the Web portal. The user configuration script can overwrite some default values in the templates, but the templates define the rules and can validate the user's configuration to avoid errors and conflicts.

DVM is only one of the components of a VE. To model a sophisticated and complete VE for applications, we also need to (1) model multiple DVMs, network connections, storages, and other components in a computing environment, (2) design and implement the VM Composition Service, which interacts with users to compose a VE, (3) design and implement the VM Reconfiguration and Adaptation Service, (4) design and implement the DVM Web Portal, and (5) enrich the Knowledge Base with templates for more hardware and software resources and their configurations.

16.3.2 VM Incarnation and Deployment

The system also needs to be deployed or incarnated based on the fundamental model explained in the above section. Figure 16.5 illustrates the workflow of the DVM instantiation service to do this task. The DVM instantiation service first accepts the input VMD script. The input VMD script, together with the predefined templates, is then compiled into a

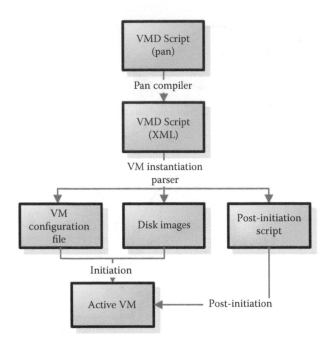

FIGURE 16.5　Virtual machine instantiation service.

VMD script in XML format. The VM instantiation parser then parses the XML-VMD script and generates the DVM configuration file, which is used by the virtualization platform as the configuration file; one or several disk image(s), each corresponding to a hard disk or a disk partition; and a post-initiation script that performs necessary subsequent configuration. The format and the contents of the configuration file depend on the resource virtualization platform used by the DVM hosting server. The resource virtualization platforms, such as VMware [2] and Xen [3] are usually called virtual machine monitors (VMM). In a heterogeneous environment, the contents and format of the configuration file are different for each virtualization platform but the structure of VMD scripts are kept in a platform-independent format. The disk images are mapped to the disk drives and the mapping is designated in the configuration file. The DVM configuration and disk image(s) are then deployed and initiated on a DVM hosting server. After the initiation, the postinitiation script is executed to complete the configuration process.

Parser fetches the information from the XML and postinstantiation scripts written in Perl to instantiate a DVM based on a Xen virtualization platform. The XML parser parses the VMD script in XML and generates a configuration file, VM disk images, and a postinitiation script for Xen. Xen is selected because it is an open-source software and has plenty of useful tools and an active discussion forum.

A VE contains more components than a DVM. We have implemented the DVM parser that can incarnate and deploy the DVM according to its model. We need to establish the VM Incarnation and Deployment Service and further extend current parsers to other components described in a VM model, such as network connections and storages.

16.3.3 Virtual Machine Mobility

As described previously, the mobility challenges are addressed in three major steps: reconstruction, runtime state migration, and communication state transfer. The reconstruction request is fulfilled by the VM Incarnation and Deployment Service. In this section, we address the runtime state migration and communication state transfer issues.

16.3.3.1 Runtime State Transfer

High-Performance Computing Mobility (HPCM) is a user-level middleware supporting process migration of legacy codes written in C, FORTRAN or other stack-based programming languages via denoting the source code [24]. It consists of several subsystems to support the main functionalities of heterogeneous process migration including source code pre-compiling, execution state collection and restoration, memory state collection and restoration, and I/O state redirection. We have developed several optimization mechanisms to reduce the migration cost, including communication/execution pipelining, and live variable analysis.

The input of HPCM is the source code of an application. The precompiler or the user chooses some points (called poll-points) in the source code. A poll-point is a point where a migration can occur. The precompiler annotates the source code and outputs the migration capable code, namely the annotated code. The annotated code is preinitialized on the destination machine before a migration. When a migration is demanded, the migrating process first transfers the execution state, I/O state, communication state, and partial memory state to the initialized destination. The preinitialized process resumes execution while the remaining memory state is still in transmission. That is, the process states are transferred in a pipelined manner. The concurrency saves significant time in a networked environment, especially when a large amount of data needs to be transmitted.

To migrate an application over heterogeneous systems, we represent the application's memory space by a memory space representation (MSR) model [25], which is a machine-independent logical representation of memory space. The snapshot of an application's memory space is modeled as an MSR-directed graph. Each vertex in the graph represents a memory block. Each edge represents a relationship between two blocks when one of them contains a pointer, which points to a memory location within another memory block. MSRLT (MSR lookup table) is a global mapping table between application memory space and the conceptual MSR model. Each memory block that may be referenced in the MSR, including a dynamic memory block, has an entry in the table. To represent a pointer, which contains a machine-specific address, the MSRLT is searched for the memory block that contains the address. The pointer is then represented in MSR by an edge to the referenced memory block. The preinitialized process restores the pointer to the correct address allocated to the referenced memory block.

16.3.3.2 Communication State Transfer

State transfer of point-to-point communication is supported by HPCM [26]. Recently, group communication protocols to handle the state transfer of group communications during a migration and implemented under MPI have been designed and developed [27].

The difficulty of migration with group communications is that all the processes in a group need to be synchronized for a new process or an updating of the group membership information. The basic idea of our protocols is to divide the synchronization process into two phases: collective synchronization and point-to-point synchronization. We define a superstep as the execution block between any two collective operations. Within a superstep, processes can send messages only through point-to-point (pt2pt) communication channels. After receiving a migration signal, collective synchronization protocol brings all the processes in a group to the same superstep. The pt2pt synchronization protocol wakes up all processes waiting for messages through the old communication group, drains the communication buffer, and preserves the messages in transmission. After synchronization, all the processes coordinate to spawn a new process, create a new group, or update the group information. The preserved communication state, together with local process states, is transmitted to the new process for continuous execution.

In these protocols, processes are allowed to execute asynchronously for better performance. Processes are synchronized only when a process in the group is migrating. After broadcasting the migration signal, the migrating process checks the current superstep of each process and determines the maximum step as the global superstep. The global superstep is sent to each process and all processes keep execution until all of them reach the global superstep. After all processes are within the same global superstep, the migrating process initiates the distributed pt2pt synchronization protocol. After all processes are synchronized, the local communication channels are drained and all pending messages are stored as the local communication state. Then, all processes collectively spawn a new process and update their local group information. The migrating process transmits memory and communication states to the new process, and finalizes its communication channels. The new process then resumes execution as P_j in a new group. If a process is woken up from a blocking pt2pt operation, this operation is repeated in the new group. After migration, the processes should first lookup the local communication state for corresponding messages before they actually receive messages from their communication channel. We only highlight the idea here. Interested readers may refer to [27] for the formal description of the protocols and the details of the system implementation.

Although we have addressed some of the major technical difficulties, VE and DVM mobility demand more than the integration of process state transfer and VM reconstruction. For a complete VE mobility service, the motility service needs to coordinate VM migrations; identifying the processes to be migrated; acquiring and remapping resources; identifying and requesting a reconfiguration and an adaptation service. The system development of VE mobility as a whole is largely untried at this time.

16.3.3.3 Experimental Tests and Results

The concept of a DVM and its associated mechanisms are new. In the following, through some case studies, we present our current implementation and some experimental results. They further illustrate the design and implementation consideration and show the potential of a DVM.

16.3.4 A Case Study of VM Modeling and Incarnation

VM modeling and incarnation is the "core" for the development of a DVM. In this case study, we show the process of VM modeling and incarnation step by step, including the establishing DVM prototype and templates, modeling a DVM with user configuration, compiling, and validating VMD scripts, generating platform-independent XML script, and instantiating and initiating a VM. We apply the proposed modeling and incarnation method to a virtual environment running SPEC CPU2000. The hardware and system configuration of the virtual computing environment is summarized as: (1) a VM running Debian Linux on IA32 architecture; (2) a CD-ROM drive mapping to the CD image of CPU2000; (3) 256M of RAM; (4) GNU gcc 3.3.5 that provides a C and a f77 compiler; (5) Intel Fortran compiler for Linux. The ifort compiler is used as f90 compiler. Because GNU GCC does not support the f90 compiller we use the FORTRAN compiler from Intel instead. Figure 16.6 illustrates the basic type structures of a DVM. The configuration is defined by the configuration tree (shown in Figure 16.7) constructed during the compiling time. Each node in the configuration tree corresponds to a resource (inner node) or property (leaf node) of the a DVM. Each node corresponds to a component or parameter such as a software package, a disk, a mounting point, and so on. The statements that define some low-level components are pruned for brevity in the figures. The DVM-type structure categorizes the software and hardware

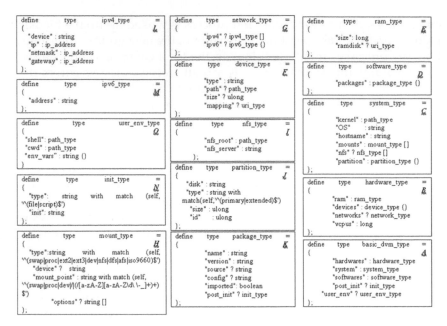

FIGURE 16.6   A DVM-type structure.

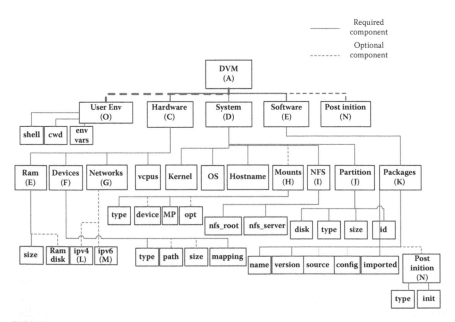

FIGURE 16.7   A DVM *configuration tree.*

resources, properties, and their values, and defines them with various types. The DVM types also define the rules to validate the user configuration. After compilation, the configuration tree is validated according to the structure and property constraints defined by the *basic_dvm_type*.

As defined in its type structure, a DVM has three major components and two optional components. The major components, hardware, system, and software, must be defined in high-level templates or in the user script. The optional components are the postinitiation file or script, and the user environment. Users may choose to define their tasks or follow-up actions in the DVM initiation, which can be executed automatically, or perform these actions manually after the DVM initiation. The *package_type* defines the semantic of a software package. *Name, version,* and *source* are used to identify a software package. The *config* defines the software configuration file, which is specific to the software. The *post-init* defines additional actions, which are performed after software installation such as starting a service. Figure 16.8 illustrates the user configuration script, which is the only script provided by the user. At the beginning of the template, the *dvm_cpu2000* is defined as an object template that can be instantiated into a VM. The *dvm_prototype* and the *dvm_functions* are two templates declaring the framework and functions to compute or validate a resource property. Several templates are then loaded into the configuration tree. The user script is not required to reference all these templates. The values of the resource properties defined in the templates are used by default. They may be overruled by a user script. Figure 16.9 shows the compiled and validated VMD script in XML.

Experimental testing is conducted to verify the feasibility of the design and measure the effectiveness of the implementation. The expectation is to have a working system that achieves minimum time to instantiate (or install) the client machines, reduces the time required to download and upload hence reducing bandwidth requirement, and supports maximum flexibility.

As all the measures of minimum and maximum are relative, we compare our approach with other existing solutions. Experimental testing is

```
# This script define a DVM supporting CPU2000 benchmark
object template dvm_cpu2000;
# include the pre-defined types and functions.
include dvm_prototype;
include dvm_functions;
# dvm_cpu2000 is derived from "basic_dvm_type" template
type "/" = basic_dvm_type;
# start the hardware configuration with a simple DVM template.
# This template define a virtual CPU, 256M memory, a hard disk
named hda.
# This is the basic configuration, no ramdisk
"/hardwares" = create ("hardware_hda_cdrom_256_1");
# There is a cdrom disk which is mapped to a iso.
"/hardwares/devices/hdc/mapping" = "file:///images/cpu2000.iso";
# create a system with default configuration
"/system" = create ("default_system_hda");
```
```
# overrule the default configurations of "system" template
"/system/hostname" = "dvm1";
# create a root mounting point, keeping some default settings
# create a cdrom point
include automount_cdrom;
# import the default configuration of "system" template
#include softwares;
"/softwares/packages/gcc" = create ("gcc_3_3_5");
"/softwares/packages/gcc/imported" = true;
"/softwares/packages/ifort" = create ("ifort");
"/softwares/packages/ifort/imported" = true;
"/softwares/packages/cpu2000" = create ("cpu2000");
"/softwares/packages/cpu2000/imported" = false;
"/softwares/packages/cpu2000/source" = "/mnt/cdrom";
#import user environment setup
include default_user_env;
```

FIGURE 16.8   A DVM user script for the CPU-2000 benchmark package.

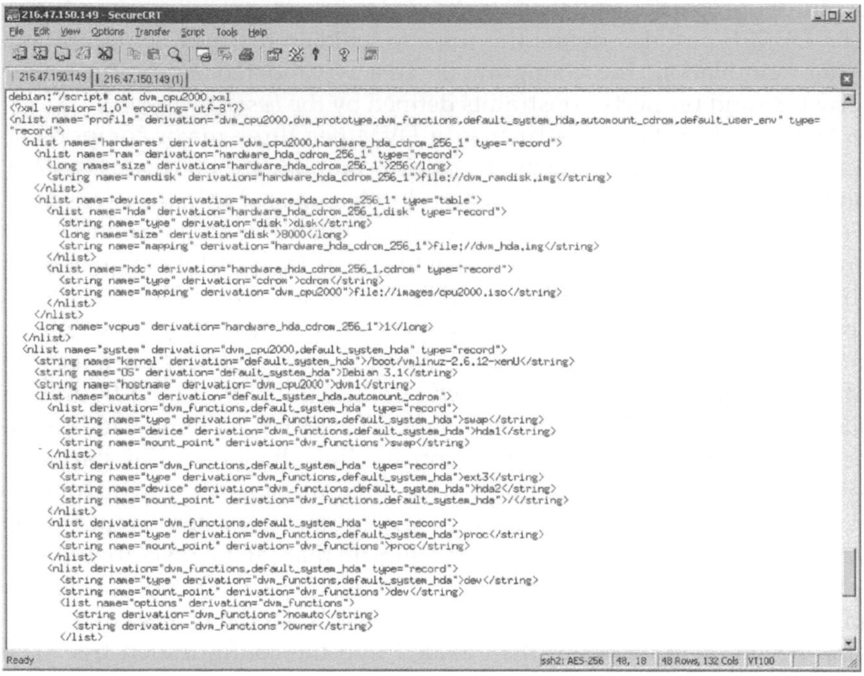

FIGURE 16.9   A VMD script in XML format.

conducted to measure the cost of migration of a DVM over a wide-area network and compare the results with that of the existing suspension/ resume method and shared global storage approach. In our method, we establish a model for a DVM, reconstruct a DVM skeleton according to the model, and then migrate the user task, where the underlying VM is first migrated via a DVM skeleton and incarnation. The migration performance test is performed on a Dell Dimension 3000 with 1 P4 2.8 GB CPU, 1 MB cache, 1 GB memory, and 160 G hard disk, and a Dell Precision WorkStation 360 with 1 P4 2.6 GB CPU, 512 K, 1 G memory, and 80 G hard disk. The effective communication capacity between the two nodes is 83.2 Mbps. Xen is used as the virtualization platform. We have tested four migration-enabled applications: the linpack sequential program that solves a dense system of linear equations with Gaussian elimination; the bitonic program, which builds a random binary tree and then sorts it; the test_tree program, which creates binary trees, assigns each node a random weight, sorts them, and then sums the weights; the gzip program, which compresses or decompresses files. We use the linpack as the representative of computational intensive applications, the bitonic and the test_tree as the representatives of memory intensive applications, and the gzip as the representative of I/O intensive applications. The VM used in all these methods are 1 VCPU, 256 MB memory, and one loop-back file virtual disk.

As shown in Table 16.1, a test is performed on three different VM migration mechanisms. Suspension and resume is the most popular approach implemented by several projects [3,20,21]. We use the suspension and restore functions of Xen to simulate this general approach. With this method, first the VM is suspended or checkpointed to a hard disk; the VM image is then moved to and restored on the destination machine. In our experiments, we transmit the VM image in its original format and in a compressed format. The former is for platforms with high-speed communication channels and the latter is preferred for wide-area networks with low communication speed. The second is for live migration with shared global storage, such as the migration supported by Xen live migration [10]. In this kind of live migration, the source and destination virtualization platform share the networked storage. The VM image is accessible from both machines. The VM is migrated with optimized memory state collection and restoration. The DVM method is different. The VM skeleton is reconstructed on the destination platform according to the DVM model, and only the VM memory state of user applications is migrated. The measured response time is the time between the virtualization platforms received the suspension or migration request and the

TABLE 16.1

VM Migration Comparison

		Suspension and Resume	Suspension and Resume (compressed)	Shared Storage	DVM
Linpack	Response (secs)	449.2	431.8	38.25	198.5
	OOS (secs)	449.2	431.8	6.4	1.99
	Comm (M)	4698	570.6	382.25	0.359
	Perf degradation (%)	0	0	23.5%	0
Bitonic	Response (secs)	449.9	432.9	29.3	198.6
	OOS (secs)	449.9	432.9	0.6	0.03
	Comm (M)	4698	566.1	275.73	0.011
	Perf degradation (%)	0	0	20.8%	0
Test tree	Response (secs)	448.9	430.9	39.21	198.4
	OOS (secs)	448.9	430.9	6.48	6.38
	Comm (M)	4698	565.0	382.25	2.08
	Perf degradation (%)	0	0	25.6%	0
Gzip	Response (secs)	449.1	438.9	67.3	201.1
	OOS (secs)	449.1	438.9	4.78	0.10
	Comm (M)	4698	644.65	664.19	0.055
	Perf degradation (%)	0	0	28.1%	0

VM is restored or migrated on the destination virtualization platform. The OOS is the time period in which the DVM is out of service. The Comm gives the amount of DVM data transmitted over the network. The perf degradation measures how much the source machine is degraded during the migration. Table 16.1 shows that our approach is much better than the other three approaches in OOS time and the amount of data transferred. Although researches in suspension and resume [20,21,24] have proposed several optimization mechanisms to optimize the migration speed, the suspension and resume mechanism inherently requires the moving of the VM image over the network, which incurs much more communication cost. The shared storage live migration performs best in response time because it does not move the VM image. However, because the live memory collection and restoration is performed on both the source and destination platforms, experiments show that the performance of the source VM is degraded 20% to 29% during migration. With the DVM method, since the VM reconstruction is performed only on the destination platform, the source VM is not affected. The DVM approach provides a significant improvement of overall performance over the existing methods. Also, notice that the live migration approach is limited to cluster computers only and cannot be extended to general network environments such as grid environments.

The above experiments demonstrate the capability of a DVM to provide support for migration. Not only does it provides support for various migration techniques but also improves their performance by utilizing its novel feature of reconstructing the VM image based on the skeleton data, particularly in wide-area image migration scene. Experiment results show the same.

16.3.5 DVM Mobility and Communication State Transfer

The HPCM process migration system is also successfully built, which can migrate a legacy code from one computer to another. Several critical mechanisms and components have been developed, including the execution, memory, and communication state transfer mechanisms [25–27], a precompiler [24], and an automatic monitoring and triggering runtime system [28] to support automatic migration. These mechanisms and components have been tested under MPI-2 [29] and a PVM environment with different applications. As shown in Figure 16.10, for the linpack benchmark, the overheads of homogeneous migration usually ranges from 0.08% to 0.5%, which is very low.

The communication state transfer protocols are tested with the IS benchmark from NAS Parallel Benchmark 3.1 [30] and mpptest [31]. We have developed a portable communication library, called MPI-Mitten, to support migration with group communication under MPI [27]. Figure 16.11 shows the MPI-Mitten overhead during normal execution when no migration is conducted.

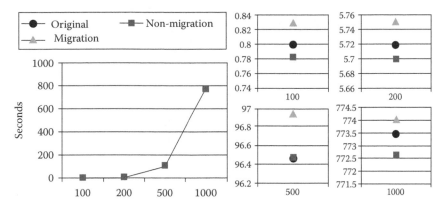

FIGURE 16.10 Migration overhead of homogeneous migration.

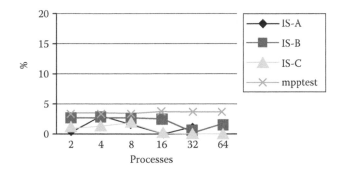

FIGURE 16.11 MPI-Mitten overhead.

The tests are performed for data sizes from A to C, where C has the maximum problem size. We performed the test on 2 to 64 processes. The overall overhead is less than 4% and average overhead is 1.22% for NAS IS and 3.55% for mpptest. The overhead is caused by the "test" operations and synchronization operations before and after communication primitives. Because the migration signal is passed through asynchronous communication primitives, it does not introduce much overhead. Figure 16.12 shows the timing result of migration. We have tested NAS IS and mpptest with 2 to 32 processes and found that the synchronization time is almost constant while the number of processes increases. The average synchronization time is 2.21 seconds for NAS IS and 2.42 for mpptest. These experiments show that our communication protocols are efficient and scalable.

These experiments show how a running process is migrated from one computing environment to another and prove the feasibility of redirecting a live communication. In a virtualized computing environment, our method to migrate the processes with their running states to a newly

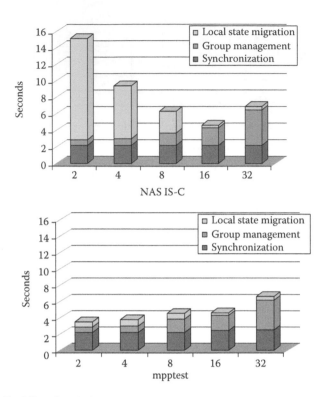

FIGURE 16.12   Migration timing.

established VM skeleton is more efficient compared with the statistics provided in [3], [10], and [21].

For DVM mobility, the most outstanding challenges are the efficiency of migration and the live communication redirection. Experiments show that our methods can address these problems. They show the great potential of a DVM in improving the migration performance of virtualized computing environments.

16.4 Conclusions and Future Work

In this chapter, we have introduced virtual machine technologies and their implementation in grid environments and discussed the development of a DVM middleware to provide a flexible and dynamic virtual computing environment. VM technologies have been identified as the able technology to support system-level virtualization so that virtual working spaces can be created and customized for users' needs in grid environments. However, applying VM technology to grids requires addressing the issues of mobility and complexity management. We have introduced the DVM concept and system for grids. The DVM system can generate a virtual environment swiftly and dynamically "on-demand." It provides dynamic virtual environments to support a wide spectrum of applications in a distributed, shared, heterogeneous, and homogeneous environment. Experimental results show that the DVM approach is feasible, efficient, and significantly better than existing methods in virtual machine deployment and virtual machine migration.

Current virtual machine technology lacks the standard and functionality to support customization flexibility. We have introduced the DVM middleware framework to support the customization flexibility and implemented it to demonstrate its correctness and effectiveness. The DVM approach can provide great support for a better standardization, a well-structured representation, and a customizable and flexible virtual resources management. While computing becomes more and more service-based, system-level virtualization becomes more and more a necessity to provide system virtualization, customization, and security.

The dynamic virtual machine or dynamic virtual space concept is not limited to grids. It can extend to any general distributed environments. For instance, it can be directly applied to support customization of a single virtual machine on sequential and parallel computers and virtual machine migration will provide an extra level of security and virtualization of mobile computing. This chapter introduced only the fundamental DVM and system-level virtualization but has revealed the need and potential of a DVM. A DVM has not been fully integrated into grid environments at this time and much work remain to be done.

References

1. I. Foster and C. Kesselman, *The Grid2: Blueprint for a New Computing Infrastructure*, Morgan-Kaufman, 2004.
2. VMware Co. Available at: http://www.vmware.com.
3. P. Barham, B. Dragovic, K. Fraser, S. Hand, T. Harris, A. Ho, R. Neugebauer, I. Pratt, and A. Warfield, "Xen and the art of virtualization," in *Proceedings of the ACM Symposium on Operating Systems Principles (SOSP)*, pp. 164–177, New York, 2003.
4. J. Smith and R. Nair, *Virtual Machines: Versatile Platforms for Systems and Processes*, Morgan Kaufmann, 2005.
5. K. Keahey, M. Ripeanu, and K. Doering. "Dynamic creation and management of runtime environments in the Grid," presented at the Workshop on Designing and Building Web Services (GGF 9), Chicago, October, 2003.
6. I. Foster and S. Tuecke, "Describing the elephant: Different faces of IT as services," ACM queue July/August 2005.

7. Amazon Elastic Cloud 2 Project, API Version 2006-06-26.
8. Available at: www.salesforce.com
9. M. Jankowski, J. Denemark, P. Wolniewicz, N. Meyer, and L. Matyska, "Virtual environments—framework for virtualized resource access in the grid," in *Proceedings of the CoreGrid Workshop on Grid Middleware in Conjunction with Euro-Par Conference*, pp. 101–111, Dresden, August 2006.
10. C. Clark, K. Fraser, S. Hand, J.G. Hansen, et al., "Live Migration of Virtual Machines," in *Proceedings of the 2nd Symposium on Networked System Design and Implementation*, pp. 273–286, Boston, May 2005.
11. Plex86 x86 Virtual Machine Project. Available at: http://plex86.sourceforge.net/.
12. IBM S/390: IBM Corporation, *Virtual Image Facility for LINUX Guide and Reference*, White Paper GC24-5930-03, February 2001.
13. J. Dike, "A user-mode port of the Linux kernel," in *Proceedings of the 2000 Linux Showcase and Conference*, pp. 63–72, October 2000.
14. A. Whitaker, M. Shaw, and S. Gribble, "Scale and performance in the Denali isolation kernel," in *Proceedings of USENIX Symposium on Operating Systems Design and Implementation*, pp. 195–210, Boston, MA, December 2002.
15. S. Hand, T. Harris, E. Kotsovinos, and I. Pratt, "Controlling the XenoServer open platform," in *Proceedings of the 6th IEEE Conference on Open Architectures and Network Programming*, pp. 3–11, San Francisco, CA, 2003.
16. P. Ruth, X. Jiang, D. Xu, and S. Goasguen, "Virtual distributed environments in a shared infrastructure," *IEEE Computer*, 38 (5), 63–69, May 2005.
17. A. Sundararaj and P. Dinda, "Toward virtual networks for virtual machine grid computing," in *Proceedings of the third Conference on Virtual Machine Research and Technology Symposium*, p. 14, San Jose, CA, May 2004.
18. R. Figueiredo, P. Dinda, and J. Fortes, "A case for grid computing on virtual machines," in *Proceedings of the 23rd IEEE International Conference on Distributed Computing Systems*, pp. 550–559, May 2003.
19. I.V. Krsul, A. Ganguly, J. Zhang, J.A.B. Fortes, and R.J. Figueiredo, "VMPlants: Providing and managing virtual machine execution environments for grid computing," in *Proceedings of ACM/IEEE Conference on Supercomputing*, pp. 7, Pittsburgh, PA, November 2004.
20. C.P. Sapuntzakis, R. Chandra, B. Pfaff, J. Chow, M.S. Lam, and M. Rosenblum. "Optimizing the migration of virtual computers," in *Proceedings of OSDI 2002, ACM Operating Systems Review*, 377–390.
21. M. Kozuch and M. Satyanarayanan, "Internet Suspend/Resume," in *Proceedings of the 4th IEEE Workshop on Mobile Computing Systems and Applications*, pp. 40–46, Calicoon, NY, June 2002.
22. J. Chase, L. Grit, D. Irwin, J. Moore, and S. Sprenkle, "Dynamic virtual clusters in a grid site manager," in *Proceedings of HPDC-12*, pp. 90–100, Seattle WA, June 2003.
23. G. Vallee, T. Naughton, and S.L. Scott, "System management software for virtual environments," in *Proceedings of the 4th International Conference on Computing Frontiers*, pp. 153–160, Ischia, Italy, May 2007.
24. C. Du, X-H. Sun, and K. Chanchio, "HPCM: A pre-compiler aided middleware for the mobility of legacy code," in *Proceedings of Cluster 2003*, pp. 180–187, Hong Kong, December 2003.
25. K. Chanchio and X-H. Sun, "Data collection and restoration for heterogeneous process migration," *Software—Practice and Experience*, 32: 1–27, April 2002.

26. K. Chanchio and X-H. Sun, "Communication state transfer for the mobility of concurrent heterogeneous computing," *IEEE Transaction on Computers*, 53 (10), 1260–1273, 2004.
27. C. Du and X.-H. Sun, "MPI-mitten: Enabling migration technology in MPI," in *Proceedings of the 6th CCGrid*, pp. 11–18, Singapore, May 2006.
28. C. Du, S. Ghosh, S. Shankar, and X.-H. Sun, "A runtime system for autonomic rescheduling of MPI 32 programs," in *Proceedings of the International Conference of Parallel Processing, Montreal*, Canada, August 2004.
29. MPICH2 homepage. Available at: http://www-unix.mcs.anl.gov/mpi/mpich2.
30. NAS Parallel Benchmarks. Available at: http://www.nas.nasa.gov/Software/NPB/.
31. MPPTEST. Available at: http://www-unix.mcs.anl.gov/mpi/mpptest.

17

High-Energy Physics Applications
on the Grid

Massimo Lamanna

CONTENTS

17.1 Introduction

In this chapter we will describe the usage of the grid in the High-Energy Physics (HEP) environment at the beginning of 2008. We will almost exclusively leverage on the experience and plans of the four major experiments at the Large Hadron Collider (LHC) at CERN [1].

This choice has multiple motivations, the most important being the fact that 2008 is the turning point year for these experiments (ALICE, ATLAS, CMS, and LHCb), which after many years of preparations are basically ready to start (first proton beams have been circulated in the LHC collider in September 2008 and the full start of data taking is scheduled for spring 2009). These experiments have played a crucial role in the evolution of grid technologies in the last several years and notably in connection with grid infrastructure projects. The most important projects are EGEE (Enabling Grid for E-sciencE) in Europe [2], OSG (Open Science Grid) in the US [3], and NDGF (Nordic Data Grid Facility) in the Nordic countries [4].

In the evolution of grid technology the HEP community and the HEP experiments have played a determinant role. The essential contribution was the enthusiastic promotion of the idea of grid computing formalized and popularized by I. Foster and K. Kesselmann in the late 1990s [5]: HEP adopted grid technology as the foundation for the entire computing in the LHC era. This led to the creation of the first production infrastructures based on grid technology.

The importance of the HEP role can be judged by following facts:

1. The HEP community had already at that time an established experience in creating long-lived collaborations across different and geographically distributed entities (universities, laboratories, etc.) funded by the coherent effort of several funding agencies. The HEP experiments were already exceeding several hundred collaborators from several tens of universities in the early 1990s (e.g., CDF experiment at Fermilab, USA). At the same time, thus still in the preparation phase, the LHC experiments were reaching an even larger scale (the largest LHC experiment, ATLAS, exceeds 2,100 physicists from 167 institutes in 37 countries). In a sense, the HEP world was proving that the collaboration scale that the grid was suggesting was attainable and even desirable when excellence and optimization of resources requires to cross existing borders (national, institutional, etc.).

2. The HEP community had already started a deep reflection about the way to provide the necessary computing power and data handling capabilities for the LHC research program. The experience of the CERN LEP experiments (active between 1989 and 2000 at CERN) and of several other HEP experiments such as CDF and D0 (Fermilab), BaBar (SLAC), NA48, and COMPASS (CERN) made very clear the importance of computing in terms of handling very large data samples (1 PB range). This was not new; from the very beginning, nuclear and particle physics were early adopters of new computing technologies. The new point was the observation that the computing infrastructure (software and hardware)

had to be planned well in advance both for reasons of cost and to master the increasing complexity of the scientific data. Along these lines, CERN set up a review of the LHC computing (the so-called Hoffmann review [6]) in 1999 to prepare and formally secure the mechanisms to build and maintain the necessary computing infrastructure. Eventually, the LHC Computing Grid (LCG) project, led by Les Robertson (CERN), was started in 2001 [7]. Notably, the LCG project was designed with the necessity to "cross" the experiment boundaries, fostering cross-experiment collaborations at the level of base tools (both in the application sector and in the infrastructure).

3. There were many examples of HEP experiments using distributed computing infrastructure well before LHC, notably on national centers like the IN2P3 Computing Centre in Lyon (France) or CINECA and CNAF in Bologna (Italy). The important point is that the grid concept suggested a complete solution to concrete issues being experienced in the HEP domain (single sign-on, role-based access and global sharing of resources). When I. Foster delivered a very inspiring talk at the CHEP (Computing in High-Energy Physics conference, the lead event for computing in the HEP community) in March 2000, the HEP community was already designing (and validating with simulation studies and prototyping work) a hierarchical model, which is still the foundation of the LCG infrastructure (MONARC project [8]).

4. The HEP community was at the heart of the European Data Grid project (EDG), led by F. Gagliardi (CERN) who then initiated the EGEE program. The HEP experience, together with innovative ideas and tools from the grid community (most notably the Globus project led by I. Foster and the Condor project led by M. Livny), initiated a number of research and development studies on the middleware necessary to provide dependable services for user communities (HEP plus Biomedical and Earth Observation applications). The software stack adopted and evolved in EDG and then in use by EGEE is the underlying base for operating the grid. In parallel, several initiatives have been undertaken by the experiments to provide high-level services to serve specific needs. All the LHC experiments developed layers on top of the base grid services. The reasons were multiple, but in general we recognize the following patterns:

- Insulate the physicists community from an infrastructure that is in fast evolution (e.g., AliEN project developed in ALICE).

- Provide a layer to optimize performances, in particular to increase efficiency, stability and minimize latency (e.g., DIRAC project developed in LHCb).

- Federate different grids, providing an effective interoperability layer for data processing and data movement (e.g., DQ project in ATLAS).

As a matter of fact all experiment-specific layers contain all the three patterns, with different level of emphasis depending on the needs of the experiments and the different phases of their evolution. All these projects were effectively a continuous stimulus to progress (in the HEP community itself and in the grid communities at large). At the same time, they allowed the maximum usage of the resources available from the different infrastructures, overcoming interoperability and instability problems observed in the early stages.

The early feedback from the user community was a decisive factor in helping the evolution of these complex technologies. The HEP community devoted significant resources (see e.g., the ARDA project described in this paper [9]) to work in close contact with the middleware communities.

The activities in (close connection with) the experiments eventually matured in a coordinated process across all the active grid communities. We are observing a sort of relay between the middleware community on the one side, the infrastructure on the other, and the applications, in particular HEP, on the third side. Over the years, three main phases have been observed. The first one had the main focus on the development of the middleware, especially prototyped in the pre-LCG phase. The second phase corresponded to the first years of LCG (and EGEE): the goal was essentially to demonstrate (by building it) a worldwide computing infrastructure. Progressively the focus went to a third phase where more and more feedback (and innovative ideas) were coming from the user communities.

I believe that either the role of the applications (HEP and others) will continue to be strengthened (via close collaboration) or the existing momentum will eventually be redistributed across national and application-specific solutions with possible loss of coherence. The HEP community, especially for the sociological strengths and its power of innovation mentioned at the beginning, is the best guarantee to keep the coherence achieved in the last few years.

In recent years, very interesting patterns of collaborations have been observed across different applications. In all major cases HEP played an important role. Initially the idea of several projects (notably EGEE) was to have the applications "validating" their services (the infrastructure, the middleware) by injecting user requirements and in using prototypes. In this perspective, a "generic" grid will be validated by exposing it to several user communities (the more the better), effectively covering more and more use cases. It is one of the main successes of these projects to demonstrate grid usage from several applications (e.g., the spectacular usage rise observed in EGEE-2). The key point is, however, different: an infrastructure at the scale of the grid should not only demonstrate its value for a large number of users like a super computing center, but also bring additional added value to its users.

On an infrastructure like the grid, the applications sit side by side and benefit from each other's experience. The fact that every activity had some specific (possibly nongeneral) use case is largely counterbalanced by the fact to find (in a sister application) colleagues sharing solutions, advising, and so on. A team of scientists (or a company) should join the grid because the balance between the advantages of the new technology is largely exceeding the aggravation to change part of their working system (which is at the base of their activity or their business). These examples and new opportunities of collaboration are the real method for attracting new applications and involving them in using grid computing.

The convergence among applications is not easy and cannot be established "by decree." There are very positive examples, even between different communities as I mentioned, but should not be the only parameter for success. The convergence on common solutions, even in the HEP community, is not automatic and has not been achieved completely. There are good reasons for this: computing is not a generic tool (at least not yet) and it is on the critical path to get faster and better results. It is therefore understandable that (as we will describe in what follows) in a few cases we can already observe full convergence. In other areas, more convergence will be eventually achieved in the near future. Ultimately, some diversity will stay.

I think it is difficult to overestimate the importance of the visionary power to the irrevocable move to the grid as the solution to all the computing of all leading-edge activities in LHC. This is something we have observed only in HEP so far, namely to commit the success of the most important scientific program to the usage of this new technology. Due to its history and computing requirements, HEP was the only science being ready, both sociologically and technically, to move to the grid to build its computing infrastructure. Often a comparison between the Web (invented at CERN during the LEP period) and grid technologies is made. The next few years will tell whether this comparison is appropriate.

17.2 The ARDA Project in LCG/EGEE

In the case of HEP, a specific effort was set up in the years 2004–2008 to investigate the usage of the grid for the so-called end-user analysis: the ARDA project. In the following we will use some of the activities of this project to guide us in the HEP usage of grid technology and of the LCG infrastructure in particular.

ARDA stands for "A Realisation of Distributed Analysis" (http://cern. ch/arda) and is jointly funded by EGEE and by CERN and with substantial contributions of several institutes such as the Russian institutes in LCG and the Taipei Academia Sinica Grid Center. With the word "analysis" in HEP we mean all computing activities, performed, almost independently, by individual physicists, sometimes organized in small teams. In general they share a common software foundation but each individual/team has a set of different executables, tailored for a specific scientific task. All analyses share part of the input data (experimental data, both raw and reconstructed plus simulation data) but often rely on private copies of "derived data." Frequent multiple passes on subsets of the data are the rule. The impact of this activity on the grid computing is relevant at least in three areas.

On the one side, the size of potential user community (in the case of the LHC experiments, several thousands physicists) is a call for a robust system, which should be reasonably user friendly and transparent. Analysis is therefore very different from the organized activities (detector simulation, raw data reconstruction, etc.) that are performed by a single expert team in a coordinated way. Realistically if a large community has to use the grid, this should not force unnecessary changes in the way of working (analysis is a day-to-day activity). With grid technologies being still in a fast-evolution phase the users should be shielded at least by nonessential changes in the internal components of the infrastructure.

The second area is again intimately connected to users' expectations. Users are interested in performing analysis on the grid only if they can get a faster turnaround time or have access to larger or more complex datasets. The potential benefit of larger resources could be reduced (or even disappear) if one needs continuous expert support as in troubleshooting activities. This observation translates into the requirement of a system, which should not only provide sheer power but also be reliable and efficient. In this case the users have to rely on the results back within dependable time limits. High efficiency implies no need for too many time-consuming operations like resubmitting jobs due to failures of the system in accepting jobs, in accessing the data or in returning the results. Simple access to relevant monitoring information is clearly the key.

The third area is data access. Data access on the grid is a field of research in itself. In the analysis use case, users should be empowered with simple but powerful tools to place, locate, and access the data. HEP is quite unique in the area of data management, as we will see in the following, due to the requirements coming from aggregated data sizes (over several PB per year for several years of functioning of the experiment and physics analysis), the need of replication and broad access (user communities of the order of several thousands scientists).

17.3 The LHC and the Grid Projects

The LHC started operation in 2008. Four major LHC experiments (ALICE, ATLAS, CMS, and LHCb) will collect roughly 15 PB of data per year, which should be processed, calibrated, and analyzed multiple times. Seamless access to the LHC data should be provided for 5000 scientists in about 500 research institutions worldwide. The lifetime of the project is estimated to be around 20 years.

The goal of the LCG (also called Worldwide LCG or WLCG) is to prepare and deploy the computing environment indispensable to perform the physics program of the LHC project. This includes the detector simulation studies to push the detectors' performance to their limit, the calibration and the monitoring of the detectors during data taking, the reconstruction of the raw data, and other selection and analysis stages. All relevant data should be accessible to all the physicists worldwide participating in an experiment.

The LCG Technical Design Report [10] estimates the computing power required for the LHC data analysis to be of the order of 100,000 CPUs (CPU available in 2004). A globally distributed model for data storage and analysis was chosen. Originally the MONARC project (Models of Networked Analysis at Regional Centers for LHC experiments) suggested a hierarchical structure of distributed computing resources (partially modified due to the emerging grid technologies). CERN and multiple computing centers worldwide are providing resources for constructing the LCG infrastructure.

The infrastructure, which has been built, has a hierarchy of tiers of computing centers. CERN is the Tier0 center of the infrastructure. The primary functions are the data recording and the permanent storage capability (tape system). The system should be capable to sustain up to 1.25 GB/s of data recording rate (ALICE experiment during heavy-ion operations) and store several tens of PB per year. Tier0 provides CPU power for data calibration and first-pass reconstruction. Tier0 distributes data to the Tier1 according to the policies agreed with each experiment.

The infrastructure has 11 Tier1's. Each Tier1 has custodial responsibility for the data received from Tier0 and for data processed in the Tier1 layer. Tier1 CPU will be used heavily in data reprocessing and in preparing big data sample for analysis. The Tier1's are: ASGC Taipei, BNL US, CNAF-INFN Italy, FNAL US, GridKa Germany, IN2P3 France, NDGF in the Nordic countries, NIKHEF/SARA Netherlands, PIC Spain, RAL UK, and TRIUMF in Canada. All Tier1s have support and data distribution (and custodial) responsibility to the next level in the hierarchy, the Tier2 centers.

So far, over 100 Tier2s are participating in LCG. At variance with the Tier0 and Tier1s, Tier2s have no long-term data storage responsibility. Ultimately they will provide the computing resources for most of the analysis activities (hence serve the majority of the users). Tier2s also have a very important role in providing the bulk of the computing power for simulation activities. Smaller facilities (Tier3) do exist, essentially to perform analysis on distilled data samples ("downloaded" from LCG centers), but are outside the scope of the LCG project and are not discussed here.

The data rates and sizes for the first two years of LHC running are summarized in Table 17.1. The luminosity is $L = 2 \times 10^{33}$ cm^{-2} s^{-1} in 2008 and 2009 and then it will reach $L = 10^{34}$ cm^{-2} s^{-1} in 2010 (event rate scales up with luminosity; event sizes can also grow due to interaction pile-up). The canonical beam time for proton–proton operations is assumed to be 10^7 seconds in 2008 and 2009. For heavy-ion running a beam time of 10^6 seconds is assumed to be $L = 5 \times 10^{26}$ cm^{-2} s^{-1}.

The column RAW corresponds to the so-called raw data, the events that have been read from the experiment readout channels, assemble and pass through a series of online filters. These data are recorded (also on tape for long-term custodial storage) at CERN and at Tier1 (normally guaranteeing at least two complete copies across the whole LCG). Raw data enter in a chain of processing steps, generating reconstructed information and analysis objects (ESD and AOD) to allow different types of physics and detector studies. The MC (MonteCarlo) columns correspond to the required simulation data. Before the LHC starts this is the dominating activity on the grid (both for the simulation and the corresponding analysis). The LCG infrastructure is built out of a collaborative effort on top of other projects and organizations like EGEE, OSG, and NDGF. All these projects have a multiscience character, particularly prominent in the case of EGEE. In all cases the HEP community is one of the major drivers.

TABLE 17.1

Event Rate and Data Sizes at LHC Start Up for the LHC Experiments

	Rate [Hz]	RAW [MB]	ESD [MB]	AOD [kB]	MC [MB/evt]	MC % of real
ALICE HI	100	12.5	2.5	250	300	100
ALICE pp	100	1	0.04	4	0.4	100
ATLAS	200	1.6	0.5	100	2	20
CMS	150	1.5	0.25	50	2	100
LHCb	2000	0.025	0.025	—	0.5	20

Note: ALICE HI refers to the heavy-ion operations. All other entries correspond to the proton–proton operations. The requirements in terms of CPU [CPU power is measured in SPECint2000, a benchmark suite maintained by the Standard Performance Evaluation Corporation (SPEC: http://www.spec.org) to measure and compare compute-intensive integer performances. This quantity has been found to scale well with typical HEP applications. As an indication, a single-core Intel Pentium 4 processor can deliver about 1700 SPECint2000. MSI2K stands for 10^6 SPECINT2000], disk and mass storage system (MSS) are given in Table 17.2.

Source: LCG Technical Design Report.

TABLE 17.2

The Requirements in Terms of CPU, Disk, and Tape Storage

	Requirements—All Experiments			
	2007	2008	2009	2010
CPU (MSI2K)				
CERN total	10.0	25.3	34.5	53.7
CERN Tier-0	6.9	17.5	22.4	32.8
CERN T1/T2	3.1	7.8	12.1	20.9
All external Tier1s	19.2	55.9	85.2	142.0
All Tier2s	23.6	61.3	90.4	136.6
Total	53	143	210	332
Disk(TB)				
CERN total	2200	6600	9200	12,600
CERN Tier0	400	1300	1400	1800
CERN T1/T2	1800	5300	7800	10,800
All external Tier1s	9300	31,200	45,400	72,100
All Tier2s	5200	18,800	32,400	49,200
Total	17,000	57,000	87,000	134,000
MSS (TB)				
CERN total	4900	18,000	31,100	45,600
CERN Tier0	3400	13,600	23,600	34,500
CERN T1/T2	1500	4400	7500	11,100
All external Tier1s	9300	34,700	60,800	92,200
Total	14,000	53,000	92,000	138,000

It is important to note that 2008 is the start-up year for LHC but is also a key year for EGEE. The year 2008 marks the end of the first part of the EU-funded project launched in 2004 as a four-year program (EGEE-1 April 2004 to March 2006 and EGEE-2 April 2006 to April 2008). A third two-year phase (EGEE-3) has started in May 2008, which will be incontestably the year where the plans for a longer-term, sustainable infrastructure will have to be clarified and unfolded. The long-term future of grid computing to the scale attained by projects like EGEE will depend on these decisions and plans.

17.4 HEP Analysis

Each experiment has prepared specific mechanisms to ease the access to the grid for their physics community. As an example we will start from the case of ATLAS and LHCb and their system called Ganga.

Ganga is a job-management system developed as an ATLAS-LHCb common project. ARDA started to collaborate with the Ganga team already in 2004 and progressively increased its contribution due to the interest and the potential of this system [11]. The basic idea is to offer a simple, efficient, and consistent user interface in a variety of heterogeneous environments: from local clusters to global grid systems. It is natural that a user develops an application on a laptop, moves to a local batch system for optimizing the analysis algorithm onto richer datasets and eventually performs full-statistics runs on the grid. Moving from one stage to another also applies in the reverse order (from the grid to the laptop), for example, when a bug-fix or an algorithm improvement should be developed and tested.

This approach responds to the fact that the physics analysis (also on the grid) is an activity performed by a large community of physicists using a variety of applications. These applications are typically built on a simulation or event reconstruction framework (foundation framework), which is experiment specific and enriched with custom code provided by each physicist. Ganga supports users using the foundation libraries by appropriate plug-ins simplifying the configuration stages for the users. On the other hand, Ganga leaves the freedom to run completely independent custom applications (or to contribute new application plug-ins).

Ganga shields users completely from the job submission details (basically the execution back-end is selected by the users by a software switch and Ganga generates the appropriate stubs to execute user code on the available resources). This is essential to allow users to execute on different back-ends in a seamless way as mentioned before. It is interesting to note that this approach also shields the users from the evolution of the middleware, hence it fully responds to the first area mentioned in the Introduction.

Ganga is written in Python. Current versions are available under the GNU Public Licence. Ganga acts as a front end for submission of computationally intensive jobs to a variety of submission back ends:

- Several batch systems including LSF, PBS, and Condor
- Grid middleware like different flavors of the LCG/EGEE middleware or NorduGrid (NDGF)
- Specialized workload management systems for the grid such as Dirac (LHCb experiment) and Panda (ATLAS experiment)

Since Ganga scripts are Python scripts, the entire power of Python is available for creating complex tasks, yet the user is not obliged to be a Python expert. In tutorials new users typically learn the necessary syntax within the first 30 minutes. In Figure 17.1 we show a basic example that is used in a typical tutorial session.

Finally, Ganga keeps track of the jobs created and submitted by the user as records in a job repository. This allows the user to manipulate Ganga

```
#

# Ganga example

# Submit 3 jobs, one local, one on batch, one to the grid

#


j=Job(backend=Interactive(),application=Executable())

j.application.exe="/bin/echo"

j.application.args=["Hello world"]

j.submit()


j2=j.copy() # make a copy of the last job

j2.backend=LSF(queue=?8nm?) # submit to LSF

j2.submit()


j3=j.copy()

j3.backend=LCG() # run on the Grid

j3.submit()
```

FIGURE 17.1 A simple example where the same job ("Hello world") is submitted to the local machine, a batch system (LSF), and the LCG grid.

jobs in between sessions. Manipulations include being able to submit, kill, resubmit, copy, and delete jobs. The repository is updated by a monitoring loop, which queries all used back ends for the status of the jobs and updates the status or triggers actions based on the state transition. For example, a job that changes into a completed state triggers the retrieval of the registered outputs from the submission back end.

Figure 17.2 illustrates the very large user basis which has been built around Ganga. It is important to note that around 25% of the user community (over 50 regular users each month) comes from non-HEP communities. As an example of the usage of Ganga outside ATLAS and LHCb, I use an example from theoretical physics. Quantum chromodyamics (QCD) describes the interaction of the constituents of the hadronic matter (quark

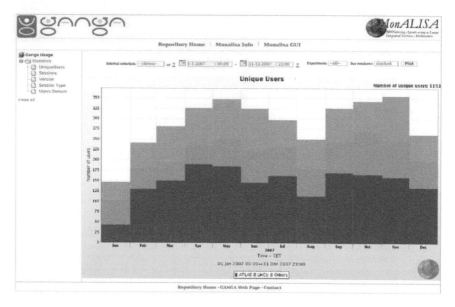

FIGURE 17.2 Ganga usage as reported by MonALISA [12]. In 2007 over 1000 distinct users (unique users) tried out Ganga. Each month, over 100 ATLAS users and about 50 LHCb users use Ganga for their activities. An additional 50 users (25% of the total) are coming from other communities (mainly outside HEP).

and gluons) and ultimately the structure of nuclei. When QCD is studied on discrete systems (Lattice QCD), it requires nontrivial computing resources. The application that we present here is a study of phase transitions in a quark–gluon plasma [13].

The interest of the example from the computing point of view lies mainly in the fact that Ganga allows a very fast porting of an application onto the grid. The clear scientific advantage is that, with an investment of about 1 week during summer 2007 for porting and running on the EGEE infrastructure, the available statistics have been multiplied by four compared to the one collected over several months on dedicated resources.

The application performs a series of iterations, descriptive of the space–time lattice to be investigated. Of these lattices, 21 different versions exist, all describing slightly different physical conditions. Independent (from a random number generation point of view) programs running on the different lattice configurations produce results that can be statistically added to study the behaviors of the quark–gluon plasma. Since the result improves with the number of iterations performed and the result is saved in the space–time lattice, it makes sense to run the application for as long as possible (ideally until the batch queue time is reached). Therefore the decision was taken to run in an infinite loop and to regularly send back the results to a simple server. This allows the script, which runs on the worker node, to be very simple and to make sure that if a job crashes or gets killed, the latest result is still available. Since results were sent back every hour, on average a job would waste one hour at most (out of several days of running).

We have exploited the natural parallelism (the 21 space–time lattice files) together with the free parameters in the configuration file. With this strategy, around 450 jobs were submitted using Ganga to both the EGEE grid and to the CERN LSF batch system. This resulted in about 9500 CPU cores to be used. The jobs ran for about one week after which they were terminated (via Ganga). Within this week the results from more than 30 CPU years could be harvested. A subset of these results have been used for presentation in conferences as Lattice 2007. The jobs ran on more than 50 sites, with a majority of jobs running on fast Intel Xeon processors (see Figure 17.3).

This example is a neat demonstration of the power of Ganga as a tool to facilitate the usage of the grid. The original goal to isolate HEP users from the details of the execution back end led to the development of Ganga, which is attracting users from different activities. Often new users discover the tool by themselves and then start using it.

Within the EGEE context, we have observed the value of Ganga also as a tutorial tool. The choice of the Python language (its flexibility and the availability of powerful extension modules) helps to guide the new users into realistic scenarios without unnecessary technicalities. The final result is that users end a three-hour tutorial and are in a position to continue experimenting and preparing to use the EGEE production infrastructure without further dedicated support effort. Ganga is used in ATLAS and LHCb. ALICE and CMS designed their own strategies to support users on the grid.

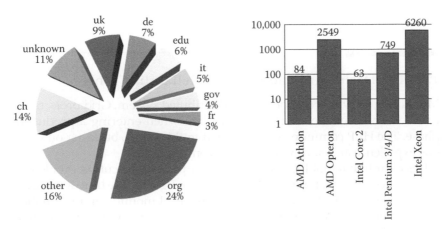

FIGURE 17.3 Distribution of top-level domains of the sites and the distribution of processors used for the lattice QCD application. Note the log scale in the processor distribution plot.

In the case of ALICE, the system conveniently couples their grid back end (AliEn) with the ROOT framework [14] (at the basis of their C++ framework ALIRoot). The key component is a very efficient gateway (a service used by multiple users) to deal with user commands. This service caches the authentication state of the clients in order to provide efficient access for interactive users. This pattern (described in the paper for the original ALICE implementation [15]) is actually used more and more in different areas of the grid middleware since it couples the strict security standards needed by the grid (basically the usage of X.509 security) with the responsiveness needed by any interactive application. This mechanism avoids excessive load generated by security at the server level since the server does not authenticate all the clients at each interaction but it basically delegates this to (a set of) trusted services. In particular, the searches in the ALICE (AliEn) file catalog can be done in a transparent way from the user prompt and from ROOT with high efficiency (also implementing features like filename completion, etc.). Again, the complexity of the sophisticated solution to provide simple and efficient access is hidden.

In the case of CMS they developed CRAB (CMS Remote Analysis Builder) [16] an application which is somewhat similar to Ganga. In the original form it was basically a client tool helping the user to submit and control jobs on the grid via a convenient set of commands and tools. More recently the usage of an optional server has been introduced allowing "disconnected operations" like, for example, automatic intelligent resubmission while the user is actually not connected.

CRAB is a also very successful application in terms of user response. In 2007, 20,000 jobs per day (with efficiency exceeding 90%) have been executed by CMS users, making CRAB the most intensively used tool in the HEP grid environment. In the next chapter, we display a snapshot of usage of CRAB in Figures 17.6 and 17.7.

17.5 The Dashboard and the Grid Reliability Tools

Monitoring is a vital component in a distributed system. Grid projects had to invest considerable effort, in particular, when entering a production phase. The HEP community contributed to this effort, building on previous experience and adding innovative contributions.

It is clear that a tool like Ganga does not prevent execution problems if these are connected, for example, to a misconfigured site or to a failure in the middleware stack. Such investigations need monitoring information. As a matter of fact, all the different actors in the grid world (operation support, middleware developers, individual users, application managers) need easy access to the available information.

A special role is being played by the Service Availability Monitor (SAM) developed at CERN within the EGEE and LCG projects [17]. SAM is capable to schedule tests on the grid infrastructure (as grid jobs and as commands from grid user interfaces) in order to collect operational data. In Figure 17.4 the status as seen by SAM for a part of the EGEE infrastructure is shown. Computer-centers' statuses are indicated by a color code. One can then drill down to the status of individual services/sites to spot operational problems or calculate the availability of the different computer centers. In the case of LCG, monthly reports compare the expected and available resources at each participating site.

SAM is clearly an essential tool to operate the grid. In addition, it is important to correlate these data with the actual user activity (usage and efficiency seen by the different types of jobs). The correlation is not always very simple due to the different ways in which different jobs (and different user communities) use the grid services offered by the computer centers. A complementary view is needed and the applications should also be involved. In practice this generated a collaboration between the HEP user communities and the operation team (at the origin of SAM and other infrastructure-oriented monitoring systems).

The combination of the experience of the monitoring system of CDF (FNAL) and the user monitor of an early ARDA analysis prototype was used to start the CMS Dashboard project [later renamed (Experiment)

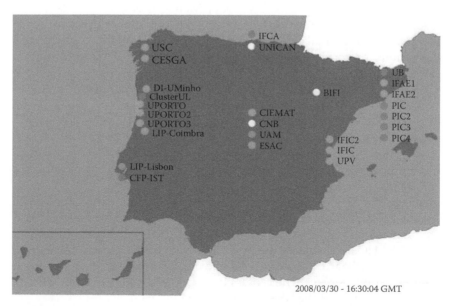

FIGURE 17.4 SAM status for a part of the EGEE infrastructure (http://www.egee.cesga.es/EGEE-SA1-SWE/monitoring/sam.sel.shtml?voselect=atlas.)

Dashboard since the same foundation is used by all four LHC experiments [18]]. The project thus started as a collaboration between ARDA and the CMS experiment.

The strategy was to give to all grid actors the right tool to manipulate and display the available data. The grid operation support, for example, could use the Dashboard to isolate site-specific troubles and use the statistics from error messages to fix the problem. Middleware development teams could collect large statistics of error conditions, concentrating on the most common (hence the most annoying for the users) factoring out site or application problems. Users are clearly interested to follow the execution (including error conditions) of their own jobs while the activity managers are interested in global figures like resource usage.

In the development of the project, the emphasis was given to the aggregation of existing information and no special effort was devoted in the development of new sensors or protocols. The main components of the Dashboard are then information collectors, the data storage (an Oracle database), and the services responsible for data retrieval and information presentation (command-line tools, Web pages, etc.).

The Dashboard is using multiple sources of information. In addition to SAM, it collects information from other grid monitoring systems like R-GMA (Relational Grid Monitoring Architecture) [19], GridIce (Monitoring tool for Grid Systems) [20], and ICRTM (Imperial College Real Time Monitoring of the Resource Brokers) [21].

The Dashboard also aggregates information from experiment-specific sources. Examples are the ATLAS Data Management, central databases (like the ATLAS Production database), job submission systems (like Ganga), and individual jobs. Information is transported to the Dashboard via various protocols (depending on the capability of the information providers).

The collection of input information implies regular access to the information sources. They are retrieved and stored in the Dashboard database. To provide a reliable monitoring system, data collectors should be resilient to run permanently and recover any missing data in case of failures. The Dashboard framework provides all the necessary tools to manage and monitor these agents, each focusing on a specific subset of the required tasks.

In Figure 17.5 we present one of the main views of the Dashboard, namely the Job Monitor. We display as an example the summary of CMS production jobs (one week at the beginning of 2008). It is worth noting that since the LHC experiments use as a rule more than one grid infrastructure, the Dashboard has been designed in order to collect information from all used resources. The centers listed in the display belong to EGEE with the exception of the US sites (belonging to OSG).

In Figure 17.6 we present also an alternative view from the Job Monitor. The dashboard database provides here the view of the analysis jobs

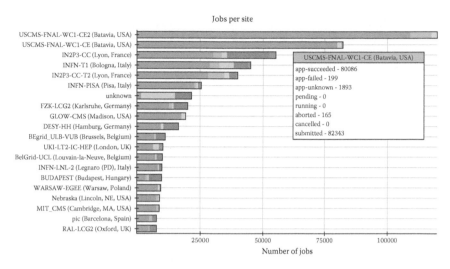

FIGURE 17.5 Dashboard Job Monitor. Summary of CMS production jobs (October 2007). The Experiment Dashboard accounts for all CMS jobs on both the infrastructures used by the experiment (EGEE and OSG).

(submitted by the CMS tools CRAB). These summary views are interesting for both the resource managers at the participating sites and the ones responsible for the computing of the experiment as a whole.

On the other hand, users are clearly more interested in concentrating on their own work, in particular to pin down problems in their activity. In Figure 17.7 we drilled down to the view provided for a given user. It is

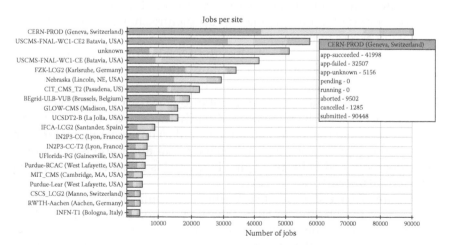

FIGURE 17.6 Dashboard Job Monitor. Summary of CMS analysis jobs (October 2007). As in Figure 17.5 the Experiment Dashboard accounts for all CMS jobs (submitted with the CRAB system) on both the EGEE and OSG infrastructures used by the experiment.

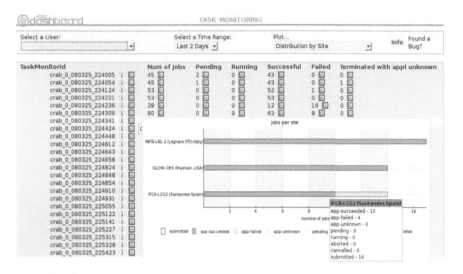

FIGURE 17.7   Dashboard Task Monitor. A snapshot of a user page is shown. There is the possibility to have a breakdown of each task (normally a set of jobs sharing the same executable running independently on a coherent dataset, i.e., a set of files).

important to know that a user rarely submits single jobs. Due to the data quantities to be analyzed, data are often organized in *datasets*; in general, collections of files containing a coherent collection of data. In this case the action to analyze a single dataset generates (in this case within CRAB) a set of jobs (e.g., one job per data file). Jobs from a single user request may be executed on different sites since data are replicated across the LCG infrastructure (in Figure 17.7, the pop-up picture shows the execution of one dataset across three sites).

The importance of an activity like the Dashboard is clear and documented by the interest in the HEP community (usage by the four LHC experiments): the Dashboard provides unbiased views of the delivered performances to specific user communities by measuring the efficiency of the user's application by monitoring directly the activity of all users. All of the project (and the Job Monitor in particular) has generated interest in several other applications in EGEE. Biomedical applications (VL-eMed) have adopted it and Diligent (Digital Libraries) is considering evaluating it in their infrastructure.

In Figure 17.8 we show another Dashboard application: Site Efficiency. In this case, the Dashboard shows the installation in use for VL-eMed (the same application runs for the HEP communities as well). In this application, job attempts are identified and the grid failures are categorized and associated with a given grid resource in a site. In case a job is automatically resubmitted multiple times due to failures, each job

SiteName (click on any site)	Successful jobs	Failed jobs	Efficiency
unknown	0	4	0.00%
SARA-MATRIX	8660	9908	46.64%
ce.gina.sara.nl:2119/jobmanager-pbs-express	4	0	100.00%
mu6.matrix.sara.nl:2119/jobmanager-pbs-long	1	0	100.00%
ce.gina.sara.nl:2119/jobmanager-pbs-medium	3878	4678	45.32%
mu6.matrix.sara.nl:2119/jobmanager-pbs-express	13	6	68.42%

JobIds	# jobs	Successful?	Error message
See all the jobids...	6	No	Failure while executing job wrapper
See all the jobids...	7	Yes	user retrieved output sandbox
See all the jobids...	3	Yes	Job terminated successfully
See all the jobids...	1	Yes	unknown
See all the jobids...	1	Yes	unknown
See all the jobids...	1	Yes	unknown

mu6.matrix.sara.nl:2119/jobmanager-pbs-medium	7	4	63.64%
ce.gina.sara.nl:2119/jobmanager-pbs-short	4748	5216	47.65%
mu6.matrix.sara.nl:2119/jobmanager-pbs-short	9	4	69.23%
NIKHEF-ELPROD	11250	1967	85.12%
tbn20.nikhef.nl:2119/jobmanager-pbs-qshort	6333	994	86.43%
tbn20.nikhef.nl:2119/jobmanager-pbs-qlong	4917	973	83.48%
LSG-AMC	18368	2058	89.92%

FIGURE 17.8 The Site Efficiency Dashboard application at work for VL-eMed. Job attempts are identified and the grid failures are categorized and associated to computing resources of the sites. The application permits the very quickly identification of a specific error pattern.

attempt is taken into account to test all available grid sites. The main difference with the Job Monitor application (Figures 17.5 through 17.7) is that in that case only the final execution of a job is considered. Site Efficiency permits very quick identification of error patterns, typically connected to a site misconfiguration. In the case of common errors the tool points to a list of explanations/solutions that are accessible via the drill-down functionality of the tool.

The future of this activity is that it will continue to grow. The availability of more data allows more sophisticated studies. Very important developments are going on to propose a unified mechanism to exchange data (e.g., using ActiveMQ http://activemq.apache.org/) and to better interface with the different systems used in the grid computer center (e.g., using Nagios http://www.nagios.org/). Here, the idea is to feedback monitoring data (like grid efficiency at a site) into the monitoring system of the site itself, allowing seamless integration between local established operational procedures and the newly available information.

17.6 Data Management

Data management is particularly interesting in the case of HEP. In this case the quantity of data (every year several PB of data have to be *added* to the data store), the replication strategies (multiple complete copies should coexist over the LCG infrastructure to provide redundant storage and data access), and the complex access patterns (especially at the level of end-user analysis) make data management a very interesting problem. ARDA invested a lot in this field, starting from middleware tests to monitoring activities. For example, a very important part of the Dashboard monitors data transfers at the level of the infrastructure services and at the level of experiment-specific steering systems.

17.6.1 Storage Resource Manager

Due to HEP-specific requirements (actually much older than the grid idea) the definition of a standard to interface to mass storage has a long history. In recent years this problem has been discussed in the context of the Open Grid Forum (OGF), which led to the definition of SRM (Storage Resource Manager). The adoption of SRM within LCG considerably accelerated the convergence on a workable standard implementation. The deployment of a nontrivial infrastructure of SRM and the operational experience will in turn be essential in the further evolution of the SRM concept.

The complexity does not only depend on the difficulty of the performance required (data size, number of files, etc.) but also because SRM is effectively an interface to be implemented by the different mass storage systems supported and in use in the grid computer centers. LCG sites use four systems, namely CASTOR (notably working at the Tier0 and in three Tier1s), dCache (in use in most Tier1s), StoRM (at the Italian Tier1 and under consideration in other centers), and DPM (essentially deployed at Tier2s). Details of the different implementations can be found in [22].

The experiments' requirements are satisfied with the SRM version 2.2 which is being deployed and now (beginning of 2008) over 160 endpoints are becoming available for the last round of readiness tests before the data taking (CCRC'08). Very much like the operations of the first services in LCG back in 2003, this is a proof-of-existence of the viability of the SRM solution to deliver the base data storage layer for LCG. It is clearly a start, since this area is in constant evolution, but the fact that this infrastructure can be actually operated by shift crews and a good service is delivered to users is clearly very encouraging.

17.6.2 File Transfer Service

An example of a high-level service built on the existing data infrastructure (and developed in close connection with the HEP community within the

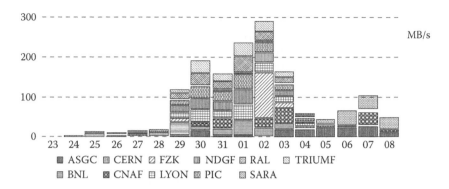

FIGURE 17.9 ATLAS cosmics data acquisition (August 23–September 8, 2007). The snapshot of the Dashboard shows the data distribution from CERN to the main regional centers supporting the ATLAS experiments.

EGEE project) is the file transfer service (FTS) [23]. FTS is a layer on top of storage (essentially SRM) and transfer protocols (globusFTP). Its main goal is to provide a dependable service namely a layer hiding short interruptions of the underlying services (essentially by retrying) and avoiding congestions by scheduling data transfer taking into account of the network capacity and shares across users and virtual organizations.

The experiments typically contact this service to schedule a transfer and poll it to see the status. By its nature the service collects book-keeping information, which is also essential for the operation teams maintaining it. In 2007, over 10 PB have been transferred. Although at this moment these massive data movements are at the heart of the HEP applications only, I believe that in the near future more applications will depend on it to distribute files across vast infrastructures of storage elements.

In Figure 17.9, we show the data transfer of one of the first tests of the full chain of data acquisition in late 2007. During a week, the ATLAS detector collected cosmic-ray events following the schema expected in normal LHC operations. In this test, ATLAS distributed the raw data and the centrally reconstructed data onto the full infrastructure (down to Tier2s). Eventually end-users performed data analysis at the remote sites.

17.6.3 Grid Catalogs

The EGEE/LCG project has developed a very successful product called LFC (LCG file catalog). The LFC is a secure, lightweight, and highly scalable POSIX-like file catalog serving a variety of communities. LFC stores catalog entries on a database back-end: supported back-ends are Oracle and MySQL.

In HEP, ATLAS uses the LFC for the local file catalogues located at Tier0 and Tier1; these LFCs control the location of files at each Tier1 (and related

Tier2s), while the ATLAS-specific catalogs orchestrate the overall data distribution and bookkeeping. LHCb uses LFC as a global file catalog. In this case several Tier1s have a full read-only replica, synchronized with the central one using Oracle data streaming functionality (Oracle Streams: the replication is performed at the back-end level).

Globally (including non-HEP applications) over 100 LFC instances are in use on the EGEE infrastructure. The largest installations have more than 10 million entries. The evolution of this successful product was always driven by HEP use cases, although inputs from other user communities have been taken into account. During this evolution the product included more and more sophisticated features both to boost performance (like bulk operations for inserting and deleting entries) and to cover security needs (integration with the EGEE security infrastructure, data encryption, etc.).

Other catalogs exist developed by different experiments. One example is the AliEn catalog, which is at the center of the AliEn system (the ALICE distributed system) [24]. In this case the catalog not only keeps location information for data files (and corresponding metadata attributes) but is also used by several components of the system. The catalog also contains the information of software installations available at different sites and the output of all the jobs.

As the final example of the fruitful collaboration between HEP and other sciences on catalogs, I choose the AMGA metadata catalogue (AMGA stands for ARDA Metadata Grid Access [25]). This system, originally developed by ARDA as a tool to validate the metadata interface in the EGEE middleware, was used as a laboratory to investigate efficient techniques to provide robust and efficient access to databases in a grid context. AMGA is the basis of a few systems in the HEP world (most notably the LHCb bookkeeping catalogue).

The AMGA system has been adopted by several applications in completely different domains (see, e.g., the Book of Abstracts of the 2nd User Forum organized by EGEE in 2007 [26]). Applications range from Climatology to Multimedia. The application we use here as an example is High-Throughput Screening in Drug Discovery. The first application in this field is WISDOM [27], active on the EGEE infrastructure since 2005. In 2006, a new phase was started with the arrival of new collaborators (most notably by Academia Sinica Taipei [28]) and with the start of a set of campaigns against the H5N1 virus (bird flu).

The basic idea is to use the grid to perform collaborative screening of potentially active chemical compounds (called ligands). This activity, called docking, can be executed on the grid by assigning single combinations of proteins and ligands to independent execution units. In order to scale up this activity, a central repository is needed (to assign the protein–ligand pairs, to store and display the results, and to implement more complex workflows). The choice for this system has been AMGA (Figure 17.10). The decisive arguments in the choice were the performance

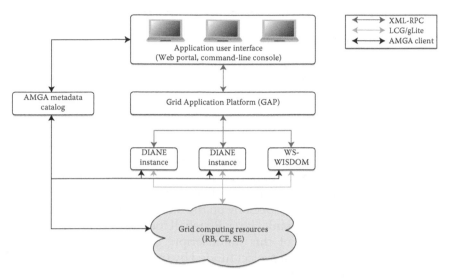

FIGURE 17.10 The system in use in the most recent challenges against H5N1 (bird flu) showing the integration of the AMGA system. The layer with the DIANE and WS-WISDOM is the component which controls the execution of the jobs on the grid. DIANE is actually a component of the Ganga system.

and robustness in supporting multiple concurrent clients and its support for grid security.

In the case of H5N1, the idea is to prepare for a fast-response system in case of the appearance of dangerous mutation for humans. In 2007, the system was demonstrated to perform as expected (delivering interesting candidates to be validated in the laboratory). A typical challenge scans several million ligands using hundreds of CPU-years in a month real time. The result is a handful of promising preselected candidates for validation in the laboratory.

17.7 Conclusions

As mentioned in the Introduction, the choice of grid technologies for computing in the LHC program is a major milestone. The actual implementation of a production grid is making possible the spectacular growth in usage also outside the HEP communities, as demonstrated by the EGEE project. Close and successful collaboration of the HEP community with other sciences in grid computing (in particular the adoption of solutions in new areas) is a promising sign of the level of maturity and the potential future of this technology.

Acknowledgments

I would like to thank all members of the team I coordinated, the so-called ARDA team (2004–2008), which was built around the initial core team started in the framework of the LCG and EGEE projects. The team and its activities grew constantly, due to the continuous support of LCG and EGEE plus fruitful collaborations with other institutes, most notably ASGC and the Russian LCG collaborators. I would like to thank especially Simon Lin and Eric Yang (ASGC); Slava Ilyin (SINP Moscow) and Vladimir Korenkov (JINR Dubna) for their support and excellent collaboration. A special thank goes to Iosif Legrand (Caltech) for the fruitful collaboration and support especially on the monitoring (MonALISA project). I would also like to thank Harry Renshall for interesting discussions during the preparation of this manuscript. This work was partially funded by EGEE. EGEE is a project funded by the European Union under contract INFSO-RI-031688.

References

1. General updated information on the LHC programme can be found on the CERN web site (http://www.cern.ch). A recent review article on the first 2 years of LHC is: Fabiola Gianotti "Physics during the first two years of the LHC," *New Journal of Physics* 9 (2007), 332.
2. Enabling Grid for E-sciencE (EGEE). Available at: http://www.eu-egee.org.
3. Open Science Grid (OSG). Available at: http://www.opensciencegrid.org.
4. Nordic Data Grid Facility (NDGF). Available at: http://www.ndgf.org.
5. Ian Foster and Carl Kesselman, *The GRID: Blueprint for a New Computing Infrastructure*, Morgan Kaufmann Publishers, Inc., San Francisco, CA, 1998.
6. S. Bethke, M. Calvetti, H.F. Hoffmann, D. Jacobs, M. Kasemann, and D. Linglin, "Report of the steering group of the LHC computing review," CERN/LHC/2001-004, CERN/RRB-D 2001-3, 22 February 2001.
7. LHC Computing Grid (LCG) home page: http://cern.ch/lcg.
8. Models of Networked Analysis at Regional Centers for LHC Experiments (MONARC) project. Available at: http://cern.ch/monarc.
9. Massimo Lamanna, "ARDA experience in collaborating with the LHC experiments," in *Proceedings of the Computing in High Energy and Nuclear Physics (CHEP06)*, Mumbai, February 2006, vol. I, p. 1081.
10. LCG Editorial Board, *LHC Computing Grid Technical Design Report*, LCG-TDR-001, CERN-LHCC-2005-024, June 2005.
11. Andrew Maier, "Ganga: a job management and optimisation tool," in *Proceedings of the Computing in High Energy and Nuclear Physics CHEP07*

Conference, Victoria, Canada, September 2007. The Ganga project home page is http://cern.ch/ganga.

12. Monitoring Agents Using a Large Integrated Services (MonALISA) project. Available at: http://monalisa.cern.ch/monalisa.html.

13. Philippe de Forcrand, Seyong Kim, and Owe Philipsen, "A QCD critical point at small chemical potential: is it there or not?," in *Proceedings of the Lattice 2007 Conference*, the XXV International Symposium on Lattice Field Theory, Regensburg (Germany), July 30/August 4, 2007, p. 178.

14. ROOT is an object-oriented data analysis framework. Available at: http://root.cern.ch.

15. Derek Feichtinger and Andreas J. Peters, "Authorization of data access in distributed storage systems," in *Proceedings of the 6th IEEE/ACM International Workshop on Grid Computing 2005*, Seattle, Washington, DC, November 13–14, 2005, pp. 172–178.

16. O. Gutsche and C. Hajdu, "WLCG scale testing during CMS data challenges," in *Proceedings of the Computing in High Energy and Nuclear Physics CHEP07 Conference*, Victoria (Canada), September 2007.

17. A. Duarte, P. Nyczyk, A. Retico, and D. Vicinanza, "Monitoring the EGEE/WLCG grid services," in *Proceedings of the Computing in High Energy and Nuclear Physics CHEP07 Conference*, Victoria (Canada), September 2007. The project web page is http://sam-docs.web.cern.ch/sam-docs.

18. A nice review of the Dashboard functionality can be extracted by the following contributions at the Computing in High Energy and Nuclear Physics CHEP07 Conference, Victoria (Canada), September 2007: J. Andreeva, S. Belov, A. Berejnoj, et al., "Dashboard for the LHC experiments;" M. Branco, D. Cameron, B. Gaidioz, V. Garonne, B. Koblitz, M. Lassnig, R. Rocha, P. Salgado and T. Wenaus, "Managing ATLAS data on a petabyte-scale with DQ2;" P. Saiz, J. Andreeva, C. Cirstoiu, B. Gaidioz, J. Herrala, E.J. Maguire, G. Maier and R. Rocha, "Grid reliability."

19. R-GMA. Available at: http://www.r-gma.org.

20. GridIce. Available at: http://gridice.forge.cnaf.infn.it.

21. Imperial College Real Time Monitor. Available at: http://gridportal.hep.ph.ic.uk/rtm.

22. F. Donno and M. Litmaath, "Dynamic storage management," in *Scientific Data Management*. CRC Press/Taylor and Francis Books, London.

23. M. Schulz et al., "Tools for the management of stored data and transfer of data: DPM and FTS," in *Proceedings of the Computing in High Energy and Nuclear Physics CHEP07 Conference*, Victoria (Canada), September 2007.

24. S. Bagnasco, L. Betev, P. Buncic, F. Carminati, C. Cirstoiu, C. Grigoras, A. Hayrapetyan, A. Harutyunyan, A.J. Peters and P. Saiz, "AliEn: ALICE environment of the Grid," in *Proceeding of the Computing in High Energy and Nuclear Physics CHEP07 Conference*, Victoria (Canada), September 2007.

25. B. Koblitz, N. Santos, and V. Pose, "The AMGA metadata service," *Journal of Grid Computing* 6 (1) (2008), 61–76. The AMGA web site is http://cern.ch/amga.

26. EGEE User Forum Book of Abstracts. EGEE User Forum, Manchester, May 9–11, 2007, EGEE-TR-2007-002.

27. N. Jacq, J. Salzemann, F. Jacq, et al., "Grid-enabled virtual screening against malaria," *Journal of Grid Computing* 6 (1) (2008), 29–43.

28. Hurng-Chun Lee, Jean Salzemann, Nicolas Jacq, Hsin-Yen Chen, Li-Yung Ho, Ivan Merelli, Luciano Milanesi, Vincent Breton, Simon C. Lin, and Ying-Ta Wu, "Grid-enabled high-throughput in silico screening against influenza A neuraminidase," *IEEE Transactions in Nanobioscience* 5 (4) (2006), 288–295. See also the ASGC Taipei website: http://www.twgrid.org/Application/Bioinformatics/AvainFlu-GAP.

18

Design and Performance Evaluation of a Service-Oriented HLA RTI on the Grid

Ke Pan, Stephen John Turner, Wentong Cai, and Zengxiang Li

CONTENTS

18.1 Introduction

Simulation is a low-cost and safe alternative for solving complex problems in various areas such as production, business, education, science, and engineering. With increased problem size and the advance of parallel computing, parallel simulation was introduced to increase the simulation speed. With the worldwide prevalence of networked computers, distributed simulation was introduced to promote the interoperability and reusability of simulation applications and to link geographically dispersed simulation components. The high-level architecture (HLA) has been standardized as IEEE 1516 for distributed simulation in September 2000 [1]. While HLA defines the rules, interface specification and object model template (OMT), a runtime infrastructure (RTI), such as the Defense Modelling and Simulation Office (DMSO) RTI [2], provides the actual implementation of the HLA standard. In HLA terminology, a distributed simulation application is called a federation that comprises several simulation components called federates; the RTI is a communication middleware through which federates in the same federation can communicate with each other.

Traditionally, HLA-based distributed simulations are conducted using a vendor-specific RTI software and federates with different RTI versions cannot cooperate with each other. To run a distributed simulation over a WAN, the required software and hardware resource arrangements and security settings must be made before the actual simulation execution. Because of this inflexibility, it is not easy to run HLA-based distributed simulations across administrative domains. To address these inflexibility issues and leverage globally pervasive resources for distributed simulations, the grid is naturally considered as a solution.

Grid computing was proposed by Foster as a flexible, secure, and coordinated resource shared among dynamic collections of individuals, institutions, and resources in different virtual organizations [3]. Among the various available grid middlewares, Globus Toolkit [4] is the *de facto* standard middleware for grid computing. Based on open grid services architecture (OGSA) [5], its latest version GT4 contains five components, namely common runtime, security, data management, information services, and execution management, to facilitate heterogeneous resource sharing. Merged with Web service standards, GT4 implements a group of related standards such as Web services resource framework (WSRF) [6] and Web services addressing (WS-Addressing) [7].

We have previously proposed a service-oriented HLA RTI (SOHR) framework that implements an HLA RTI using grid services [8]. The various grid services of SOHR cooperate with each other to provide the functionalities of an HLA RTI as services. Federates participate in federations by invoking specific grid services without the installation of a heavyweight vendor-specific RTI software at the local site. Since grid services are used for communications firewalls can be pierced, and distributed simulations can be conducted across administrative domains. Moreover, the various grid service components can be dynamically deployed, discovered, and undeployed on demand. All these features of SOHR enable convenient execution of HLA-based distributed simulations on a WAN. The performance test of SOHR in [8] was based on GT4.0.2, which does not support persistent TCP/IP connections. With the newer version GT4.1.1, which supports persistent TCP/IP connections, the performance is expected to be improved significantly. This chapter elaborates on the design of SOHR and evaluates its performance after its migration to GT4.1.1.

The rest of the chapter is organized as follows: Section 18.2 introduces some related work on HLA-based distributed simulation on the grid. Section 18.3 describes the framework overview of SOHR. After laborating on the design details of the various framework components in Section 18.4, Section 18.5 describes the experiments conducted and their results. Section 18.6 summarizes the major benefits of SOHR. Finally, Section 18.7 concludes the whole chapter and overviews future work.

18.2 Related Work

Three approaches can be defined for an HLA-based distributed simulation on the grid, namely a grid-facilitated approach, a grid-enabled approach, and a grid-oriented approach.

18.2.1 Grid-Facilitated Approach

In the grid-facilitated approach, grid services are defined to facilitate the execution of HLA-based distributed simulations while the actual simulation communications are through a vendor-specific RTI. Several projects using this approach are introduced in this subsection.

The Grid-HLA Management System (G-HLAM) was proposed by Rycerz for efficient execution of HLA-based distributed simulations on the grid [9]. HLA interfacing services are defined for interfacing and managing federates or RTIExec processes on each site, such as saving and restoring a federate for migration. Broker support services are defined and they reside on each federate site providing necessary information about the local federate such as the current host load and application behavior. The migration service is defined to coordinate federate migration between source and destination sites by communicating with their interfacing services. To decide when and where federate migration should be performed, collective services are defined including the application monitoring main service manager, the performance decision service and the broker service. They communicate with the broker support services of each federate for global performance monitoring, decision-making, and migration triggering. Although the various grid services cooperate for an optimized configuration of federation execution, the actual simulation communications are through a vendor-specific RTI.

Wu et al. [10] proposed a data grid called Aegis for large-scale distributed simulation applications. It offers data resource management services and computing services for HLA-based simulation applications while the simulation execution still uses a vendor-specific RTI. Choi et al. [11] proposed an RTI execution environment called RTI-G based on the OGSA. In order to achieve high performance, RTI-G utilizes various GT3 services, such as MDS, GRAM, and GridFTP, to enable dynamic resource allocation and automatic execution of HLA-based distributed simulations.

A simulation grid is an infrastructure that integrates grid technology and the HLA to realize dynamic and secure resource sharing, optimized resource utilization, collaborative activities and fault tolerance for distributed simulation applications [12,13]. In a simulation grid, HLA-based distributed simulations are grid-facilitated in various ways. For example, registration services are defined for dynamic model resource discovery and scheduling services are defined to generate proper simulation deployment schemes according to monitored computing resource status.

Our group proposed a load-management system for HLA-based distributed simulations in [14]. The load-management system and its federate migration mechanism are based on grid services while communications between federates are through a vendor-specific RTI. It was later extended with an efficient migration protocol in [15].

We also proposed a grid service-based framework for flexible execution of large-scale HLA-based distributed simulations [16]. RTI executive processes and federate models are encapsulated in grid services for dynamic federation creation and management. An index service is provided for registration and dynamic discovery of the various grid services. Simulation communications are also through a vendor-specific RTI.

As the grid-facilitated approach still replies on a vendor-specific RTI for simulation communication, it requires cross-domain trust and particular prior security setup for HLA-based distributed simulations to be conducted across administrative domains, which is very cumbersome.

18.2.2 Grid-Enabled Approach

In the grid-enabled approach, grid (or Web) service interfaces are provided to enable HLA-based distributed simulations to be conducted in a grid (or Web) environment. There are various forms of this approach. One form is that a client federate communicates with a federate server using grid (or Web) service communications and the federate server representing the client federate joins an HLA-based distributed simulation using a vendor-specific RTI. Another form is that different federations are executed using a vendor-specific RTI at their own local sites and grid service interfaces are defined to enable communications between the federations so that a larger federation community can be formed over the grid.

An example of this approach is the work done by the XMSF group [17] to integrate simulations with other applications using Web services [18]. Web service interfaces are provided for a particular simulation application so that other applications on the Web can interact with the simulation application. For example, a remote passive visualizer can periodically probe the simulation progress and display it accordingly; when the progress is not satisfactory, an army C4I (command, control, communications, computers, and intelligence) system can send a command to the simulation application for adjustment. The communications between the simulation application and the other applications are through Web service invocations while the communications inside the simulation application are through a vendor-specific RTI.

As another example of this approach, new Web service APIs have recently been proposed for the HLA IEEE 1516 standard [19]. With the Web service APIs, a client federate on the Web communicates with a Web Service Provider RTI Component (WSPRC), through which the federate joins a federation.

Zhu et al. [20] proposed a grid-based distributed simulation architecture (GDSA), in which grid service interfaces are defined for a federation so that multiple federations can be integrated through the defined interfaces. This enables HLA-based distributed simulations to be carried out over the grid. Xu and Peng developed a grid-based simulation service bus (SSB) [21], through which a remote federate on the grid can join a local federation in a LAN and multiple federations are able to exchange data over the grid. This effort follows the grid-enabled approach.

In a simulation grid [12], RTI grid service components are deployed to provide non-real-time communication services between federations and thus enable HLA-based distributed simulations to be executed over the grid. To run HLA-based distributed simulations on the grid, our group proposed a federate-proxy-RTI-based framework [22]. Proxies are provided so that a remote federate on the grid is able to join a local federation through grid service communications with a proxy. All proxies and local federates reside in a LAN and communicate with each other through a vendor-specific RTI. This effort also follows the grid-enabled approach. Recently, Chen et al. [23] integrated the federate-proxy-RTI based framework with the HLA RePast middleware system and therefore proposed an HLA grid RePast platform for executing large-scale agent-based distributed simulation on the grid [24]. The major drawback of the grid-enabled approach is that vendor-specific RTI execution environments and communication servers have to be set up beforehand, which makes this approach not so flexible.

18.2.3 Grid-Oriented Approach

In the grid-oriented approach, the RTI is implemented using grid services according to the HLA specification. The six HLA service groups should be mapped to different grid services in order to create a service-oriented architecture. This approach was raised in Fox's keynote at the Distributed Simulation and Real Time Applications Conference in 2005 [25]. Our SOHR framework follows the grid-oriented approach and its design is discussed in the next two sections.

18.3 Framework Overview of SOHR

The architecture of our SOHR framework is illustrated in Figure 18.1. It contains seven key grid services, namely the RTI index service, the local service (LS), and five management services, all of which are implemented based on GT4. All services except the RTI index service follow the WS-Resource factory design pattern [26].

In the WS-Resource factory design pattern, information is organized into resource instances: a resource home keeps track of the multiple resource instances: a factory service is defined to create new resource instances; an instance service uses the resource home to find the client-specified resource instance and operates on it.

The RTI index service provides a system-level registry so that all other services are able to register their end point references (EPRs) here and dynamically discover each other. It also provides services to create and destroy federations in the system. The five management services correspond to the six HLA service groups. Each of them consists of a factory service, an instance service, and multiple resource instances, with one resource instance for each federation. For example, the Federation Management service (FMS) provides functionalities of the HLA Federation Management service group as a grid service and consists of the factory service FMFS, the instance service FMIS and multiple resource instances FMRIs, with one FMRI for each federation.

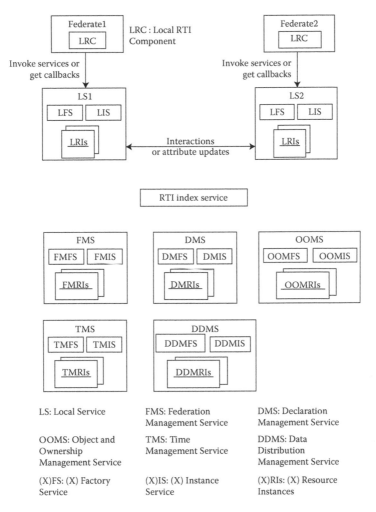

FIGURE 18.1   Framework overview.

The LS is used as a messaging broker of federates and consists of the factory service LFS, the instance service LIS, and multiple resource instances LRIs, with one LRI for each federate. A federate communicates with the outside world through its LRI by invoking services and getting callbacks. As shown in Figure 18.2, the LRI is structured into six modules. Each of the modules except the callback module corresponds to one of the management services. The callback module is used to buffer callbacks for a federate.

The objective of separating the HLA service groups into different grid services and using a modular structure for the LRI is to create a plug-and-play paradigm for an HLA RTI implementation so as to build an extensible

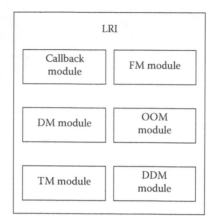

FIGURE 18.2   LRI structure.

SOHR framework. There may be multiple algorithms for the implementation of an HLA service group, and these can generally be classified as centralized algorithms and distributed algorithms. In a centralized algorithm, the major processing is done by the corresponding management service while the corresponding module in the LRI simply keeps some necessary information related to its federate. In a distributed algorithm, the major processing is done by the corresponding module in the LRI while the corresponding management service simply keeps relevant centralized information of the federation.

A particular module in the LRI and its corresponding management service cooperate to provide the services of the HLA service group. Based on different algorithms for the HLA service group, multiple combinations of a particular module of the LRI and its corresponding management service can be implemented and plugged into SOHR. A user is able to choose the most suitable algorithm based on a specific federation scenario to optimize the performance. This plug-and-play paradigm has been demonstrated by the implementation of three time management algorithms in SOHR [27]. It may be thought that the separation of the HLA service groups into various grid services may impact the overall performance of SOHR. However, this is not the case as most of the communications, which are generally attribute updates and interactions, are directly conducted between LRIs in a peer-to-peer manner.

The local RTI component (LRC) is a federate's local library that implements the HLA service interfaces [1] and does the translation between HLA service interfaces and the corresponding LIS grid service invocations. It makes legacy HLA-based federates work in SOHR. A decoupled design is used between a federate and its LRI so that the LRC passes all requests by the federate to the underlying LRI and the LRI buffers call backs for the federate in its callback module. Both HLA 1.3 and HLA 1516 service interfaces are supported in two different LRC libraries. The LRC is lightweight, so a federate can be run on resource-limited platforms such as PDAs and cellphones. Most of the local RTI processing, such as those related to DDM and TM, is offloaded to the LRI in a similar way as offloading it to network processors in [28], so the federate has more computation time for its real simulation work. Since the callbacks are buffered by the LRI, federate migration is simplified [29]. The migration protocol proposed in [15] requires the source federate to flush all messages pending delivery and insert them in a queue. Then the queue is encoded and transferred to the migration destination where it gets decoded. Additionally, a complex algorithm is used in the protocol to prevent event loss. SOHR does not have these problems when migrating federates. A source federate just transfers the federate code and its status. The migrated federate can then connect to the same LRI to fetch the buffered callbacks.

18.4 Detailed Design of SOHR

18.4.1 Local RTI Component

The LRC is a federate side library that implements the HLA service interfaces and makes legacy HLA-based federates work in SOHR. It simply does the translation between the HLA standard interfaces and LIS grid service invocations, so it is quite lightweight. Both HLA 1.3 and HLA 1516 service interfaces are supported in two different LRC libraries. When referring to a specific HLA service interface, its HLA 1.3 version is used for the description in this chapter.

The major component in LRC is the RTIambassador class, which implements various HLA service interfaces such as createFederationExecution and tick. The other components are supporting classes and exception classes such as the SuppliedParameters class used in the sendInteraction service and the InteractionParameterNotKnown class that defines a type of exception thrown when an interaction parameter is not valid. The RTIambassador class translates HLA service requests to corresponding grid service invocations of the LIS, for example an RTIambassador. A createFederationExecution request by the local federate will be translated to a createFederationExecutionOut grid service invocation of the LIS, which will further contact the RTI index service to create a new federation. A pulling mechanism is used for callbacks to the local federate. When the local federate requests for callbacks by calling the RTIambassador.tick (or evoke multiple callbacks in HLA 1516) function, the LRC pulls callbacks from the LRI by invoking the getCallbacks service of the LIS. If any callback is received, the LRC delivers it to the local federate by calling the corresponding callback function provided by the local federate.

18.4.2 Local Service

The LS operates as a messaging broker for federates. On initialization, the RTIambassador instance in each federate calls LFS's createResource function to create a new LRI for communication purposes. A federate's LRC communicates with the outside world through its corresponding LRI and the LIS defines miscellaneous grid service interfaces for accessing the LRI, such as an interaction-sending operation invoked by other federates' LRIs to exchange an interaction and an unsubscription notification operation invoked by the DMS to notify an unsubscription made by another federate.

As introduced previously, the LRI has six modules and each of them except the callback module corresponds to one of the management services. The FM module communicates with the FMS to provide HLA federation management services to the federate. The DM module communicates with the Declaration Management Service (DMS) to provide HLA declaration management services to the federate. It keeps subscription and publication information of the LRI's corresponding federate as well as relevant subscription and publication information of other federates. This subscription and publication information ensures correct message exchanges between federates' LRIs. The OOM module communicates with the OOMS to provide HLA object and ownership management services to the federate. It is responsible for directly exchanging messages with other federates' LRIs with the help of the DM and DDM modules. The TM module communicates with the TMS to provide HLA time management services to the federate. The DDM module communicates with the Data Distribution Management Service (DDMS) to provide HLA data distribution management services to the federate. It keeps the relevant overlapping information of regions to ensure correct message exchanges between federates' LRIs.

The callback module maintains two callback queues, namely the TSO queue and non-TSO queue. The TSO queue keeps time-stamp-order messages while the non-TSO queue keeps receive-order messages as well as system callbacks such as timeRegulationEnabled and timeAdvanceGrant. Other modules insert callbacks in the corresponding queues of the callback module. The callback module works closely with the TM module to ensure the correct delivery sequence of TSO messages to the federate.

Originally, we were considering whether each of the modules could be implemented as a separate grid service. Realizing that it will make the framework performance too low because of the close communications between modules, we decided to bundle all modules in the LS and implement each module as a separate Java class. An LIS grid service invocation is translated to a method invocation of the corresponding module class that will take over the processing work. For example, a getCallbacks Grid service invocation of the LIS is translated to a CallbackModule. getCallbacks Java method invocation that will scan the callback queues and return the available callbacks as a list.

18.4.3 Federation Management Service

The FMS is a grid service related to the HLA federation management service group. The FMFS defines a createResource operation invoked by the RTI index service to create an FMRI for a new federation. The FMIS defines some operations corresponding to the HLA federation management service group such as joinFederationExecution and resignFederationExecution. It also defines operations for fetching some information kept in FMRIs. When a federation is created, a new FMRI will be created and it is the FMRI's responsibility to create a new DMRI, OOMRI, TMRI, and DDMRI for the new federation. The FMRI operates as a federation-wide registry. It keeps the EPRs of the DMRI, OOMRI, TMRI, and DDMRI created for the federation. It also keeps the EPRs of federates' LRIs for the federation. It generates federate handles for new joining federates. When a federate resigns from the federation, it is responsible to notify the whole federation about this resign for the deletion of any information related to the federate.

18.4.4 Declaration Management Service

The DMS is a grid service related to the HLA Declaration Management service group. The DMFS defines a createResource operation invoked by the FMRI to create a DMRI for a new federation. The DMIS defines some operations corresponding to the HLA declaration management service group such as publishInteractionClass and subscribeInteractionClass. It also defines operations for fetching some information kept in DMRIs. The DMRI keeps the subscription and publication information of all federates in the federation. When a DMRI is created and initialized, it reads the federation execution data (FED) in HLA 1.3 or FOM document data (FDD) in HLA 1516 file and generates handles for the interaction classes, interaction class parameters, object classes, and object class attributes defined in the federation. All these handles are kept in the DMRI and the DMIS defines operations for fetching the handle information. Since these federation object model (FOM) [1] data are centralized in the DMRI, the DMIS can define some operations to enable dynamic change of the FOM which is not supported in most of the traditional RTIs. For example, the DMIS can define an addObjectClass operation for adding a new object class in the federation. Similar to the HLA evolved modular FOMs approach [30], the DMIS can define some operations to enable a FOM module to be dynamically added to or deleted from the federation FOM.

18.4.5 Object and Ownership Management Service

The OOMS is a combined grid service related to the HLA object management and ownership management service groups. The OOMFS defines a createResource operation invoked by the FMRI to create an OOMRI for a new federation. The OOMIS defines some operations corresponding to the HLA object management and ownership management service groups such as registerObjectInstance and attributeOwnershipAcquisition. It also defines operations for fetching some information kept in OOMRIs. The OOMRI keeps all object instances in the federation and their handles. It also keeps the attribute ownership information for each object instance. Since the attribute ownership information is specified per object instance, it is convenient to include both the HLA object management and ownership management service groups in the OOMS.

18.4.6 Time Management Service

The TMS is a grid service related to the HLA TMS group. The TMFS defines a createResource operation invoked by the FMRI to create a TMRI for a new federation. The TMIS defines some operations corresponding to the HLA TMS group. The TMRI keeps relevant information for time management in the federation. The operations defined by the TMIS and data kept in the TMRI depend on the time-advancing algorithm used in our SOHR framework.

If a centralized algorithm is used, the TMS does the major processing of time management, such as the recalculation of lower bound time stamp (LBTS) in HLA 1.3 or greatest available logical time (GALT) in HLA 1516 for each constrained federate based on time information of regulating federates. Since time information is frequently exchanged, the TMS may be easily overloaded with increasing federation size. So a distributed time-advancing algorithm is more attractive to SOHR.

A distributed algorithm, such as the one in RTI version F.0 [31], can be easily adapted to our SOHR framework. The TMIS defines operations for changing of a federate's regulating or constrained status such as enableTimeRegulating and disableTimeConstrained. The TMRI keeps the regulating and constrained status of all federates in the federation. When a federate enables/disables its constrained status, the TMRI notifies all regulating federates' LRIs to update their TM modules. In this way, a list of constrained federates in the TM module of each regulating federate's LRI is always kept updated. With the always updated list of constrained federates, a regulating federate's LRI directly sends its time information to each of the constrained federates' LRIs. The TM module of a constrained federate's LRI may recalculate the constrained federate's LBTS (or GALT in HLA 1516) when time information of a regulating federate is received. In this way, the major processing of time management is done by the TM module of each federate's LRI, which makes SOHR more scalable.

A new time-advancing algorithm can be incorporated in to SOHR by adding a TM module of the LRI and a TMS based on the new algorithm, which makes SOHR an extensible framework. This has been demonstrated by the implementation of three distributed TM algorithms in SOHR [27].

18.4.7 Data Distribution Management Service

The DDMS is a grid service related to the HLA data distribution management service group. The DDMFS defines a createResource operation invoked by the FMRI to create a DDMRI for a new federation. The DDMIS defines some operations corresponding to the HLA data distribution management service group. The DDMRI keeps relevant information for data distribution management in the federation. Similar to the case of TMS, the operations defined by the DDMIS and data kept in the DDMRI depend on the DDM algorithm used in our SOHR framework. Basically a DDM algorithm can be classified as a centralized algorithm or a distributed algorithm. In a centralized DDM algorithm, a DDM coordinator calculates the overlaps between regions (a process called matching). In a distributed DDM algorithm, the matching is done at each federate's local site.

If a centralized DDM algorithm is utilized in SOHR, the DDMRI acts as the DDM coordinator for its federation. It keeps information of all regions and calculates their overlaps. It sends the updated overlapping information to relevant federates' LRIs to update their DDM modules. The DDMIS defines relevant operations for updating region information kept in the DDMRI, such as a createRegion operation for creating a new region and an associateRegionForUpdates operation for associating attributes of an object instance with a region. If a distributed DDM algorithm is utilized in SOHR, the matching is done by the DDM module of each LRI whereas the DDMRI may simply keep some centralized information for data distribution management, such as all regions and their respective creators.

There are many DDM matching algorithms such as the region-based approach [32], grid-based approach [33], hybrid approach [34], and sort-based approaches [35,36]. Each algorithm has different performance, but there is no algorithm that fits all federation scenarios [37]. Multiple combinations of a DDM module of the LRI and a DDMS can be incorporated into SOHR based on different algorithms so that the combination with the best performance can be used for a specific federation scenario.

18.4.8 RTI Index Service

The RTI index service provides a system-level registry so that all other services are able to register their EPRs here and dynamically discover each other. A load balancing mechanism is employed in the RTI index service to improve the overall system efficiency. It keeps the load information of each registered service in terms of the number of created resource instances and defines operations for returning the least loaded services of different types. For example, it keeps the load of each registered DMS in terms of the number of created DMRIs and defines an operation for returning the least loaded DMS. It also defines operations for creation and destruction of federations and keeps the mapping between each federation and its FMRI EPR.

18.5 Experiments and Results

We have previously implemented a SOHR prototype [8]. It was based on GT4.0.2 that does not support persistent TCP/IP connections. Now, the prototype has been migrated to GT4.1.1 that supports persistent TCP/IP connections and its performance is compared with the one based on GT4.0.2. The implemented HLA service groups now include federation management, declaration management, object management, and time management. The implementation of the object management service group supports both interactions and object attributes. The ownership management and data distribution management service groups have not been implemented yet.

18.5.1 Ping-Pong Experiment Design

To compare the performance of SOHR based on GT4.0.2 and GT4.1.1 and the performance of the DMSO RTI 1.3NG [2], a ping-pong experiment was designed. A federate sends an interaction to another federate and then waits to receive an interaction from that federate before repeating the same process. The average round trip time is measured and the one-way latency is calculated as half of the round trip time.

18.5.2 Ping-Pong Experiment Trace on SOHR

The trace of the initialization stage of the ping-pong experiment on SOHR is shown in Figure 18.3. The communications between a federate and its LRC are through HLA service calls and callbacks, while all the other communications are through grid service invocations. Before the actual simulation execution, the RTI index service, FMSs, DMSs, OOMSs, TMSs, and LSs have to be deployed and started on respective hosts. Each FMFS, DMFS, OOMFS, TMFS, or LFS registers its EPR with the RTI index service (step 0). When LRC1 (Federate1's RTIambassador instance) is created, it asks the RTI index service to return the LFS with the lowest load (LFS1), creates and initializes a new LRI (LRI1) using the returned LFS (steps 1, 2, and 3). Since a new LRI is created with LFS1, LRC1 updates the load information of LFS1

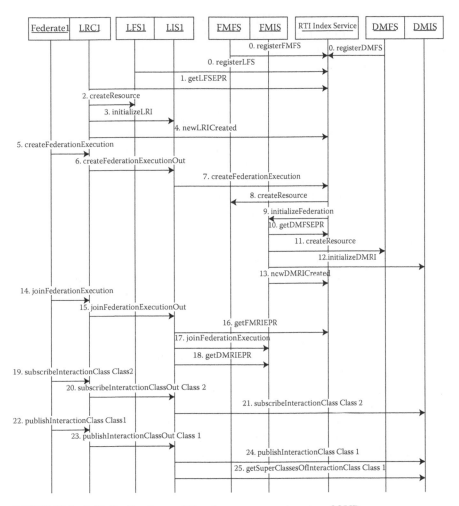

FIGURE 18.3   Initialization trace of the ping-pong experiment on SOHR.

in the RTI Index Service (step 4). The creation of a new federation starts with Federate1's RTIambassador.createFederationExecution request (step 5). LRC1 translates this request to a createFederationExecutionOut service invocation of LIS1 (step 6), which next forwards the invocation to the RTI index service (step 7). After the RTI index service receives the request, it chooses the FMFS with the lowest load, creates and initializes an FMRI for the new federation (steps 8 and 9). The initialization of the FMRI creates and initializes a DMRI (steps 10, 11, 12, and 13), an OOMRI, and a TMRI for the new federation. After the federation is created, Federate1 calls an RTIambassador.joinFederationExecution request (step 14), which is translated to a joinFederationExecutionOut service invocation of LIS1 by LRC1 (step 15). The LIS1 asks the RTI index service for the FMRI EPR of the federation (step 16) and invokes a joinFederationExecution service invocation of the corresponding FMIS (step 17). The DMRI EPR of the federation is fetched and stored in LRI1's DM module (step 18). Similarly, the OOMRI EPR and TMRI EPR of the federation are fetched and, stored, respectively, in LRI1's OOM module and TM module. Next, Federate1 requests to subscribe to interactions of Class2 (step 19) and publish interactions of Class1 (step 22). Each of the two requests is forwarded to LIS1 by LRC1 (steps 20 and 23) and further forwarded to DMIS (steps 21 and 24). The getSuper-ClassesOf InteractionClass service invocation of DMIS (step 25) is to deal with an interaction class hierarchy defined in the FOM. Since no interaction class hierarchy exists in this ping-pong experiment, it simply returns an empty list. After Federate1 finishes its initialization, Federate2 joins the created federation, subscribes to interactions of Class1, and publishes interactions of Class2 in a similar way.

After the initialization stage, interactions are exchanged between Federate1 and Federate2 as shown in Figure 18.4. Federate1 keeps calling RTIambassador.tick until an interaction is received from Federate2 (step 29). Federate2 requests to send an interaction of Class2 (step 26) and LRC2 translates this request to a sendInteractionOut service invocation of LIS2 (step 27). The LRI2's DM module keeps the list of subscribers and their LRI EPRs. Because Federate1 is subscribing to interactions of Class2 (step 19), it should be included in the list. This causes a sendInteraction service invocation of LIS1 by LIS2 (step 28). Then LRC1 receives this interaction (step 30) when Federate1 calls an RTIambassador.tick (step 29) and delivers this interaction to Federate1 (step 31). After Federate1 receives the interaction, it requests to send an interaction of Class1 and this interaction is received by Federate2 in the reverse way (steps 32, 33, 34, 35, 36, and 37).

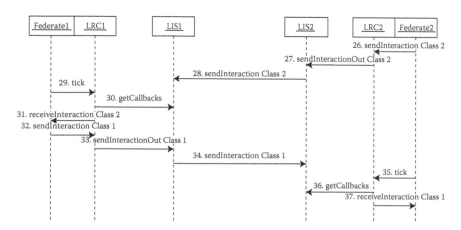

FIGURE 18.4   Interaction trace of the ping-pong experiment on SOHR.

This is one round of the ping-pong style of communication. After multiple iterations, the two federates resign from the federation and the federation is destroyed by Federate2. After the whole simulation execution, the FMRI, DMRI, OOMRI, TMRI, and LRIs created for the federation have all been destroyed.

18.5.3 Experimental Configurations and Results

Our experimental testbed consists of a cluster in Nanyang Technological University (NTU), Singapore, and two hosts in the E-Science Centre at the University of Birmingham, UK, as shown in Figure 18.5. The cluster has a 10 Gb/s infiniBand connection. Each node of the cluster is installed with two dual core Xeon 3.0 GHz CPUs, 4 GB RAM, and Redhat Enterprise Linux 4 OS. The two hosts in the E-Science Centre are connected by an ethernet and each of them is installed with two Xeon 3 GHz CPUs, 2 GB RAM and Redhat Enterprise Linux AS 3 OS. The cluster has only one externally accessible node which is its main node (pdcbl), and our experiments were configured based on this restriction. The performance of the DMSO RTI and our SOHR framework was measured both inside the cluster (LAN) and across continents (WAN). The RTI index service and all management services were deployed in the cluster for all experimental configurations of SOHR. The experimental configurations and latency results are shown in Table 18.1.

The experimental results show that SOHR's performance both on the LAN and on the WAN is improved significantly after its migration to GT4.1.1. This is due to the support of persistent TCP/IP connections by GT4.1.1. Without persistent TCP/IP connection support as in GT4.0.2, each service invocation creates a new TCP/IP connection for communication and this brings the overhead of connection setup and tearing down per service invocation. With persistent TCP/IP connection support as in GT4.1.1, multiple service invocations between the same client and server

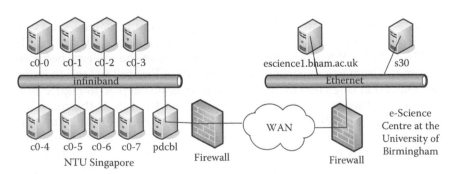

FIGURE 18.5 Experimental testbed.

TABLE 18.1

Experimental Configurations and Latency Results in ms

DMSO RTI					
Test	Federate1	Federate2	rtiexec + fedex	Result	
LAN	c0-6	c0-7	c0-0	5.3	
WAN	pdcb1	Escience1	pdcb1	116.5	

SOHR					
Test	Federate1	LS1	Federate2	LS2	GT4.0.2	GT4.1.1
LAN	c0-6	c0-4	c0-7	c0-5	85.6	19.6
WAN A	Escience1	pdcb1	c0-6	c0-4	1026.8	241.6
WAN B	c0-6	pdcb1	s30	Escience1	916.7	154.7

share the same TCP/IP connection for communication, so the overhead of the connection setup and tearing down is amortized among the multiple service invocations. Since SOHR's performance on GT4.1.1 is much better than its performance on GT4.0.2, its performance results on GT4.1.1 are used for comparison with the DMSO RTI in the following paragraphs.

On the LAN, SOHR's latency is 19.6 ms, which is 3.7 times that of the DMSO RTI. SOHR's latency is larger than that of the DMSO RTI due to the overheads of grid service invocations. Basically, grid service invocations have two kinds of overheads, processing cost and communication cost. As discussed in [22], the size of a SOAP message is around 1–2 kB per grid service request/response, which is much larger than a DMSO RTI message. This increases both processing and communication costs in SOHR. To further analyze this, we carried out a test of a simple add(int a) service invocation on the LAN using GT4.1.1, and found that service invocations by clients executed on the same machine as the service and also on a different machine take almost the same time of 8 ms inside the cluster. This means that the major overhead of the LAN experiment is the processing cost while the communication cost is not high due to the fast intra-cluster network connection.

Our SOHR was tested on the WAN using two configurations, with two LSs deployed at the NTU side (WAN A) and one LS deployed at each side (WAN B). WAN B has better performance, because the interval between the time an interaction reaches an LRI (step 28 in Figure 18.4) and the time the interaction is fetched by the receiving federate (step 30 in Figure 18.4) depends on the service invocation sequence of the LIS. If the receiving federate tries to fetch the interaction (step 30 in Figure 18.4) before the interaction reaches the LRI (step 28 in Figure 18.4), this incurs extra time and the effect is more prominent if a remote LS is used as in WAN A. Sleeping time can be used by the receiving federate to delay the interaction fetching (step 30 in Figure 18.4) until the interaction has reached the LRI (step 28 in Figure 18.4). This generates an appropriate service invocation sequence of the LIS so that WAN A may have the same performance as WAN B. WAN B's latency is 154.7 ms, which is around 1.3 times that of the DMSO RTI. The relative performance of SOHR to the DMSO RTI on the WAN is improved compared with the performance on the LAN. The communication cost becomes the dominant overhead on the WAN because of the long distance.

18.6 Major Benefits of SOHR

The major benefits of the SOHR framework are summarized as follows:

- The functionalities of an RTI are provided as grid services so that distributed simulations can be conducted without any vendor-specific RTI software.

- Grid services, which can pierce firewalls, are used as the communication infrastructure so that distributed simulations can be conveniently conducted across administrative domains on a WAN.

- The six HLA service groups are mapped to different modules in the LRI structure and different management services so that multiple combinations of a particular module of the LRI and its corresponding management service can be implemented based on different algorithms for an HLA service group. This plug-and-play paradigm makes SOHR an extensible framework. A user is able to choose the most suitable algorithm based on a specific federation scenario to optimize the performance. New algorithms can be implemented and studied on SOHR, which makes SOHR a good experimental environment for distributed simulation research.

- The decoupled design between a federate and its LRI makes the LRC very lightweight so that a federate can be run on resource-limited platforms such as PDAs and cellphones. Both HLA 1.3 and HLA 1516 service interfaces are supported in two different LRC libraries. As the major local RTI processing is offloaded to the LRI, the federate has more computation time for its real simulation work. The LRI buffers callbacks for its federate, so federate migration is simplified.

- Handles of interaction classes, object classes, parameters, and attributes are centralized in the DMRI, and grid service interfaces can be provided by the DMIS to enable dynamic configuration of the FOM, which is not supported by most of the traditional RTIs.

- The various grid services can be deployed and undeployed on demand. The RTI index service operates as a system-wide registry so that the other services can dynamically discover each other. The RTI index service also employs a load balancing mechanism to improve the overall system resource efficiency.

- Other grid services can be easily integrated into SOHR for more functionalities such as checkpointing services for fault-tolerance purposes and messaging services for optimization of cross-cluster communications.

- Based on the grid, SOHR can utilize the underlying grid infrastructure, such as the grid security infrastructure, to evolve a secure, scalable, and coordinated large-scale distributed simulation environment.

18.7 Conclusions and Future Work

HLA-based distributed simulations on the grid can be categorized into three approaches, namely a grid-facilitated approach, a grid-enabled approach, and a grid-oriented approach. We have previously proposed a SOHR framework, which is so far the only work following the grid-oriented approach. The functionalities of an RTI are provided as grid services so that distributed simulations can be conducted without any vendor-specific RTI software. Grid services are used as the communication infrastructure, so distributed simulations can be conveniently conducted across administrative domains. SOHR creates a plug-and-play paradigm for an HLA RTI implementation. A new algorithm for an HLA service group can be easily incorporated into SOHR, which makes SOHR an extensible framework. Other benefits of SOHR were described in the previous section. This chapter elaborates on the design of SOHR and evaluates its performance after its migration to GT4.1.1. Experimental results show that, after the migration, SOHR's performance is improved significantly and is now comparable with the performance of the DMSO RTI due to the support of persistent TCP/IP connections by GT4.1.1. In the future, the DDM service group will be added to the SOHR prototype to make it a more complete HLA RTI. Our sort-based DDM matching algorithm proposed in [35] will be implemented. Some realistic case studies will be developed to further evaluate the SOHR prototype.

Acknowledgments

The authors would like to thank Dr. Dan Chen and the E-Science Centre at the University of Birmingham, UK, for their help in the WAN experimental configurations and testing of this project.

References

1. IEEE Standard 1516 (HLA Rules), 1516.1 (Interface Specification) and 1516.2 (Object Model Template), 2000.
2. Defense Modeling and Simulation Office (DMSO), "High Level Architecture RTI 1.3NG Programmer's Guide," Version 5, 2002.
3. I. Foster, C. Kesselman, and S. Tuecke, "The anatomy of the grid: Enabling scalable virtual organizations." *International Journal of High Performance Computing Applications*, 15(3), 2001, 200–222.
4. Globus Toolkit Version 4. Available at: http://www.globus.org/.
5. I. Foster, C. Kesselman, J. Nick, and S. Tuecke, "The physiology of the grid: An open grid services architecture for distributed systems integration," Open Grid Services Infrastructure WG, Global Grid Forum, Edinburgh, Scotland, 2002. Available at: http://www.globus.org/alliance/ publications/papers/ ogsa.pdf.
6. WSRF (Web Service Resource Framework)—Primer. Available at: http:// docs.oasis-open.org/wsrf/wsrf-primer-1.2-primer-cd-01.pdf.
7. WS-Addressing (Web Service Addressing). Version 1.0. Available at: http:// www.w3.org/TR/2005/CR-ws-addr-core-20050817/.
8. K. Pan, S.J. Turner, W. Cai, and Z. Li, "A service oriented HLA RTI on the grid," in *Proceedings of the IEEE International Conference on Web Services*, Salt Lake City, UT, 2007, pp. 984–992.
9. K. Rycerz, "Grid-based HLA simulation support," PhD thesis, University van Amsterdam and AGH Krakow, 2006.
10. W. Wu, Z. Zhou, S. Wang, and Q. Zhao, "Aegis: A simulation grid oriented to large-scale distributed simulation," in *Proceedings of the 3rd International Conference on Grid and Cooperative Computing*, Wuhan, China, 2004, pp. 413–422.
11. K. Choi, T. Lee, and C. Jeong, "RTI execution environment using open grid service architecture," in *Proceedings of the 5th International Conference on Computational Science*, Atlanta, 2005, pp. 866–869.
12. X. Chai, H. Yu, Z. Du, B. Hou, and B. Li, "Research and application on service oriented infrastructure for networkitized M&S," in *Proceedings of the 6th IEEE International Symposium on Cluster Computing and the Grid Workshops*, Singapore, 2006, Volume 2, p. 67.
13. N. Li, Z. Xiao, L. Xu, and X. Peng, "Research and realization of collaborative M&S services in simulation grid," in *Proceedings of the 6th IEEE International*

Symposium on Cluster Computing and the Grid Workshops, Singapore, 2006, Volume 2, p. 69.

14. W. Cai, S.J. Turner, and H. Zhao, "A load management system for running HLA-based distributed simulations over the grid," in *Proceedings of the 6th IEEE International Workshop on Distributed Simulation and Real Time Applications*, Fort Worth, TX, 2002, pp. 7–14.

15. W. Cai, Z. Yuan, M.Y.H. Low, and S.J. Turner, "Federate migration in HLA-based simulation." *Future Generation Computer Systems*, 21(1), 2005, 87–95.

16. W. Zong, Y. Wang, W. Cai, and S.J. Turner, "Grid services and service discovery for HLA-based distributed simulation," in *Proceedings of the 8th IEEE International Symposium on Distributed Simulation and Real Time Applications*, Budapest, Hungary, 2004, pp. 116–124.

17. Extensible Modeling and Simulation Framework (XMSF) Project. Available at: https://www.movesinstitute.org/xmsf/xmsf.html.

18. J.M. Pullen, R. Brunton, D. Brutzman, D. Drake, M. Hieb, K.L. Morse, and A. Tolk, "Using web services to integrate heterogeneous simulations in a grid environment," *Future Generation Computer Systems*, 21(1), 2005, 97–106.

19. B. Moller, "A first look at the HLA evolved Web service API," in *Proceedings of the IEEE 2006 European Simulation Interoperability Workshop*, Stockholm, Sweden, 2006, paper no. 06E-SIW-061.

20. S. Zhu, Z. Du, and X. Chai, "GDSA: A grid-based distributed simulation architecture," in *Proceedings of the 6th IEEE International Symposium on Cluster Computing and the Grid Workshops*, Singapore, 2006, Volume 2, pp. 66.

21. L. Xu and X. Peng, "SSB: A grid-based infrastructure for HLA systems," in *Proceedings of the 6th IEEE International Symposium on Cluster Computing and the Grid Workshops*, Singapore, 2006, Volume 2, pp. 68.

22. Y. Xie, Y.M. Teo, W. Cai, and S.J. Turner, "Service provisioning for HLA-based distributed simulation on the grid," in *Proceedings of the 19th ACM/IEEE/SCS Workshop on Principles of Advanced and Distributed Simulation*, Monterey, CA, 2005, pp. 282–291.

23. R. Minson and G.K. Theodoropoulos, "Distributing RePast agent-based simulations with HLA," in *Proceedings of 2004 European Simulation Interoperability Workshop*, Edinburgh, Scotland, 2004, paper no. 04E-SIW-046.

24. D. Chen, G.K. Theodoropoulos, S.J. Turner, W. Cai, R. Minson, and Y. Zhang, "Large scale agent-based simulation on the grid." *Future Generation Computer Systems*, 24 (7), 2008.

25. G. Fox, A. Ho, S. Pallickara, M. Pierce, and W. Wu, "Grids for the GiG and real time simulations," in *Proceedings of the 9th IEEE International Symposium on Distributed Simulation and Real Time Applications*, Montreal, Quebec, Canada, 2005, pp. 129–138.

26. B. Sotomayor, "The Globus Toolkit 4 Programmer's Tutorial." Available at: http://gdp.globus.org/gt4-tutorial/.

27. K. Pan, S.J. Turner, W. Cai, and Z. Li, "A hybrid HLA time management algorithm based on both conditional and unconditional information," in *Proceedings of the 22nd ACM/IEEE/SCS Workshop on Principles of Advanced and Distributed Simulation*, Rome, Italy, 2008.

28. A. Santoro and R. M. Fujimoto, "Offloading data distribution management to network processors in HLA-based distributed simulations," *IEEE Transactions on Parallel and Distributed Systems*, 19(3), March 2008, 289–298.

29. Z. Li, W. Cai, S.J. Turner, and K. Pan, "Federate migration in a service oriented HLA RTI," in *Proceedings of the 11th IEEE International Symposium on Distributed Simulation and Real-Time Applications*, Chania, Crete Island, Greece, 2007, pp. 113–121.

30. B. Moller, B. Lofstrand, and M. Karlsson, "An overview of the HLA evolved modular FOMs," in *Proceedings of the 2007 Spring Simulation Interoperability Workshop*, Norfolk, VA, 2007, paper no. 07S-SIW-108.

31. C.D. Carothers, R.M. Fujimoto, R.M. Weatherly, and A.L. Wilson, "Design and implementation of HLA time management in the RTI version F.0," in *Proceedings of the 29th Winter Simulation Conference*, Atlanta, GA, 1997, pp. 373–380.

32. D.J. Van Hook and J.O. Calvin, "Data distribution management in RTI 1.3," in *Proceedings of the 1998 Spring Simulation Interoperability WorkShop*, Orlando, FL, March 1998, paper no. 98S-SIW-206.

33. S.J. Rak and D.J. Van Hook, "Evaluation of grid based relevance filtering for multicast group assignment," in *Proceedings of the 14th DIS Workshop on Standards for the Interoperability of Distributed Simulation*, Orlando, FL, March 1996, pp. 739–747.

34. G. Tan, Y. Zhang, and R. Ayani, "A hybrid approach to data distribution management," in *Proceedings of the 4th IEEE International Workshop on Distributed Simulation and Real-Time Applications*, San Francisco, CA, August 2000, pp. 55–61.

35. K. Pan, S.J. Turner, W. Cai, and Z. Li, "An efficient sort-based DDM matching algorithm for HLA applications with a large spatial environment," in *Proceedings of the 21st ACM/IEEE/SCS Workshop on Principles of Advanced and Distributed Simulation*, San Diego, CA, 2007, pp. 70–82.

36. C. Raczy, G. Tan, and J. Yu, "A sort-based DMM matching algorithm for HLA," *ACM Transactions on Modeling and Computer Simulation*, 15(1), January 2005, pp. 14–38.

37. C. Raczy, J. Yu, G. Tan, T. S. Chuan, and R. Ayani, "Adaptive data distribution management for HLA RTI," in *Proceedings of the IEEE 2002 European Simulation Interoperability Workshop*, London, UK, June 2002, paper no. 02E-SIW-043.

Index